MARCH PAST

MARCH PAST

A MEMOIR BY

LORD LOVAT

WITH AN INTRODUCTION BY
SIR IAIN MONCREIFFE

WEIDENFELD AND NICOLSON
LONDON

This book is dedicated to the officers, non-commissioned officers and men of No. 4 Commando.

*Take these men for your example.
Like them, remember that prosperity
can only be for the free, that freedom
is the sure possession of those
alone who have the courage
to defend it.*
PERICLES, 430 BC

CONTENTS

ILLUSTRATIONS

MAPS

(*both maps were drawn by John Payne*)

INTRODUCTION

I count it an honour to have been invited by Shimi Lovat to write this introduction to his book.

Many people may now think of him as the Damascus blade of commando leaders. Indeed, the war record speaks for itself. But Lovat is more than that. He is, in the best sense of that currently misused word, a splendid anachronism: a Highland Chief and territorial magnate surviving and functioning perfectly in our drab egalitarian age.

I recently took part in a debate at a well-known university union, where it was pointed out that the word anachronism means 'out of its time' and that our times are sadly out of joint. The motion 'This House regards the British aristocracy as a frivolous anachronism' was soundly defeated. Lord Lovat is of the aristocratic mould that once made this country the greatest in the world and may one day help to pull it together again.

Meanwhile, a lone wolf by nature, and quintessentially a perfectionist, he has got on with the job for which he was born and bred. He started with heavy responsibilities. Born the Master of Lovat, heir to an historic estate and world-wide clan, he has the inherited advantages of an ice-cold brain, Fraser cunning and courage, and also being what a television commentator recently called 'by far the best looking of all the Highland chiefs'.

His father, the sixteenth Lord Lovat, KT, was one of the best-loved figures in the Highlands and the inspiration of his own patriarchal life. Shimi's indrawn quality, impeccable artistic taste and love of natural beauty come from his mother's father, the last Lord Ribblesdale. Each was a *grand seigneur* in the finest sense. From both the author has derived his independent spirit, backed by the traditions which have strengthened independent judgement in all his actions.

This book is not a complete autobiography, but a memoir set more in the form of essays, describing a bygone era, and written in the style of that day. It ends at the moment when Lovat was severely wounded in battle. We hear about many facets of an unusually interesting and varied life. Lovat could have played an ambitious rôle in Parliament on the political stage. When he was only thirty-three, Winston Churchill wished to make him Captain of the Gentlemen-at-Arms as a preliminary to such a career, writing, 'It is a political post and its holder sits on the Government Front Bench in the House of Lords and can take part in the defence of Government policy as desired by the Leader of the House. It would give me great pleasure if you felt able to join the Government at this time.'

Lovat declined Churchill's offer and, although he later accepted temporary office as Under-Secretary of State for Foreign Affairs, becoming responsible for the functions of the Ministry of Economic Warfare when these were taken over by the Foreign Office, he resigned on the defeat of the 'Caretaker' Government in July 1945. (His Commons colleague as Joint Parliamentary Under-Secretary was Lord Dunglass, afterwards Prime Minister as Sir Alec Douglas-Home.) He has since preferred to leave national politics to his brother, Hugh Fraser, MP, who was the popular President of the Union at Oxford when we were undergraduates together before the war. Hugh later rose to be Secretary of State for Air and a Privy Councillor, but the Chief of the Clan felt his duty lay at home.

Lovat was offered the Lord Lieutenancy of Inverness-shire by the late Lochiel. But this too was declined in order to concentrate on estate activities and travel abroad on their behalf. He retained, however, a political platform at the highest level as a speaker in the Lords on Highland affairs, and in his local power base by serving for a record forty-two years as an elected member of Inverness County Council: the county is the largest in Scotland, stretching from the Moray Firth to the Isle of Skye and beyond into the Hebrides. Among his many achievements in this field, to give one example, was urging the modernization of Mallaig Harbour and the creation of the Mallaig Harbour Authority, which has transformed a small West Highland village into the greatest herring port in Europe. The new A9 road bridge construction across the Beauly Firth, cutting distance to the North, was largely due to Lovat's prompting. But the splendid Commando War Memorial at Spean Bridge is probably his greatest pride.

The Lovat estates themselves once extended from coast to coast, supporting several hundred families, and were then considered among the largest in Europe. When his father succeeded to Beaufort Castle (we are told in Sir Francis Lindley's biography *Lord Lovat*):

the property itself comprised some 250,000 acres, of which the larger part consisted of heather-clad hills.... The property stretched up the Beauly Valley and Glenstrathfarrar to the top of Scurr na Lapich: further south and west the lands embraced the old Fraser country of Stratherrick and Fort Augustus and the isolated property of Morar on the west coast. There was also much good farmland in the fertile Aird district round Beauly and extensive stretches of woodland.... Another source of revenue was the salmon fishing which comprised the whole Beauly river and all its affluents.... It was an honourable but onerous inheritance: and all his life it was Lovat's principal preoccupation to fulfil his duty by it.

When he died in 1933, the text of Scripture was recalled: 'Know ye not that there is a Prince and a Great Man fallen this day in Israel?' The great task which Shimi Lovat himself has carried out, bringing the Fraser territory into the mid-twentieth century and then protecting it from the pitfalls dug by urban rulers in the shifting sands of our present time, is, in a sense, the repeat performance of a two-man team. Neither Shimi nor his father ever raised the rent of a *bona-fide* crofter living on the property. The sixteenth Lord Lovat was the apostle and first Chairman of the Forestry Commission and leader of Highland reafforestation: his son has been the apostle of a cattle revival as well.

He has become the pioneer *par excellence* in commercial hill farming, developing the idea of New World cattle ranching in the Beauly Valley, using the winter low ground reclaimed and reseeded to support high summer range. It has transformed the Lovat estates and at one stage included 35,000 acres in hand: the biggest single farming unit of the kind in Britain. As a leading breeder of pedigree flocks and herds of Beef Shorthorns, Aberdeen Angus cattle and black-faced sheep, he proved that it is possible to make an economic success of hill country in his north-east part of the Highlands.

For many years Chairman of the Anglo-Scottish Cattle Company, Lovat is much in demand at home and abroad as an international judge. He travels widely as a lecturer and agronomist. His interests overseas expanded. His life has been a full one in other spheres. There have been travels in South America, for he knew the remote regions of Brazil and the Argentine during and before going up to Oxford. He refers all too briefly to big-game shooting and other adventures overseas, and to mining exploration in Canada – where in a lighter vein Lovat became a connoisseur of redskin lore and 'Western' movies. Lovat has handed over his territory, transformed and modernized, to his sons. No man is better qualified to describe the life of a Highland Chief and the varied complexities of running a great estate in the fluid world of today.

Expeditions can prove an exhausting business. The year 1951 was a particularly heavy one. Beginning as a cattle judge at the Sydney show, Lovat made an extensive tour of Australia and New Zealand after joining the board of a big pastoral company where his advice proved valuable. He then went straight to Calgary in Alberta to open the famous Stampede: the greatest festival in the Canadian West, opened in Jubilee year by Prince Charles last summer. Thence to Nova Scotia to act as Chieftain of the Highland Gathering there, and then on to receive doctoral degrees *honoris causa* from two different Canadian universities. Returning that summer, he may have strained his heart when his three-year-old daughter, Tessa, was swept away on a Li-Lo, while playing on a beach at Cambus Darroch, which looks over five miles of sea to Skye, and across which he vainly pursued her. Father and daughter were picked up by a rowing boat far out in the Sound of Sleat. He was hunting moose in the North Woods during the Canadian fall. That December, while judging the Smithfield Show, Lovat collapsed. This was the beginning of some enforced leisure, and eventually the writing of this book.

Tessa Fraser is now Lady Reay, wife of the Chief of the Clan Mackay, so Lovat not only has a son who is called the Master of Lovat but also a grandson – the Master of Reay, both among the rare holders of that uniquely Scottish title. Known in medieval Gaelic as 'Mac Shimidh' (son of Simon), Lovat is the twenty-fourth Chief of Clan Fraser of Lovat.

The rôle of a Highland Chief has taken on an increasing importance during our lifetime, and with the advent of air travel is now perhaps at its most significant since the downfall of the original clan system. It is to the Highlands that we owe not only the word 'clan' (which is the Gaelic for 'children') but also all that it connotes. Originally it signified the numerous descendants of the earliest forebear, but very soon came to include their followers and dependants as well. Many followers of Lords Lovat took the surname of Fraser. But over the centuries, inter-marriage within their own district made them blood relations in any case. Scotsmen at home perhaps tend to take their traditions for granted, but, as I have written elsewhere, 'Frasers far afield in Toronto or Brisbane, Chicago or Wellington, can find in their history those roots that give them identity and distinguish them from being merely numbered ants in some enormous, urban ant-hill.' The greatness of this emotive pride, to quote the late Dame Flora MacLeod of MacLeod, is that 'it is beyond and outside and above divisions between nations, countries and continents. It takes no note of age or sex, rank or wealth, success or failure. The spiritual link of clanship embraces them all.'

These international bonds give Scotland a unique position in the whole comity of nations. Clansmen of long descent look to their home country and their Chief, however many generations or even centuries their particular families may have been settled as citizens of other lands. Wherever I roam abroad, from Japan to Brazil, the first thing I get asked about is tartan and the kilt and who has the right to wear it. This is a matter on which Frasers are not in any doubt. Clan societies have been formed everywhere in North America, such as the Frasers of California, and I was told in Nova Scotia, that second home of emigrant Highlanders, that Fraser is the most common name there. Latent clan feeling has received a fillip from the mobility of modern communications.

It is as 'Mac Shimidh' that Lovat's indefinable star quality makes him stand out among his fellow Chiefs, remarkable though some are. No man commands greater loyalty, and thousands attended the Clan Gathering of Frasers which he held at Beaufort after the war. Lovat does not believe in attending gatherings in the Lowlands, which he feels is more the province of his kinsman Alexander Fraser, Lord Saltoun, as Head of the whole Name. But he welcomes descendants of Gaelic-speaking Frasers returning to visit their ancestral Highland homeland, which remains his territory. Just as he visits them in new homes overseas. For the Chief is the focal pivot of the Clan throughout the world.

Historically the Frasers came to Scotland as mailed knights during our country's formative period, when the best and most adventurous blood of many warlike races went into the foundations of what were to become our Clans and Names. Originally dynamic Angevins from the Plantagenet county of Anjou, their surname derived from the nickname Frezel, which had been borne by the *seigneurs* of La Frezelière as early as 1030, before the dawn of heraldry. The first settler, Simon Frisel or Fraser, was a Scottish landowner in 1160. Other Frasers appear soon afterwards, and early in the following century Sir Bernard Fraser was Scottish Ambassador to England. When heraldry began, their shields in battle and tournament bore the 'fraises', or strawberry flowers, a pun on their name, that still appear in the Lovat Arms. In 1288, at the end of our peaceful Golden Age, William Fraser, Bishop of St Andrews and brother of Lovat's ancestor, was so prominent a statesman as to be Regent of all Scotland north of the Forth for that infant queen, the tragic Maid of Norway.

By this time plate armour was coming into use, and the medieval Frasers took a leading part in the War of Independence against England. Sir Simon Fraser of Oliver Castle, who held great estates in Tweeddale, fought for both Wallace and Bruce and defeated the English three times in one day at the Battle of Rosslyn; but he was later captured and by King Edward I's personal command suffered in London the ghastly public death of being hanged, cut down while still alive and then castrated and his bowels drawn out before he was beheaded and quartered. Sir Alexander Fraser, Chamberlain of Scotland, married King Robert Bruce's sister: she too had been a prisoner of the English who kept her in a cage in public. They were the ancestors of the Frasers of Philorth, now Lords Saltoun, to whom King James VI in 1592 granted the unique privilege of having their own University of Fraserburgh, with free education.

Curiously enough, the Frasers are connected with another university, Simon Fraser University in British Columbia, of which the present Lovat is Hon. LLD. It is named after the intrepid Canadian explorer, who was descended from a younger son of the second Lord Lovat. For the Chamberlain's younger brother, another Sir Simon Fraser, who (according to the former chronicler Froissart) 'chased the Englishmen three days after the Battle of Bannockburn' was the fore-father from whom the Lovat Chiefs derive their Gaelic title of 'Mac Shimidh'.

Hugh Fraser, Lord of Lovat, already held the nucleus of the future clan territory in the Highlands by 1367, and his grandson the third Chief was made a Peer of Scotland before 1464. A generation later the Frasers appear to have come up against a well-known local problem, when we are told that in the year 1500 'Huchone Frissell in Glenconie – that is, Hugh Fraser, tenant of Glenconvint – the best and most in estimation of the Lord Lovat's kin', slew a nine-foot dragon in the heather: reminiscent of the Loch Ness phenomenon of which the eagle-eyed author of this memoir once caught a brief glimpse in the water off the Stratherrick shore below Glendoe in the Fraser country.

The disastrous clan battle against Clanranald, when the third Lord Lovat and the Master were slain with nearly all the main force of their clansmen in 1544, was a setback for both sides unparalleled in tribal warfare. By 1562, however, Frasers had recovered sufficiently in numbers for the young fifth Lord to wait upon Mary Queen of Scots in Inverness at the head of four hundred of his armed clansmen. Indeed, Mac Shimidh's power became so great that a later Lord Lovat was able to write to his neighbours on his succession, warning them that if a single cow was stolen from any of his clansmen he would 'pursue it himself in person to their utmost bounds and borders'; he also hung up a golden chain at the 'Stock Ford of Ross' (Beauly River) and left it there, and nobody durst steal it.

The celebrated Simon Fraser, eleventh Lord Lovat, had abducted my own ancestress, widow of the ninth Lord, in 1697, and forcibly married her while a piper drowned her screams and clansmen cut off her stays with their dirks and so thrust her into bed. One of the most remarkable Chiefs in Highland history, he was created Duke of Fraser by 'the King over the Water' but was beheaded as a Jacobite on Tower Hill in 1747: the last nobleman to die under the axe. His eldest son, General Simon Fraser of Lovat, raised the 78th Fraser Highlanders and commanded them at the capture of Quebec, when they scaled the Heights of Abraham. Later he raised two more battalions of Fraser Highlanders, renumbered the 71st, for further campaigns. The younger son, Colonel Archibald Fraser of Lovat, MP for Inverness, who raised the Fraser Fencibles when Napoleon's invasion was expected, took the lead (with the Marquis of Graham) in Parliament to obtain the repeal of the laws banning the Highland dress, and was thus responsible for the restoration of the kilt.

The next Chief, Thomas Fraser, fourteenth Lord Lovat, was restored to the forfeited peerage by a special Act of Parliament approved by Queen Victoria. Through his mother, a Leslie of Balquhain (the family of Count Leslie who successfully plotted Wallenstein's assassination), he was a devout Catholic, the Faith that has so moulded the family ever since. A great improving agriculturalist, he was horrified by the notorious Clearances, and sheltered a hundred crofter families evicted from a neighbouring estate, providing them with lands first in Glenstrathfarrar and then on the more fertile braes above Beauly.

The late Lord Lovat carried on the Fraser tradition by raising the Lovat Scouts, often called the 'last of the clan regiments', which he commanded with distinction in the field during the Boer War. Raising special troops for battle is in the Lovats' blood. In the Western Desert during the Second World War, his first cousin David Stirling, known as 'the Phantom Major', personally raised and commanded the SAS, still so dreaded by our country's foes. For their exploits, 'dead or alive', both David and Shimi had a high-priced reward put on their heads by Hitler's personal directive.* On his mother's side, too, Lovat descends from Colonel Thomas Lister, afterwards created first Lord Ribblesdale in 1797, who during the American War raised, at his own expense, a regiment of Horse called 'Lister's Light Dragoons'.

Although Shimi Lovat was four years ahead of me at Oxford, his name was a legend there in my own happy pre-war undergraduate days. When I became an officer in the Scots Guards, once again Shimi

* '*Nacht und Nebel, Rückkehrung erwünscht*' – 'Dangerous terrorist to be exterminated'.

had been ahead of me, and was a byword in the regiment. My soldier-servant Guardsman Fraser, a loyal clansman, was told by the old hands that the two most unpopular peacetime Scots Guards' Officers were a pair (now dead) who shall be nameless, while the best liked were the Master of Glamis and Lord Lovat.

Those had been the days when my present wife was born in the Commandant's House at the Guards' Depot at Caterham, described by Rudyard Kipling as 'Little Sparta'; and Lovat was among the officers who had toasted the health of the Commandant's new-born babe. I never knew her father, Colonel 'Faulks' Faulkner, but he was famous as a soldier, and I well remember the shock that ran through the Brigade of Guards when we learnt of his death commanding the Irish Guards in Norway. The character given Shimi by the ordinary guardsman was borne out by Colonel Faulks' confidential report on his work at the Guards' Depot in 1936–37: 'Lovat has exerted a powerful influence on K Company and all Caterham recruits, not only on the playing fields where he has captained both the Depot's rugger and soccer teams, but by good example to young soldiers who have gained in confidence and pride in their regiment.' Coming from Colonel Faulks, anybody who served in the Household Brigade in those days will realize that this was praise indeed.

My first meeting with him can be dated precisely from my Game Book, kept meticulously since childhood and then in the care of Guardsman Fraser, a Perthshire gamekeeper in private life, who was well aware of his Chief's prowess as a big-game hunter. For Lovat is the Chairman of the Shikar Club, which provides an annual dinner for those who have sought adventure in lonely places. Over the years, Shimi has become the staunchest of friends, and by no means a fair-weather one.

The war came soon after our first meeting, and Lovat was distinguishing himself in a new field. Not for nothing was he a descendant of Henry of Navarre. His talent for converting men's skills and disciplining the excitement of the chase to forge a delicate weapon of war was making him an exacting and brilliant leader of commandos. The Lofoten raid had given a boost to army and civilian morale. The enemy could, nowhere, feel safe from British soldiers along the Atlantic seaboard: other forays were to follow.

The Dieppe raid is recounted at some length in this book, but I feel brief extracts are worth quoting from Terence Robertson's *The Shame and the Glory* (London, 1963):

Even allowing for the relative magnitude of the tasks and forces involved, the economy of Lovat's military plan, which he code-named 'Cauldron', was in startling contrast to the welter of detail in the military orders for Jubilee. [The main plan of the full-scale operation, by great masters of the art of War, which was a tragically unsuccessful shambles.] In outline it consisted of twelve paragraphs on four pages: in detail it was expanded by only two pages. Furthermore, it budgeted for such unforeseen eventualities as late landings, wrong landings, lack of surprise, heavier opposition than expected, and for failure on one of the two beaches to be used. The Jubilee plan provided alternatives only in the event of some infantry landing-ships being sunk on passage.... Lovat's Commandos touched down precisely on time at precisely the right place, and once again tactical surprise was achieved.... The raiders were re-embarked at 7.30 am precisely as laid down in Lovat's timetable.

A Canadian officer once told me that, *for his rank and rôle*, Lovat probably had the best military brain on either side in the Second World War.

Many readers will have seen the film *The Longest Day*, that depicts his part in the historic D-Day battle, for which Dieppe had been a mere rehearsal. At Dieppe he astonished superiors by electing to assault the enemy stronghold with only two-thirds of the forces available to him. In Normandy his Commando Brigade fought six miles through the German defences on the first day. Lovat's achievements in the war were those for which he was bred: he helped to raise, train and finally command a brigade of shock troops on D-Day. Decorated many times for gallantry, though he doesn't tell us so, this book ends when he was dangerously wounded. As soon as he had sufficiently recovered, he was sent as one of a Parliamentary delegation to the Kremlin. The Americans were being driven back by Rundstedt in the Ardennes. The Russians had deliberately halted their advance to allow the Nazis time to liquidate the Polish Resistance rising in Warsaw. The Allies were in dangerous discord. The delegation was intended to bolster the Soviet alliance with reassurances. Quoting Byron, Churchill wrote to Stalin that he was sending him in Lord Lovat 'the mildest-mannered man that ever scuttled ship or cut a throat'.

Easter Moncreiffe   IAIN MONCREIFFE OF THAT ILK
Perthshire

FOR A' THAT

The conviction that a preface is seldom read rarely discourages the aspiring author from putting pen to paper. In similar vein the writer's childhood – the perambulator experiences (sure target for the critic) – are best left to summary execution by a friend.

The reader wants to get on. No guest coming to dinner expects to be kept waiting in the cold, pressing at the doorbell. Iain Moncreiffe has done his best to soften the blows: I have stepped into obvious pitfalls with the excuse that this memoir requires an explanation to be understood.

Years ago I was asked to write a book, but the offer was declined. There were many duties to perform in Scotland, nor did I approve of war stories suiting the mood of the day which were rushed into print before the dust had settled over Europe.

I made a half-promise, however, to put something together without committing myself to a subject or the date of completion, but also determined to avoid the blood and thunder currently in demand.

Time's winged chariot has caught up with excuses: if readers now seek adventures in less familiar fields I have missed the mark. A variety of subject-matter has been added to counterbalance more objective views which should put this memoir into better perspective. If the war chapters appear too long or the background too short, the events described are interlocking and, I hope, relate to form a pattern.

It has been said 'the child is father of the man'. Perhaps this contention will explain the far away and long ago build-up of a story which by accident ends in the climacteric of war. For clarification, I will trace the outline of the tale.

March Past began slowly with a series of sporting asides, scribbled in a hurry after long periods of stagnation. But at least it made a start. Most of the early efforts have been discarded or used elsewhere. Next came childhood, school and university; then the army. Though the war chapters are important the intention has been to cast back into the past and recapture a way of life set in more leisurely surroundings.

The exercise of presenting a fragmented career as a co-ordinated whole to give it shape and meaning has not appealed to me – the discipline required to write a serious book would have spoilt the fun. Concentration never figured in the scheme of things and this contribution is offered for quite different reasons.

Present fashion seems to enjoy and certainly expects confessions of doubt and insecurity. My efforts are not concerned with self-analysis or the subconscious mind: there have been few traumatic experiences. I have sought highlights and milestones under clear skies, and the ensuing sketches make no attempt to raise aesthetic or philosophical issues. Self-criticism may occur, but those who seek a dash of eroticism or indications of repression will be disappointed. I have escaped them all and it can be assumed that what follows will be healthy if not profound.

There are faults: the book has a beginning but no end, and is inclined to wander about in the middle. Various anecdotes have been cribbed from friends and relations who handed down much folklore describing situations which will not be seen again.

But what of myself? Sir Iain has done his best to provide the answers. I fear the author will prove an unimaginative fellow, devoid of ambition: self-sufficient, yet hostile to altered circumstance. I have no defence against these charges. Change has not come easily. My youth positively reeked of privilege and opportunities no longer enjoyed by, or indeed available to, the citizens of today: life was uncomplicated and the way secure.

Then came the war. It changed the world as I knew it and put a full stop to fluent writing. Have we won the Peace? I mistrust bureaucracy – rubber stamps – officialdom and forms in triplicate: the simplicity of an age that ended in my youth is no longer possible. Yet in that rustic virtue lies an attribute which is shared both by genius and numbskull.

Today I find less generosity of spirit; reducing every standard to a common denominator is difficult to understand. Self-interest has taken the place of good manners. It will be no surprise to learn my politics remain strictly feudal!

In recent years hard things have been said about the ownership of land, stirring up passion and provoking a real or imaginary sense of injustice. My 'withers are unwrung.' From childhood, a stern but respected father taught his sons the sound principle that the possession of broad acres involved more duties than rights, and as many responsibilities as pleasures. Excuses were not accepted for absentees who gambled away estates and became generally disliked in the process.

There was much amiss with a family who could not fend for itself in a faster spinning society. On this advice I grew to manhood.

'Chary of praise and prodigal in counsel' – my father was an uncompromising man, and he was usually right. The economic depression of the early thirties took him unawares. He died after my coming-of-age celebrations, within a year of my joining the army.

I remained in the Scots Guards, trying to economize during the time it took to pay off death duties; then sent in my papers and went on holiday before proposing to the girl I had fallen in love with. We got married and settled down at Beaufort to lead a quiet life – so we hoped. But it did not work out like that. War, which is not a good time for young wives, was declared the week our son was born.

> When did hearts so careless beat,
> When was grief so far.
> For love is a child in the days of peace,
> But a man in the days of War.

The subsequent upheaval and five more years in uniform is the burthen of this story. It is a large canvas, multiplied by varied experiences. The perspectives must be finely balanced. I have concentrated on local colour and anecdotes that may appeal to modern thinking: commonplace stock sheets have been cast aside.

Camp-fire tales and blood sports interest few people, but there are no barriers between fishermen. Foxhunting and the pursuit of game were reluctantly discarded, but time was when Whyte Melville's lines, 'I freely admit that the best of my fun I owe to horse and hound', seemed rather important. Wounds put an end to riding horses. The war was a sad and bloody business in which I lost my friends: a light-hearted approach can best serve their memory.

Public duties and the preoccupations of running an estate and raising a large family came later. In the best traditions of wardroom and mess, women, politics and religion have been avoided.

Pompous individuals and the 'unco guid' are also left behind. Instead, I have attempted to provide some aspects of the war, record a few legendary figures, and make the most of a cheerful approach to life. There has been one serious drawback: I find it difficult to write in the first person. Pages on oneself must sparkle, or the manuscript is better consigned to the dustbin. I had reached this painful conclusion somewhere about the sixth chapter, when the door of a two-seater aeroplane heading for Bimini in the Gulf Stream suddenly blew open, and the suitcase containing this book and all I owned plummeted from on high into the swamps and everglades of Florida. The vision of

Seminole Indians prodding among alligators and water hyacinths for lost property turned the MS into a work of art only comparable with the Unfinished Symphony. It has spurred on and possibly improved my latest literary efforts. Although references were lost, the set-back has done nothing to impair a certain licence which highlights happy memories.

There is a danger of re-living the past in the light of a golden sunset, but is that such a bad thing? Sun is preferable to shade outside the bullring.

More questionable is the psychological change from the boy to the man. Thus the path that ran over the moor to the loch, which in memory was five miles long, is in reality less than three. To the boy's short legs it seemed like five miles. Conversely—in youth the salmon pool below the castle, when fished before breakfast, could be reached in no time, running all the way; the grown man takes ten minutes to get there, walking at a more leisurely pace.

Some of these things will come out in the text, but now the reader must imagine he is drowsing by an open fire, listening to a veteran who once led an active life, and finds pleasure in recounting the tales he has been told.

Should repeated reference to my native land offend the reader, he must accept that Scotland may be the true heroine in this story. It only remains to thank old friends for the fun we have had together.

Here then is a budget of adventures. I hope it will interest the kind of people to whose support Sir Walter Scott once attributed his popularity: 'Soldiers and sailors and young persons of both sexes of bold and active dispositions'.

Grateful acknowledgements are due to my daughter Fiona, my daughter-in-law Drusilla and Ina Maclean who have deciphered hand-writing, attended to punctuation, and inserted most of the capitals. Also to various men of letters who spent much time at Beaufort. Quotations from their works and poetry, admired during my formative years, freely embellish this memoir, adding the flavour of an anthology to an otherwise pedestrian excursion into literary fields. Iain Moncreiffe supplies a charitable introduction.

Further thanks go to the authors Peter Kemp (*No Colours No Crest*, Cassell); Cornelius Ryan (*The Longest Day*, Gollancz) and James Leasor (*Green Beach*, Heinemann); extracts from the books quoted have been extremely useful. The co-operation of these authors is much appreciated. Equally important are the works of friends and relations: Derek Mills-Roberts's *Clash by Night*, John Durnford-Slater's *Commando* and

Peter Young's *Storm from the Sea* (all published by William Kimber),
and Murdoch McDougall's *Swiftly They Struck* (Hamlyn). Our shared
experiences illuminate this story. I have also consulted *Impressions and
Memories* and *The Queen's Hounds*, by my grandfather, Lord Ribblesdale
(Cassell); *Dear Youth* by Aunt Barbara Wilson (Macmillan), *Lord Lovat*
by Sir Francis Lindley, *Maurice Baring – A Postscript* by Laura Lovat
(Hollis and Carter), *The Arches of the Years* by Halliday Sutherland
(Geoffrey Bles), *Ronald Knox* by Evelyn Waugh (Chapman and Hall,
reprinted by permission of A. D. Peters & Co. Ltd), *Tartans* by
Christian Hesketh and *How Very English* by Kenyon Goode. I am in-
debted to Renate von Stern for the war photographs of the German
army obtained from various sources.

Donald Gilchrist's *Castle Commando* (Oliver & Boyd) adds some
vivid descriptions to the heat of battle. Other officer friends who served
together have contributed to *March Past* by frequent communications.
At the risk of missing someone, I wish to thank Robert Dawson, Bill
Boucher-Myers, 'Doc' Patterson, Fairy Veasey and Tony Smith (who,
like Derek and Donald, were all in No. 4 Commando), also my cousins
Andrew Maxwell, Bill and David Stirling, and Lieutenant-Commander Rupert Curtis, DSC, RNVR, who took us across the Channel
in the last adventure. Max Harper Gow and Bobbie Holmes, two
matchless members of Brigade Headquarters, have helped to jog my
memory.

Last but by no means least, two splendid helpers – Henry Brown
and Pauline Rawlinson of the Commando Association, who have, in
their own free time, constantly retyped, cross-checked and revised the
accuracy of dates, personnel, place-names and troop movements.

Margaret Lindsay completed the fair copy.

Written at intervals: SIMON LOVAT
Hay River, Great Slave Lake,
North West Territories, 1964;
completed Goat Cay, Great Exuma, 1977

PART ONE

EARLY DAYS

CHAPTER ONE

WILLIE THE MOON:
TRIBUTE TO A CLANSMAN

> He loved the brook's soft sound,
> The swallow skimming by.
> He loved the daisy-covered ground,
> The cloud-bedappled sky.
> And everything his eye surveyed,
> The insects in the brake,
> Were creatures God almighty made;
> He loved them for His sake.
>
> JOHN CLARE

The tall old man with the walrus moustache, washed-out blue eyes and hesitant manner usually wore a baize apron. He spent the morning shining shoes and turning the mangle for the laundry maids. In August 1914 the task of cleaning the field-boots of senior officers recently mobilized for the First World War was added to these chores. He spent the rest of the day, like his father before him, trimming wicks and mantles and polishing the chimneys and brass on the vast array of oil lamps needed to light the castle. When lit, the translucent globes identified our retainer with this curious lunar sobriquet. In Scotland, men of the same name are identified by their calling: Dan the Ferry, Coffin John, Allie the Toll and Johnnie Deaf Ears (the last-named, a nightwatchman) being some examples of local humour. Peterhead, who took round the plate on Sundays, had served his time as a warden in the east-coast prison of that town; Tulloch the Toes was a chiropodist. In the stately homes of England Willie Fraser – the Moon – might well have been described as 'The Boots'. No such thoughts were entertained at Beaufort for a friend who bore the same name as his chief and whose forebears, two fearless brothers, had fought and died in the Highland centre when the clans charged at Culloden.

Judged by material values 'the Moon's' life had not been a success. He had emigrated, against advice, to Canada with a wife and young

family, only to find the bitter cold of a prairie province overmuch for
a delicate chest, and the work too heavy for a man better suited to be
a poet than a pioneer.

Today a bard is something of a rarity, even in the Western Isles:
gramophone, radio and television have done their worst to Celtic culture. But it was not so at the turn of the century when Gaelic was a
living language, and the *seannachie** in demand at every *ceilidh*† and
winter fireside. The Moon, though too shy to recite his own epic sagas to
advantage, was also a musician of note who played both the violin and
squeeze-box with equal facility at dances, where his services were ever
in demand, on schoolhouse floors around Beauly, or in the straw barns
of Saskatchewan. In fact, Old Willie was something of a native genius.
My father's decision to pay the whole family's passage home was a
popular one, and caused quite a stir in the district.

The first Gaelic lessons, received at an age when the infant mind is
supposed to readily adapt itself, clashed with the war. Sadly, I never
grasped the language, for Nelly Cameron, nurserymaid and the
bilingual daughter of the Struy *grieve*‡, was called up to work in an
ammunition factory. Nor, in the great upheaval, did I turn my hand
to the fiddle of the Bard (as he became increasingly known, distinguishing him from his brother Hugh, who worked the croft and never
married). Certain poems were repeated at the old man's knee, to the
annoyance of the schoolroom authorities. Translated into English,
they concerned the family misfortunes in Canada. The lines were in
blank verse: had they scanned, I would have remembered them. Down
the long years I can still conjure up the isolation of the log cabin
with its turf roof, a winter wolf howling at the stars, snow-shoes
crackling, the frozen stillness, and the Northern Dancers flickering in
the intense cold. The moment of climax came when all Willie's children
were put in the oven to prevent them from freezing to death! These
stirring tales, never forgotten, implanted a growing interest in the New
World, and subsequently a love for Canada and its vast prairies and
backwoods. After reading Jack London's *White Fang* and *The Call of the
Wild*, such twaddle as *Les Malheurs de Sophie* (beloved by French
educational crones) seemed very cold porridge.

Willie, the returned prodigal, was too old for National Service. Now
safely harboured after storm, he radiated an essential happiness that
was touching to behold. In a fold of the Park, his home, Brae Cottage,
sat tucked away beside a giant Wellingtonia on the bank of the Beau-

* Story-teller.
† Musical get-together with the neighbours.
‡ A bailiff or foreman.

fort Burn. A bubbling spring rising between flagstones down a short footpath provided the water supply. A sunny, happy spot, without either flowers or garden – just a potato patch, a few skinny chickens, some bee-hives and a large family. Three of the more adventurous offspring returned to Canada, but, with one exception, none of the Fraser children lived to make old bones. Yet the bees thrived greatly there in home-made skeps of plaited straw, before Isle of Wight disease played havoc in the North. The Moon's clover honey off the improved leys on the Home Farm was famous in the Aird district. Out of reach of heather the same colonies – an appropriate term – found their way up to the lime-trees in the forecourt, which kept them busy until early August. They topped off a good season – and summers seemed hotter then – in the herbaceous borders of my mother's autumn garden.

That was the scene of Willie's dwelling, long since pulled down, when I first went out to tea more than sixty years ago. And what a meal it was! Piles of soda scones, pancakes, shortbread, oatcakes, butter and honey were consumed in the kitchen, preceded by an odd jesting verse, delivered in lieu of grace, by the Moon, who was something of an autocrat in the family circle.

> Now pass round the cookies
> The buns, the baps, the gingersnaps.

This formula never varied. Neither did the cookies (a New World curiosity) ever appear in person – nor for that matter the buns, the baps and gingersnaps. There was sometimes an alternative to the feast, already heaped on a bare wooden table. For a centrepiece, couched in isolation, there wobbled on occasion a red jelly. Its fluted sides caught the light when tipped quivering from a castellated mould. This launching ceremony was watched in expectant silence. Mrs Fraser, who came, I think, from Bridge of Weir and spoke the unfamiliar Doric of Renfrewshire, then urged the family on in a stage whisper: 'Press the jeel – it'll nae keep.' These directives remain a jest whenever a strange dish appears in the dining room. The illusion created by that curious cookie rhyme appealed greatly to a child's imagination. Although learned parrot-wise, it has survived where much else has been forgotten. R.L.S. was right in saying that words should be treated like musical instruments and made to play in tune.

The walk home was a leisurely affair. People in the summer of 1914 took their time about everything. As in the East, haste was very properly considered to be an attribute of the Devil. A day was twenty-four hours long. The countryside rose early and went to bed with the sun. There was no hurry and nowhere to go, except – maybe – a shinty match or the cattle market in Inverness, the County Show, or a church bazaar. Weddings and funerals were treated as major events. The village supplied the daily needs of a prosperous community. My own legs just carried to the Home Farm; the kennels, the smithy and the stables provided other important walks.

Then came the war. Up in the castle old Willie performed a Canadian parlour trick to reward the morning march from the nursery to the lamp room. There, at eleven o'clock, the Moon took off the baize apron, produced a rawhide pouch of tobacco from one pocket, tissue-thin paper from another, flint from a third, and, in what seemed one and the same movement, proceeded to roll and light a cigarette, all timed to twenty seconds. Even Jack Vickers, my father's hard-swearing English valet (by then in uniform), who smoked Player's, gave me cigarette cards, and, rumour had it, got through a case of whisky every week, nipping with other batmen and orderlies in the house-keeper's room, had to admit that this was quite a performance. The 'locals' – keepers, fishing ghillies and grooms – who reported each morning to the gun room for orders or to lend a helping hand, smoked, or sometimes chewed, Black Twist (plug). The contrast in styles is worth recording – Old World and New. The Highlander of that period spent a fair part of the day getting his pipe alight and drawing to satisfaction. The Black Twist was first cut up in chunks with a clasp knife. Much crumbling and drying of the treacly mess in horny palms was necessary before the matches gave out or the fire took hold. To this end the pipe was loaded by degrees, and with immense concentration. Once started, a tin lid with airholes was fitted over the top, which kept the bowl smouldering all day – even in a waistcoat pocket.

I remember mobilization brought much noise, and a strange variety of sounds and smells. The wartime bustle of a country sprung to arms had a confusing effect on a child's Arcadia. At three-and-a-half my nose was able to identify the changed appearances of various members of the family by the smell and touch of such things as field-boots and breeches when standing, or brass buttons and khaki, laced with moth-balls, or cigar smoke when sitting in a chair. But the faces are forgotten: perhaps they stood too tall.

The cavalry horses seemed to fill the park. The squadrons were watered, splashing knee-deep in the Beauly, and with the day's work done munched their hay standing roped in endless rows on both sides of the river.

Outside the castle, pipers played all day in the tented camp: 'Hey Johnnie Cope, are you waukin' yet?' at Reveille, and 'Brose and Butter' for the men's dinners. The drums and mounted band practised every afternoon. As dusk fell, a piper of repute – and the best of them came from Barra and South Uist – played Retreat down the lines, always choosing a classic air associated with clouds of glory. Willie the Moon was my mentor on the names and meanings of those tunes. I learned one of the favourites the men enjoyed: 'Cogadh na Sith' (the challenging title signifies 'Peace or War') had rallied the Highlanders at Waterloo, when a piper from the Reay Forest stepped unbidden from a hard-pressed British square and played round it, between repeated charges of French cavalry. The nursery was asleep long before Lights Out, but awake long enough to hear a trumpeter from the Fife and Forfar Yeomanry blowing the Officers' Mess call that translates into the words:

> Officers' wives get puddings and pies,
> Soldiers' wives get skillee.

Cavalry regiments have qualities that are all their own. Upwards of two thousand mounted men arrayed in the panoply of war provided a stirring spectacle, once seen never forgotten. But soon it was all over. The Highland Mounted Brigade,* with father in command, left Beaufort for Blairgowrie, leaving only the Assistant Quartermaster for administrative details, and a rear party to clear up the mess.

My small world sighed, turned over and went to sleep again. The euphoria sweeping through London was not apparent in the North of Scotland. Those figures of Vaudeville's imagination – 'Kaiser Bill', 'Little Willie' and 'Berlin-or-bust-before-Christmas' – never reached a countryside where native caution and common sense prevailed. In the Highlands war was a serious business, which for young men then amounted almost to a profession. The Camerons (the County regiment), and likewise the Seaforths, were both to raise more than twelve fighting battalions to supplement the regulars in the Highland Divisions.

In Lovat's Scouts there were many veterans of the South African campaign. Raised by my father in 1899, and recruited from the best type of mountaineers – stalkers, crofters, hill shepherds and ghillies – these part-time soldiers were trained to standards of high efficiency. Hardy, and independent to a fault, unless led by the right man, the Scouts had distinguished themselves against the Boers, but the availability and preparedness of Territorials was sadly misread by Kitchener, who preferred reservists as first choice for the Expeditionary Force, confronting seven German armies. Recalled to the colours at

* The First and Second Regiments of Lovat's Scouts, the Fife and Forfar Yeomanry, the Inverness Battery of the Royal Horse Artillery, Signals, and Field Ambulance.

short notice, many from sedentary occupations with no time to get fit, the reservists' feet were destined to suffer badly in the retreat from Mons. The setbacks of 1914 caused surprise, with worse to come, but the Scouts, who trained for a fortnight each summer, and were described in the Army List as Mounted Yeomanry, remained waiting in Home Forces. After a triumphant return from the Cape, General Hector MacDonald, a local hero and contemporary on the Veldt, had said of them, 'As Scouts or Guides – on pony or on foot: as individual marksmen or as a collective body in the fighting line, they are all specialists and picked men. They are a splendid band of Scotsmen, which is the highest compliment that I can pay them.'

Well before mobilization they had, in a clan sense, gained an enviable reputation, scoring a notable success over their Perthshire rivals, the Scottish Horse, commanded by Tullibardine, the Duke of Atholl and Chief of all the Murrays. The epic tale is told round Beauly to this day. It happened thus when light cavalry played the part that armoured cars or recce vehicles do today. In the army manœuvres of 1909 between North and South Land, opposing forces were planned to converge during the advance from Perth and Inverness, and join battle in Badenoch. Towards evening on the second day, father's cavalry patrols contacted South Land and vedettes at Kingussie. In an outflanking movement he had swum the regiment's ponies across the Spey, 'where ford there was none', below Kinrara, and passing through Glen Tromie, crossed into Gaich and climbed up to the Forest Pass, where the horses were abandoned. Then, after a night march of over twenty miles on the track that leads southward through the high hills, the Scouts fell on the Scottish Horse camp at Loch Ordie near Blair Atholl. The cookhouse was seized in time for breakfast and the rival CO (guess who? – the Duke himself) was captured sleeping in his tent. A great day for local patriotism!

The following autumn there was more inter-clan horseplay: my father, Uncle Hugh and Uncle Alistair, who all fought in South Africa, were going south for my parents' wedding. The train stopped at Blair: the Duke of Atholl was on the platform seeing off some friends. A taunting enquiry as to his health and what time he had got up that morning drew his Grace to the carriage door. He was instantly seized and pulled bodily, like a cork from a bottle, through the window.

As the train started to puff away, Bardie, so called by his cronies, reappeared through the steam, but without his kilt, which was handed over to the station-master at Pitlochry for early return. So much for the rough good humour of the day.

Little love was lost between the Murrays and the Frasers. Old Simon

Lovat, who parted with his head after the '45, had treated Tullibardine
contemptuously when that family sought to capture him after being
outlawed in 1697 for the rape of a relation – a Murray heiress.

But disagreements went back further, to the day in 1306 when Sir
Simon Fraser, the Patriot and William Wallace's most able Commander, was hanged, drawn and quartered, at Edward 1's orders, on
London's Tower Hill, together with an earlier Atholl leader. Because
of his rank the belted earl was given a higher ladder gibbet for execution. Sir Simon objected on the grounds that he was the most senior
officer on parade! A good story and less apocryphal than the jesting
tale of Noah, who offered MacNeill of Barra a place in the Ark. That
chieftain replied that he had a boat of his own. So much for Highland
pride.

Now once again men rode away to fight on foreign shores, in a mood
and from a setting (if years had rolled away) not dissimilar to that of
Agamemnon's host, when the Greeks left High Town in Mycenae for
the Trojan War. Strangely – and no one knew it at the time – the
regiment was destined for the Dardanelles, those same narrow waters
where Asia Minor and Priam's flattened walls look across the sea to
Gallipoli.

The martial scene made a deep impression on Old Willie. In
winter time the lighting of lamps on passages and staircases began at
four o'clock. On formal occasions, with visitors in the house, the Bard
changed out of his Lindsay Wolsey trousers into a claw hammer frock
coat, and then advanced – flanked by the pantry boy and the 'orra
man', who normally carried coal and firewood into the living rooms
upstairs. The recruited colour party were now taught to march in step;
it must have impressed the Assistant Quartermaster. In the nursery
wing, with only a girl in attendance to open doors, the timely appearance
of the Moon bearing lights had the same sort of effect on my sister
Magdalen as Leary passing by in Stevenson's charming glimpse of the
Edinburgh street scene.

That was the end of the nursery era. In August 1915 my mother,
dressed in black and seemingly a different person (for unbeknown I
had already lost three uncles), took us south to say goodbye before
the Scouts sailed for Egypt and the Dardanelles. Beaufort, handy for
Invergordon and the Fleet Anchorage, was offered as a naval hospital.
We did not return until the end of the war. It was to prove a long
and unsettling upheaval in a child's life.

More than three years later Willie was waiting in the forecourt to carry
in the luggage. His face was the only one I could properly remember,

and this winter homecoming to a freezing house without heating or electricity was not an unqualified success. The stark size of the castle, set high above the river, and the brooding silence of the woods and hills, was a little frightening after Lyegrove and the creature comforts of Gloucestershire.

Then the days began to lengthen, chilblains disappeared and almost by surprise the flowers came out again. On May Day I took off shoes and stockings and ran barefoot with the local children for the summer. There is an Arabian proverb which runs, 'Behold I was without shoes and complained until I met a man who had no feet.' To run barefoot as a boy was considered the sign of a hardy lad in country districts. Spring comes late and grudgingly to the Highlands. Now suddenly it was warm, and Nature was reborn.

The morning was fine, and the Moon suggested we should go fishing. We got the worms before lunch, chosen for colour rather than size, in the kitchen garden. Then the old two-piece greenheart – or was it hickory? – with the stiff top joint was put together, a swanshot tied to the line, and lessons in casting (underarm) began on the lawn. There came a time when a well-lobbed throw fell on a spread handkerchief, but the results of that first week's fishing were disappointing. Willie blamed the light and low water. Then, snapping the barb off the worm hook, he proceeded to catch a fingerling trout and a salmon parr in as many casts, lifting them out bodily with a quick jerk on to the bank. They were reverently returned to the water. 'You must never hurt the wee ones,' he whispered, as we crept upstream to the next pool. Careless movement and a hurried approach on a sunny day are worse faults in stalking (by land or water) than inaccuracy or misjudging distance, and far more difficult to correct. It took two more fishing seasons to learn not to disturb feeding fish. Then I graduated from worm slinging to the cast fly.

In the burn brown trout over a quarter of a pound were hard to catch in clear water. In a spate anything could happen. The precious salmon parr were easy at all times. When the rod was useless and the day was warm Willie would remove his jacket and tam-o'-shanter – the only concession ever made to Highland dress – roll up his shirt-sleeves and then, face down, nose just clear of the water, would ever so gently feel under the grass banks with the tips of his fingers. Tickling fair-sized trout in deep water, where both hands have to be applied, is a far finer art than guddling under stones in a shrunken hill stream, where smaller fish can be pinned by force with no way of escape.

But fishing was a different story when the burn ran high and coloured after heavy rain. Then the 'badgeoch' – or garden fly, best known in the worm tin – came into its own. With the rains came hungry sea trout, hurrying up from the main river to compete with big, bottom-feeding brownies in most of the deeper pools. There were other kinds of fishing. After good days in the burn, with trout to spare, we set night-lines for pike, baited with the smaller fish, in the slack water of the Beauly above the Cruives. The three salmon beats (known as the Castle water, Home Falls and Downie) were let to different tenants throughout the season, but encouragement was given at all times to destroy vermin. Thus the river-keepers were responsible, both winter and summer, for all forms of predators, above and below water. Their wages were supplemented with a vermin bounty, ranging from ten shillings for seals (which in high water sometimes pursued the running salmon as far up river as the Red Bridge) to a sixpence for cormorants, herons and mergansers. Gill nets and basket traps were set in every backwater for spawning pike or eels in season, and, if that was not enough, the jack-pike were shot whenever they ventured to sun themselves in shallow water. In face of this competition we felt superior craftsmen when our marker floats – or more often tightly-corked wine bottles painted white – started first to bob, then move slowly away as the boat approached the backwater lair. Sometimes a hooked fish swam away and hid itself in reeds or undergrowth; there was no knowing the size of the pike at the end of the short length of wire trace. Sometimes a good jack would charge out into deep water with a bow wave like a destroyer, to disappear, bottle and all, as it swam into the river. This kind of fish provided an exciting chase. On one occasion I casually picked up what seemed a rejected offering – my hand was actually round the bottle, which had not stirred – when the boat was deluged with a shower of spray from a monster under the bows. Ochon! Ochon! – the codline tore off the bottle neck and that big one got away.

The first setting in the spring provided the best scores: with our men at the front, the war years helped to increase the pike population. On a record morning Willie nailed eight heads from twelve bottles on the oak tree by the gallows where the salmon nets were dried. The weights ranged from three to fifteen pounds. Highland pike in cold running water seem to be more voracious than the land-locked pond dweller, though they do not grow to such great size. They spell death to salmon smolts, returning all silvery in easy stages to the sea, and I have even taken the undigested carcass of a drake teal from a fish that was less than six pounds in weight. (A cormorant will swallow a three-pound grilse or sea trout, but can be knocked on the head while he digests it.) Chasing bottles in a boat is better sport than hauling in fixed lines pegged to the bank. With the floating bait, one is less likely to catch eels.

While Willie knew all the best pools inside the Park, and through the beech wood and Home Farm, he seemed reluctant to go beyond the road bridge into Boblainy Forest, which my father said had never been fished. At first I thought that the walk was too much for him, then, that he was embarrassed, as I was myself, by the inmates of a German prisoner-of-war camp – embittered men still felling timber in 1919 round the Hundred Steps.

Where the valley narrowed into a mysterious tunnel, below the tall, singing fir-trees, Willie had some reason to turn home. Bull's-eyes were bought as a poor alternative at the post office, or a visit paid to the miller, who came out white and floury to tell of big fish seen in the mill stream that ground his corn. After the dear old man had retired, Willie confessed that fear of disturbing the fairies had held him back, for any trespassing in this private realm of enchantment might annoy the 'Little People'. The green colour of the water – quite different from the brown sherry of the peat stream – proved that they washed their garments there; it was clear evidence that we had ventured too far from the beaten track; like Kilmeny, we could have lost our wits or been spirited away.

Back in the schoolroom it was good to read Hogg's poem about how the Ettrick shepherd, a man who could write immortal verse, once wept because, at the age of twenty, he had had no schooling. Hogg taught himself to write with carved horn letters an inch high, as he watched his master's sheep on the hills and well-heads above Tweed. If Burns is the poet of humanity, James Hogg was the man best acquainted with the supernatural. Willie Fraser, steeped in the legends of the Gael, shared with Hogg that sense of the unseen.

> When seven lang years had come and fled,
> When grief was calm, and hope was dead,
> When scarce was remember'd Kilmeny's name,
> Late, late in a gloamin' Kilmeny came hame!

And so the Highland summers passed leisurely by as I grew in size and experienced the adventures dear to a country boy – some at Beaufort, some further afield, with friends of my own age. Up at Oxford, I recaptured a fishing experience in the following essay, which won a prize in a magazine.

THE BURN

It is on the Balmacaan march that the burn rises, on the moorland plateau called Fan Blair in the Gaelic, which divides Glen Urquhart from the Aird and Strathglass. It is a high place and a windswept one – for there is little shelter – but on a clear day there is no finer view in all Scotland.

There is no lack of rushy springs. Crystal clear and the colour of brown sherry; Loch Bruiach is the main provider, but little waters gather and flow from all the recesses of the hills.

From green well-heads the burn passes swiftly down slopes over long miles of heather, juniper and bog myrtle, singing as it goes and sweetening the close-cropped pastures of bleak uplands till it tumbles precipitously over a series of waterfalls at Culnaskiach and the Hundred Steps, into the valley below. Then a great wood of larches, fir and spruce trees hides everything from prying eyes: bracken grows high along the banks and the pools are dark and deep. The old people will tell you the water is tinged with green, for at dawn the fairies cleanse green garments there: be that as it may, the burn, when it emerges where the public road runs from Kiltarlity to Eskadale, has lost some of the high spirits of its early march.

Now it slips decorously between nodding cornfields and wide pastures, under bridges and over water-wheels, that grind the oats in farm steadings; past shoal and shallow where cattle splash in summertime and the gean tree sheds white blossoms on the water on through the village sawmill with current to spare for the flour mill beyond. It crosses the park below the castle wall, then disappears in the Beechwood where the hurrying stream enters the broad bosom of the Beauly, and so at last to the sea. Its course is run, but the burn goes on for ever. It waters a happy valley and a dearly loved one. Since the misty days of Fingal its inhabitants have prospered in a manner uncommon to the Highlands. Generations of Fraser children have climbed about its banks, looked for birds' nests, and swum or fished there on hot summer days.

At an early age, I knew every hole that held a trout right up to the edge of the moor, the best places for night lines and where to look for otters, or to set vermin traps for wild cat or fox and find the tracks of big wood stags – seldom seen and exceedingly cunning – that lay up near the water and lived off the crops and arable land. Early memories, however trivial, stand out like milestones on a forgotten highway, and it seems only yesterday that I ran barefoot down to the Junction pool to show my father, recently demobilized, a goosander sitting tight on her nest among the driftwood of a winter flood. That was in June, and the date 1919.

Then there was an old beech tree below the castle, hanging out of a sandy bank: the roots on one side had been washed bare, and by digging out some earth and gravel, a shallow cave was cunningly fashioned, and enlarged each summer as its solitary occupant grew bigger: a curtain of plaited withies provided shelter in rough weather, forming as snug a nest as that hidden refuge of the '45, where Cluny Macpherson offered to make his Prince welcome on the slopes of Ben Alder; as with Cluny, the secret of the hole was carefully guarded. There I kept my treasures – stoatskins, lines and tackle, an eelspear, an old cooking pot into which some cousins were sick after eating a boiled oyster catcher, knives, candles, dry firewood and some matches. To a small boy allowed to run wild, this was the perfect life, and if schoolroom lessons suffered from distractions, the time spent in wood and water was not wasted. It is easy to harden up out of doors, and, more important, acquire a sixth sense, not learnt in books, which might prove useful on more serious occasions.

Ten years later I was fishing at Morar on the West Coast when the gale of wind and rain swept the Beauly district in the middle of July. I returned to find the river running bank high, with bridges washed away and trees down all over the estate; among them was the friendly old beech beside the burn. The bank had given way, and, like a full-rigged ship with leaves for sail, it had crashed at an angle across the stream, trunk and branches forming a coffer dam between the banks. The pool above, perhaps thirty feet long and nine feet deep, was still very dirty; a drowned sheep and much debris twirled slowly in the eddies, but I put up a rod, caught three fresh-run sea trout with a worm and got snarled up with a fourth which appeared to be a heavier fish.

Highland spates go down as quickly as they rise. Next day the burn was clear and almost back to summer level. Leafy branches previously submerged now poked out – the spars and halliards of the capsized galleon – above the waterline. The sun was bright in a blue sky, and all the pigeons in the wood were cooing after the rain, but in the pool itself there was no sign of life. I fished it down carefully, first with a Peter Ross and then a Zulu – not a move; the sea trout were either back in the main river or stuffed with food and waiting to do some more bottom feeding after dark. It was curiosity, or maybe a stirred memory of the water-baby days, that made me strip and run along the slippery trunk of the beech on to a limb that lay at right angles and flush with the water over the centre of the pool. At first I could see nothing in a labyrinth of shadows; waving green streamers of imprisoned leaves moved gently in the undertow of the amber water; then gradually everything swam into perspective; in the dim twilight far below little pieces of mica in the gravel made pinpoints of reflected light. A fathom deep a brown trout hung like a jewel feathering with golden fins in a patch of sunlight. Out of a black shadow a thickset fish cruised into the sunlight – it was a sea trout, and the brownie bolted. Then a cloud passed over, casting the pool into darkness – it gave an opportunity to change position and wriggle out on to another cross-tree.

When the sun came flooding back through the timber I nearly fell off the perch for, lying straight below, looking enormous on a rib of sand, was the broad back of a salmon: but wait a minute – there was something wrong here – those black spots, humped shoulders and thick tail root – could it be – yes, by all the powers it was – a sea trout, the father and mother of its kind! I slid back to land feeling a trifle awed by the close proximity of such a monster, like the boys in 'Horatius'.

> who unaware,
> Come to the mouth of the dark lair,
> Where growling low a fierce old bear
> Lies amidst bones and blood.

I had unwittingly discovered the hidy-hole of a fish, the like of which had not been seen outside a glass case! How big? That was hard to say, for weights, judged under water, are deceptive, but at a range of five feet the fish looked all of ten pounds.

The question was how to get him out. Big trout are greedy, and I spent the afternoon looking for frogs and rigging up a stiff topped spinning-rod and a wire trace. At ten o'clock the light started to fade: time to go; it would be dark under the trees in the Beechwood – the evening was warm and very still, so still that the snap of a night-jar hawking moths in the Lime Walk and the thump of rabbits gone to ground made disquieting sounds. Far away on Farley Braes a man was trying out a set of pipes. But in the wood itself the silence was complete and crushing. How the frog wriggled in the dark, but it had to be done, and 'loving him like a brother', as Izaak Walton would have it, he was placed on a treble hook and, keeping well upstream, I plopped the kicking offering into the run at the head of the pool, paying away line into the hurrying darkness. There was an immediate response: I felt rather than saw the boil and struck back hard, so hard that a half-pound trout was jerked high over my head on to the shingle. Judging distance at night is curiously uncertain: the next frog seemed to have reached the Beauly before the rod jarred in a tremendous pull. No mistaken identity this time! The big trout tore once up and down the pool in a smashing run, turned at my feet, and then drove downstream for the open river – there was a jolt and the rod point sprang back to perpendicular. I couldn't hold him and he had broken the heavy tackle in the submerged beech tree.

The pool may have given a rosy blush as I waded ashore cursing, midge-bitten and defeated. But Fortune is a fickle Jade and the trout was still there at first light next morning, swimming restlessly about the shrinking pool with a yard of glittering trace hanging from a lower jaw. Immediately action had to be taken for the fish at any moment might force a passage back to the main river; indeed, it was a surprise to find him there at all – but how was he to be caught? No rod, lure or triangle could hold him in those snags, and a gaff, a spear, or a carbide bomb seemed an unfair way to end his days. Gentle reader, if you are a member of the Stockbridge Angling Club, or a purist of high degree, do not read further! I cut a straight ash plant, four foot long, ran down to kennels for a copper snare, smoked the shine off the wire, and with some tarry twine fashioned an implement strong enough to hold an elephant.

The early sun was just finding the pool as I crept out along the beech trunk. The pleasant warmth must have affected the great trout, for he had ceased his nervous patrol and now lay doggo, head and fore-end hidden in protecting branches. It was an awkward angle and a long arm reach to get down to him: I was numb to the armpit in the cold water before the snare settled at last over the flukes of his tail, but further it would not go – the blind fin and the current caused the trouble, and I dared not strike for the base of a sea trout's tail, unlike a salmon, is too thick and rounded for a certain hold. The fish seemed to like the tickle of the noose for he stood on his head to make it more difficult. Then something went wrong; I think I misjudged the refraction of the water and prodded with the ash plant, for the trout bolted suddenly upstream. But he had seen nothing, and was back in half an hour, this time in perfect position, immediately below, on the same rib of sand where first discovered. As I drew the snare gently over his head the big trout turned sleepily on his side, and then shouldered forward to rub his tummy like a pig against the wire. I drew a deep breath, got on one knee, then shifting both hands down the ash plant, struck upwards with all my might. The fish came up in a flurry of white water: for a second I had him struggling all over the branch, then I lost my balance and we both fell into the pool. I was the first to recover. The big trout rolled on the bottom, the snare wrapped fast round his middle: he was sick, but he was free, and rapidly recovering. I am a rotten swimmer and cannot dive at all, but despair gives courage. I wallowed down into the depth, bumping through the roots and branches. There, feebly struggling to keep upright, rolled the glittering prize! He would certainly have got away but for the snare, which I seized with one hand and got to his gills with the other: then we shot to the surface and the fight was over.

The kitchen scale allowed 8 pounds 10 ounces. He remains a record for the burn: years later, a sea trout of over 10 pounds – perhaps a relative – was taken in the nets, stripping spawning salmon for the Hatchery, at the tail of Castle Hill, within shouting distance of the spot where this story is told.

A man is as old as his knees. The Moon had long since retired, and the gentleness that characterized his life brought him finally to the occupation dearest to his heart: still closer communion with God's creatures, great and small. No field naturalist in the accepted sense, he knew by instinct the hidden secrets of the Park, and in the shared silence of declining years, Nature did not find herself afraid.

When we were children, the Moon had taught us to identify birds the hard way: the snatch of unseen song, a floating feather on an empty pond, the swift silhouette in flight against the sky, the droppings in a hedge, or the lining of a last year's nest, which explained the species of the former occupant. Birds' names were all his own, and would defeat your modest ornithologist. They bore something of his old-world courtesy, and I would not have them forgotten. Thus thrush and blackbird became the 'mavis' and 'merle' of a former era. Wood pigeons to him were 'cushats'. The skylark was the 'lavrock', and all the titmouse tribe were 'bluebonnets'. Various buntings were known collectively as 'yellow yates'. Arrival dates of the summer migrants (for 'gowk' and 'whitrick' read cuckoo and sandpiper) were duly recorded. Black cap, wagtail and 'seapilot',* that called or sang along the poppling water beside the cot house, gave especial joy.

It is against that background that I remember Willie on a lazy afternoon in the Long Vac, when I went down to Brae Cottage almost for the last time. I found him in a rocking chair watching a busy pair of blue-tits with young in a nest box cleverly fashioned from the stringy bark of the Wellingtonia. I had John Buchan's *Life of Scott* under one arm, and read aloud a verse from 'The Last Minstrel', which includes the lines:

> When throstles sang in Hareheadshaw,
> And corn was green on Caterhaugh,

* Oyster catcher.

And flourish'd broad Blackandro's oak,
The aged Harper's soul awoke ...

Then would he sing achievements high,
And circumstance of chivalry,
Till the rapt traveller would stay,
Forgetful of the closing day;
And noble youths, the strain to hear,
Forsook the hunting of the deer;
And Yarrow, as he roll'd along,
Bore burden to the Minstrel's song.

To my chagrin Willie sat in silence and made no comment. But the words, with their gentle tribute, struck a chord in our own last minstrel. Here was Sir Walter at his best, and the old poet's eyes lit up. He always spoke in a low voice when his interest was aroused. 'That's a grand bit about the deer,' he whispered.

Maurice Baring was an admirer of 'this silent man in life's affairs', and made a point of visiting the Moon each summer for news of the bees and other important matters. 'The less print, education and press – there you find the greatest culture,' he remarked, after a happy discussion with the old man on the merits of Queen Victoria. 'People supposedly educated will always read what they want to read and only listen to what they wish to hear. Some who have no learning develop vision on things they know nothing about.'

Speaking of Willie's Canadian sagas, Maurice commented, 'The essence of poetry is emotion remembered in tranquillity; the essence of journalism is sensation captured on the wing.'

Willie's death, before the Second World War, followed a bad fire in the east wing at Beaufort, which destroyed the picture gallery, the ballroom, and the library, with all their treasures; also the billiard room, in which sporting trophies hung upon the walls. Jack Fraser, one of the Bard's sons (later to serve in Lovat Scouts), distinguished himself that night in the rescue operations, but not quite in the way that my mother expected. Scorning the family portraits by Allan Ramsay, Raeburn, and Sargent, all shrivelling into flames, he dashed through the fire and smoke to the stuffed heads in the billiard room, furthest from the exit, seized a forty-eight-pound salmon (I think it stood next to my own sea trout), and returning through the blaze, triumphantly presented my mother with the hard-won prize.

In 1939 Maurice Baring, living at Rottingdean, was already in an advanced state of paralysis agitans, though able to sit in his garden on sunny days where, from a wheeled chair, he picked the dead heads off the roses. By 1940 windows on the South Coast were continuously shaken by low-flying aircraft, and the sirens – if not gunfire – became intolerable to his nervous system. That summer my mother removed Maurice to Eilean Aigas, an island in the Beauly, where he remained for the duration of the war. There he wrote almost his last poem:

On the death of Willie the Moon,
who for thirty years
cleaned and lit the lamps
at Beaufort Castle

If you're wanting a job to be done well and soon
The Man who can do it is Willie the Moon;
He works all the day and he gives of his best,
But now he is surely in need of a rest;

It's no use a-calling; he's far far away,
He's trimming the lamps on the wide Milky Way,
All the pipers in Heaven will strike up a tune
Of welcome to Willie – dear Willie the Moon.

Maurice died at Beaufort on 14 December 1945, after years of suffering bravely borne. Always cheerful, his strength of mind and self-discipline was an inspiration to a younger generation in uniform. My mother wrote a moving tribute to this great European, entitled *Maurice Baring – A Postscript*, which described him perfectly. But it might be appropriate to mention that Jack Fraser, invalided out of the Scouts, and the 'new Moon' as of right, helped to lift and feed the dying man in his extremity.

CHAPTER TWO

THE LOST GENERATION

Here dead lie we because we did not choose
To live and shame the land from which we sprung.
Life, to be sure, is nothing much to lose:
But young men think it is, and we were young.

HOUSMAN

After the Scouts sailed for Gallipoli, we moved to Lyegrove, the home of my aunt (then Diana Wyndham; now the Dowager Countess of Westmorland). Her husband, Percy (Coldstream Guards), and John Manners had already been killed in confused fighting at Landrecies before the Battle of the Marne: the nearest point the Germans got to Paris and a 'close-run thing'. The memory of my generation grows dim, but there, weary and footsore, the Expeditionary Force turned at bay after the retreat from Mons. There followed a series of fluctuating battles – half-remembered by description, yet blurred like the flickering passage of a silent film. I learnt the monosyllabic names slowly – Mons, Marne, Aisne, Ypres, Loos and the Somme – helped by coloured pins which my mother stuck inaccurately into the map provided by Mr Asquith. Later, in Flanders, the pronunciation became more difficult.

By autumn the Expeditionary Force, under Sir John French, in two corps commanded respectively by Haig and Smith Dorrien, covered by Allenby's cavalry division, were augmented into three Army corps with the appearance of the Seventh Division. The character of the war changed on their arrival. Antwerp had fallen, but now, after the race to the sea which saved the Channel ports, the opposing forces turned inwards, collided and then locked horns like straining bulls. The war of movement was over and the complicated trench system built up which was eventually to stretch from Switzerland to the Channel Coast and so remained for the duration of the conflict – a slogging match maintained at dreadful cost to either side.

The foretaste of destruction (by howitzer battery fire) came in October during the First Battle of Ypres: here the small British professional army sacrificed itself as the shield and bulwark of the nation. At Ypres the Seventh Division, considered the finest in the army (it included a Guards Brigade) was virtually destroyed. The men were literally blasted out of their trenches. The battle lasted three weeks and for part of that time the new division fought unsupported and alone: the Channel ports lay open to the rear. There was massive slaughter but no result. Our infantry regiments lost 346 out of the 400 officers in front-line trenches. There, Uncle Hugh Fraser, acting Second-in-Command of a Scots Guards Battalion, was killed leading a bayonet charge in a counter-attack on the enemy. His grave was never found, but his name is on the Menin Gate and his memory is not forgotten.

My mother kept a scrapbook. It makes sad reading, consisting of obituaries, personal letters, poems and tributes to friends who had fallen. Aged five, I liked what Sir Francis Lindley* wrote about my sporting uncle, especially the part which concerned an elephant:

The early promise Hugh Fraser had given on the shinty field as a natural leader had not belied him in later life: dressing his own salmon flies, his hands on rifle, rod or gun were as dexterous as any woman's. Neither smoking nor drinking alcohol he was impervious to fatigue or cold and could spend all day on the hills – hind shooting in winter, clad in an old suit and cotton shirt, without waistcoat or vest to keep out the wind and snow.

Like so many men of formidable appearance (and especially his own uncle Henry Fraser, who had commanded the regiment in the Crimea), he had the kindest of hearts.

A much-travelled big game hunter, who had shot over two continents (as witnessed by the heads of kudu, sable, ibex and markhor in the gallery at Beaufort), Hugh had shown his strength and presence of mind one day in Delhi when the Viceroy narrowly escaped death at the hands of an assassin.

Captain Fraser was one of Lord Hardinge's ADCs, riding behind him in the procession when the bomb was thrown. Jumping off his horse, he found the vice-regal elephant so terrified by the explosion that it would not kneel to allow anyone onto its back to get the wounded man down. Tearing off his red tunic, Hugh picked up a heap of packing cases gathered from a neighbouring bazaar and, mounting the precarious ladder, lifted the Viceroy, who had fainted through loss of blood, as easily as if he were a child and carried him down to safety.

Many Scotsmen, friends and relations, fell during that first year of war. James Seafield and many neighbours died at Ypres. Ian

---

* Sir Francis Lindley, GCMB, CB, CBE. A favourite uncle who married my father's sister and in a distinguished career was Ambassador to Norway, Portugal and Japan.

Maxwell (my first cousin) was severely wounded. Here the Frasers of Ardochy both lost their lives. The London Scottish were the first Territorials to go overseas, a fine regiment with many public school men serving in the ranks, and their pipers played them over the top at Messines in a charge that cost three hundred casualties. The Germans advanced against them in massed columns as darkness settled on the field, with bands playing 'Deutschland über alles'. After the attempted breakthrough the hodden grey kilts of half the young bankers in the home counties lay strewn over the battlefield. The kilted regiments ('Die damen aus der helle': 'the ladies from hell') were not favoured by the enemy: for after fixing bayonets they took few prisoners.

As children, we stepped uncomprehendingly from daylight into darkness. The slaughter mounted, there seemed no escape from the blight that spread from home to home. The bereavements came early or late, as Britain's strength drained steadily away. Families grieved in different ways according to feelings and circumstance. The early pride was followed by despair. Perhaps the first blows seemed easier to bear when war was an act of patriotism and victory just round the corner – but I do not think so now.

Our personal losses in France, and in the less brutal theatre of the Dardanelles, were sustained before disillusionment set in. My father wrote from Gallipoli on 30 October 1915:

We had a most successful landing without losing a man: a hurry and rush to get underground before daylight. We had six casualties on the second day and seven yesterday. Everyone has worked well but we are short of entrenching tools. We are situated in a great amphitheatre. We occupy the stalls: the Turks sit in the gallery and boxes, so you can imagine who sees most and has the better time. We are getting dug in deeper every day but when the wet weather comes there will be inevitable flooding and we must move forward for winter quarters onto higher ground.

No one can have any idea of the follies committed here. I have always had a poor opinion of the soldier politician but I can hardly believe some of the stories that are apparently authentic.

Water is all important but even now, after six months, there is no way of boiling it as there is no fuel. Chlorate of lime is necessary for sterilizing what one cannot boil and there is none. The flies are bad but there is no disinfectant to keep them from the latrines, etc. Wood and corrugated iron are essential for head cover from shrapnel. For Brigade HQ I have got seven planks and one sheet. I want 500 times that amount for the Brigade lines.

I hear YMCA buildings are to be put up in far away Mudros and so will be advertised at home, no doubt. There is no mention of looking after those here, who come out of the firing line.

So much for a grumble. I have told the Principal Medical Officer either to send in a strong report on the sanitary and medical arrangements or take the next boat to Imbros and report direct to Ian Hamilton. We have had glorious weather with sea bathing every day. The Turks rarely shell the beach viciously, just an occasional round to remind us they are there. They fight wonderfully fairly, allowing parties out to bring in the wounded and never shoot at hospital ships.

Everyone speaks of Charles and his great gallantry. His men adored him and they said he simply did not know the meaning of fear.

Percy and Hugh were gone: now Charles, my third uncle, was dead. Contrary to orders he left the immunity of the Foreign Office to serve with friends in the Hood Naval Division, bound for Gallipoli. Twice wounded, the second time severely, he returned unflinchingly to duty. It was on 15 August that he received his third and fatal injury, dying in a hospital ship on the way back to Alexandria. Charles was a Balliol scholar with a double first in Honour Moderations and Greats. Earlier he had written, 'I know I shall not die: that does not mean I will not be killed.'

All this happened before we left Scotland. The attack on the Dardanelles, the Narrows across which Leander once swam to visit Hero, had proved a costly failure: heads fell, including Winston Churchill's. The bungling ended when the Expeditionary Force withdrew in January 1916. The curtain rang down on a chain of mistakes unrivalled in British history. Two cardinal errors had been compounded – loss of surprise and reinforcement of failure. The heavy casualties came from dysentery as well as Turkish bullets.

Meanwhile in France and Flanders, the Germans threw the weight of their next offensive against the French at Verdun: fierce fighting in the spring at La Bassée was followed by the Second Battle of Ypres, where poison gas was used for the first time. In May I lost my godfather, Julian Grenfell. He died of wounds after shrapnel hit him at Hooge. Billy, his brother, who had stood proxy at my christening, lost his life three weeks later attacking over the same ground. As a girl my mother had been in love with both these god-like young men. Treasured letters went back to Pop, the 4th of June, Eton and Balliol days with Uncle Charles. Also yellowing photographs which showed the boys bathing at Gisburne in the Ribble with Laura, Diana and Rosemary Leveson Gore (later Lady Ednam), three beautiful girls with their hair down and, for the period, surprisingly few clothes on. I remember Billy, who carried me piggy-back, was the gentler character: he boxed light heavyweight, ran the quarter-mile and played tennis for Oxford University and was a better scholar than Julian, the ferocious boxer, who challenged all comers in uniform, professional or amateur, and while he lived was said to have enjoyed the war. As a poet my godfather is remembered for a tender love of nature. Describing his last poem, 'Into Battle', published in *The Times* the week of his death, Maurice Baring wrote, 'It sums up all that is best of England and English things – like a Constable or a speech from Shakespeare.' Julian left his long dogs to my mother, but there was trouble getting them back to England. Catch and Hold were so nimble they could jump onto a side-saddle without loss of balance when Mama exercised the old hunters. But Toby, the greyhound in a favourite poem, was dead before my day: in the mind's eye I see the black dog galloping and hear his wild footbeat:

> Shining black in the shining light
> Inky black in the golden sun,
> Graceful as the swallow's flight
> Light as swallow, wingèd one.

Ronnie Knox, his Oxford contemporary, said of Julian, 'He had the Greek love of form but not for its own sake: strength, speed and skill were to him the assertion of the dignity of man.' To my parents' generation the early deaths of the Grenfell brothers exemplified the supreme sacrifice of the war, and the fate of those dearest to the family formed a background to the upheavals which changed our life at home.

'How beautiful upon the mountains are the feet of him who brings good news and heralds peace.' Left to our own resources, the three years of childhood's upstair-downstair existence (mostly down) in Gloucestershire were a lonely period. My parents were in France and the winters seemed interminable. To make a start I turn to England in the spring of 1916. It is of my maternal grandfather, Lord Ribblesdale, that I now wish to write.

From Lyegrove we trapped over Easton Gray, the home of great-aunt Lucy Graham Smith (née Tennant), a friendly house above the slow-flowing Avon, where Ribblesdale had sought refuge after Uncle Charles's death. Mother did her best when available: as the closest daughter she took charge of a broken heart. But there was no escape from the brooding sorrow of the man we came to visit – hard driven in the dogcart.

My grandfather must have been a remarkable person. I was too young to assess the qualities of this *fin de race* nobleman, who sat silent in a chair with legs thrust out so far they seemed to trip my sister's attempts to offer homage with cowslip posies. There in the shades, inside or out, he stayed all day, after we were led forward to receive a fleeting smile: immobilized without coming in for meals. In the stone fountain there were goldfish, and crested newts with orange stomachs dotted with black spots which kept me busy with a jam jar. These, too, were presented to grandfather.

To the modern scribe Ribblesdale would appear a colourful figure – rather larger than life, perhaps the last of the dandies – all terms he would have cordially detested, being fastidious in his use of the English language. With all his talents or despite them, life for 'the Ancestor', as he was known in society, had not been easy. His father, a compulsive gambler, had retired after the loss of great possessions (like Beau Brummell) to live in Continental obscurity, in his case at Fontainebleau, now a suburb of Paris. There had been barely sufficient funds to send the small son to Harrow. During school years, he spent the holidays alone with distant relatives, notably the statesman Lord John Russell (Finality Jack), at whose table he supped with Charles Dickens. But if the Tommy Lister boy of frenchified appearance arrived in England without cash or acquaintances, he was soon to overhaul his contemporaries. And nature, having set her hand to the task, did not look back but drove a straight furrow till the work was done. Whereas eminent Victorians were inclined to corpulence and measured their inches in girth rather than by stature, Lister grew tall and slim. After Harrow (more studies with Charles Kingsley and a military crammer), came the army, a commission in the Rifle Brigade, and service overseas.

Contemporaries agree that the period of French exile had given a grace and style to the young man that went deeper than the elegant pose later painted by Sargent, who portrayed a figure which conjures up the Regency and a more distant past. The portrait, presented to the nation, now hangs in the Tate Gallery as a memorial to his lost sons. Rosa Lewis (who knew the leading figures in Edwardian society while cooking dinners at big houses) insisted that the Ancestor resembled a thoroughbred among cart-horses. For in his patrician looks lay the essence of nobility.

For thirty years he was extremely poor. Then the improvident father died by his own hand, and for a time the sun shone. After retiring from the army, it is not surprising to find Ribblesdale married to a clever, rich and beautiful wife who brought out his talents in London society. Considered a fair artist and a wit in two languages, he became the Liberal Whip in the Upper House and a favourite in Queen Victoria's household. His duties in the Lords (1896–1907) were less arduous than they are today, for we find him sleeping under the stars in South Africa and trekking in a flying column with his son Tommy's regiment (the 10th Hussars) in the last stages of the Boer War, dressed in an old covert coat and armed with an umbrella.

Appointed Master of the Queen's Buck Hounds, he was not only admired as a supreme horseman but credited for bringing the rudiments of *haute école* and dressage to England from the Continent. Edward VII, once known as Prince Tum Tum, first nicknamed him 'the Ancestor'. My grandfather loved the saddle best of all. (At the end of this book is an extract from Ribblesdale's hunting diaries which should appeal to countrymen who enjoy a day with horse and hound. Written in his attractive style, it describes a scene in the forest of Compiègnes near Paris. See Appendix I: 'La Chasse au Cerf'.)

I shall pass over his Parliamentary duties. The political opinions that divided the ruling classes into Whigs and Tories were strong, but their opposing views were respected by both parties. My grandfather's links with the past went back to such towering figures as Disraeli and Gladstone, both of whom are described in his book *Impressions and Memories*. But I prefer to touch on the lighter task which went with his court appointment. Lady Wilson (my mother's sister) left a glimpse of her father in this other England*:

Those who saw him lead up the course during the Royal Ascot meeting have never forgotten it. He rode a bright chestnut horse called 'Curious', a conspicuous one, both for its colour and fine action. In his dark green coat, wearing the embroidered belt from which hung the gold hound couples dating back from the time of Queen Anne – with the June sky above and the emerald ribbon of the race course stretching out into the distance – he was a figure to delight the eye.

To see him ride thus in the perfection of his horsemanship at the head of the Royal Procession – one felt that here was a worthy successor to that long line of Masters who, from the time of Henry II, had held that office. Members of the Brocas family – of Gascon origin – who for three hundred years were hereditary Masters – had they seen him that day – and who knows that they did not – would have rejoiced in his gallant bearing and romantic person.

Ribblesdale's association with Ascot has not been forgotten. He has a prestige race named after him at the Royal meeting.

Then fortunes changed. A happy family life was coming to an end. The world was changing too. Sorrows now heaped upon his head. In the space of a few short years Ribblesdale lost his wife, his mother and both his sons.† The cumulative effect was too much for his nervous system, but the last blow was the worst. The Ancestor never got over the death of Charles, whose career had shown such brilliant promise. He was too old to readjust himself. Prostrated by grief, a steady decline

---

* Barbara Wilson, *Dear Youth* (London, Macmillan, 1937).

† Captain Tommy Lister, who won the DSO as a subaltern in South Africa, was killed in Somaliland in 1909 at the Battle of Jidballi.

began, described by the doctors as melancholia. He never rode again, nor did he attend another meet of foxhounds. It was in this sorry state that I remember him.

Silent stable yards epitomized the sorrows at Lyegrove and Easton Gray. All the good young ones – men and mounts – were gone. That winter, 1915–16, my mother rode what remained of the old blemished hunters, the faithful friends never to be sold; going like a bird, she hunted with the Beaufort pack from both sides of the country – getting home after dark, for there was no transport. The old Duke said she went best over the big places perched high on the men's horses. She rode no weight, and though hair came down and reins were held too loose, she knew no fear.

Married at eighteen from No. 10 Downing Street (then occupied by her aunt Margot Asquith), my mother was generally considered a delicate woman, but in the saddle the light touch and easy seat taught by the Ancestor tired neither horse nor rider. She was barely twenty-four, but it was the last season for them all. In July the guns started to thunder on the Somme. After their summer's rest (turned out to grass), all but one of the Ribblesdale and Wyndham hunters were put down.

Father and daughter had much in common: they were Europeans in the truest sense. She shared Ribblesdale's love of France; she also inherited the classic looks, slim figure and long tapering hands. But more important, my mother remained light and cheerful in adversity, which held all the family together, something that exemplified Chesterton's opinion of gentleness which can be a mark of strength, not weakness, in spiritual as in bodily things.

I never knew my grandmother, Charlotte ('Charty') Tennant. She died of consumption in 1911, the year I was born. By all accounts, she was a woman of spirit. Before the debate on Irish Home Rule, Ribblesdale, as Liberal Whip, was preparing to face the big guns carried by Lord Salisbury, when he received a telegram from the Glen:* 'Hit him hard – hit him often and always below the belt.' The Tennant sisters were ahead of their times.

Great-Aunt Margot, as a girl, kept a skull in her bedroom facing the *prie-dieu* on which she knelt for prayers. She also sported a nightdress in the racing colours of an ardent admirer; but she 'put a wartime foot in it' saying that she could never hate the enemy. (Soon after the Asquith administration gave way to Lloyd George. There is little doubt the fiery Welshman intensified civilian commitment to the war

* Glen is the home of the Tennant family in Peeblesshire above Traquhair and close to Neidpath Castle, once a Fraser stronghold.

effort.) A photograph of the two sisters – Margot short, Charty tall – shows tight curls and determined Scots chins. Anthony ('Puffin') Asquith, the film producer, together with some of Charty's grand-children, have inherited this curly coat of hair.

Rosa Lewis, who ran the Cavendish Hotel, spoilt me as I grew up. She is known to posterity as the Duchess of Jermyn Street and has many biographers, so I will not add to tributes accorded to this remarkable woman. I first met her at the tender age of, I suppose, six weeks, at my christening ceremony. Her heyday was past when I came down from Oxford; but I got to know her serving as a young officer in London District. As well as the fine head carriage and stately bearing, I recall a scathing tongue and surprisingly shrewd judgement.

Rosa was treated with polite but somewhat distant formality by my mother and her sisters – though 'Burghie' Westmorland, a breezy sailor (later to become a member of the family), tried his best to break the ice. Rosa's down-to-earth approach may have jarred the Lister sensi-tivity, but no more; for she had proved their father's faithful friend, understanding his peculiarities well enough to make a home for him, after Charty's death, in the hotel, into which he moved with some of the best furniture from Gisburne, the family home.

Daphne Weymouth has written (and I believe it to be true): 'People may have wagged their tongues when Lord Ribblesdale moved into Mrs Lewis's hotel, and no doubt this removal gave rise to a myth which persisted until quite recently that not only did he give her the hotel in the first place but was also her lover. Certainly there was a very close bond of friendship between them but there is no evidence that this ever developed into a love affair.' Rosa personally ironed his daily copy of *The Times*, cooked light meals when he was out of sorts and supervised all the laundering of the Ancestor's fine linen. The family (and Rosa) positively disliked the rich American Ava Astor (née Willing), a hard and worldly woman* who married Ribblesdale in 1920 when little was left except an empty shell. The couple had already drifted apart before his death (in 1926). One very good reason, according to Rosa, being the unfortunate New World habit of smoking in the dining room during meals. I paid my respects once in New York to find that Mrs Astor (whose former husband had gone down in the *Titanic*) played bridge all day, surrounded by small yapping dogs that got no exercise. Rosa made much play on her maiden name 'Anxious and Willing'. She died 'desperately poor' aged eighty-six – leaving a large fortune.

* Miss Willing, known in her youth as 'The Baltimore Belle', lived in Grosvenor Square until the Second World War began. Renouncing the Ribblesdale title in 1940, she returned to the United States as Ava Astor.

One post-war spring, Ribblesdale bestirred himself and took us, with Diana (whose second husband, Boy Capel, had died in a car crash) to Paris for a look at family graves. We also inspected the carriage at Compiègnes where the Armistice was signed. I best remember an all-pervading smell of garlic and the fierce appearance, with disjointed shambling gait, of French colonial troops (from Senegal I think); I had not seen such white teeth in such black faces before. The uniforms of Zouaves and Spahis, and the little grey barbs on which they rode, made a brave show clattering down the Champs Elysées.

My mother shared Ribblesdale's love of France. She was bilingual and he liked her style. We visited a hat shop in the Place Vendôme; as a favour mother was allowed to choose one of the more outrageous creations the vendeuse brought forward. (Diana, in mourning, took no part.) That hat parade was the only time I ever saw the old man smile, while the feathers were being tied on. On this visit Ribblesdale's own appearance turned heads in the street. He sported a hard hat with high flat crown – half-way to a topper – and a coaching coat with fur lining and frogged buttonholes that reached to his ankles. This garment was mantled by a double astrakhan collar and weighed a ton. I still wear it, cut to size, in cold winter weather at funerals. The Ancestor also left me his riding whips and serviceable bootjack. He approved of fine linen, chosen for durable texture. Some of the double sheets at home bear his embroidered monogram: the date of origin, stitched below the capital R, goes back to 1896.

Again I pick up events. The year of the Somme started badly: there was a short respite when father returned from Gallipoli (wasted with dysentery and pale after a long convalescence in Malta) to take command of a cyclist division in East Anglia. We moved for a short spell to a house in Norfolk. He was constantly on the move, and we hardly saw him before he left for France. My mother and Diana followed to nurse as VADs in the Sutherland Hospital at Ostend. It was another lonely winter, spent below stairs with old servants: they had their own private sorrows. We did not see our parents for six months.

I learnt to read without attempting to spell. Words came easily when interest was aroused. Enforced solitude has the advantage of heightening perception and provided me with a retentive memory.

Lyegrove's garden was small, and here again the constricted premises helped application. With bird-nesting sites exhausted, there were opportunities to memorize flowers, shrubs and the vegetation on the down beyond – but more especially, the butterflies and other insects that flitted among the blooms. Caged pets had little appeal, and I soon tired of guinea pigs, rabbits and even white mice, who at least proved themselves Bohemians – always trying to escape. Rabbits and Belgian hares were disappointments, seemingly unaware of well-ordained marriage laws, and, like barmaids of ill repute, inclined on occasion to swallow their young.

Before he was killed, Uncle Charles had presented the 'Peruvian Cavies', but the high-sounding nomenclature was the best thing about them. From personal experience, Charles warned the creatures would show little affection – indeed, exhibit the worst characteristics of sloth, greed and cowardice at every opportunity. This rather put me off guinea pigs; they proved difficult to carry around, and the strong smell added to other shortcomings. Though the Cavies chirruped engagingly and grew hairy coats that covered beady eyes, I was not sorry when they disappeared after being dumped for a summer holiday in the walled garden. The place was fox-proof, but a stoat must have got at them.

When my parents returned for a short leave we were strangers to each other. Father was noticeably a general. When he spoke his voice said at once he expected to be obeyed. He got a surprise when he saw his children: one glance at my appearance was enough. First, I was stripped to the skin (which I found embarrassing), and then examined, like a horse, for physical imperfections. Various layers of superfluous clothing (I remember velvet pantaloons and mother-of-pearl buttons) were at once destroyed or thrown away. My mother possibly shed a tear when the curls, hanging in ringlets, were shorn off and dropped in the wastepaper basket. Like Samson, bereft of strength, but in appearance more like a plucked chicken, I shivered on the carpet. Considered too skinny after inspection, a supplementary diet of raw meat juice strained through muslin and sweetened if necessary with brown sugar, was prescribed, to be gulped every morning before breakfast. This gory mixture took a lot of swallowing; the treatment was discontinued when rations were reduced and beef supplies ran out. (Lambs' tails and other unmentionables, then a local delicacy, were eaten as an alternative to spring onions for high tea that summer.) I later heard this potion traced to father's early days elephant hunting in Ethiopia and the southern Sudan. In the former country, he had been entertained by the Emperor Menelek to feasts of raw camel flesh. This revived him after fever contracted on the papyrus swamps and malarial regions of the Sud.

The best news came at the end: I was told to run barefoot in shorts for the summer, with permission to explore beyond the grounds – in fact, I could go anywhere I wished, if I kept clear of roads and got home by tea. From that moment a new life began, and with it an independence which made it difficult to comply with indoor routine and certain other provisos, for my mother imported a Welsh girl, Rolanda Hirst, to look after Magdalen and teach some rudimentary lessons.

I am grateful to father for the meat juice and bare feet. In retrospect, he might have steered the transformation from the Fauntleroy image to the grimy urchin who never surrendered the new-born freedom. But there was a war on: the grown-ups returned to France and I ran as far as my reinforced legs would carry me.

If there were children of comparable age in the neighbourhood they kept their distance for I disliked toys and party manners. There was no common ground for entertainment and, like a Red Indian boy, I was happy in silent company: not shy, on the contrary curious and possibly demanding. Like the Celt of old who waited at the crossroads, compelling the passer-by to tell of something new, I learnt the ways of Gloucestershire. Beyond the boundary wall, a crippled earth-stopper of the Beaufort Hunt, who spent the summer trapping moles, became my closest friend. Another powerful ally was the pigman at the nearest farm, tenanted by Mr Dan Iles, where Spotted Gloucesters (today an uncommon variety of porker) were raised in large numbers. The old stockman had a good turn of phrase: 'Dogs look up to you, cats look down on you, but pigs is equal.' The local rustics were good people who gladly taught a boy their ways to work the land. There are fine farms around Badminton, which, in those days, raised pigs as a sideline. Where there are pigs there are rats in wartime, and rats meant Jack Russell terriers when the wheat stacks were threshed on the surrounding lands.

After the hunting season was over (the Beaufort hounds still kill a May fox) ratting provided a social occasion. The Old Duke* put in an appearance at these events for a chat with his tenants and to see the young terriers blooded. A great man in every sense of the word, he weighed nineteen stone without a saddle and carried the horn for forty-seven seasons, hunting five days a week with his own hounds. He had suffered a recent accident. While trying to break two hunters into harness, they had misbehaved, broken a shaft and tipped the carriage over. He fell out and the wheels passed over his body breaking both his legs. But that did not stop the old man, who followed hounds in a dog-cart, with a small boy handy to jump out and open the field gates. It was said he knew better than the fox which way to run, and was there, or thereabouts, at the end of the fastest hunt. The Duke of Beaufort lived for Gloucestershire and the land. His simple tastes and unassuming

* The ninth Duke of Beaufort: born 1847, died 1930.

manner made him a king in his own country, where blood sports were the order of the day. The few pheasants around were virtually raised for the benefit of the foxes; the closed season dragged interminably.

So the ratting of long ago was something of an event. Farmers and their wives trapped over from far and wide in Sunday best to see the fun. The men wore brown bowler hats, breeches and leggings; most of them carried a terrier. Cheese and draught beer were dispensed by the tenant, while the womenfolk retired to the parlour to partake of refreshments such as cowslip wine or maybe something stronger.

The farm organization was impressive. The last rick, high among the straw detritus, stood in isolation (work on the rest had been completed that morning). It towered intact, surrounded with a ring of rabbit netting. It was the final refuge of the creatures we had come to destroy. All the rats sought shelter there after their homes had been fed into the thrashing mill. The stack literally bulged with vermin. Lambs bleated in the nearby fields. In contrast, the assembly spoke in low voices. The clean smell of bunched straw was in the air; farm servants, string tied below the knees of corduroy trousers, stood around, leaning on pitchforks. The scene was set for the Duke's arrival.

The clock struck the hour as His Grace arrived, all hale and hearty, to be welcomed with a respectful greeting. After shaking hands with his tenants, the fine old man (his jolly face was framed in bushy mutton-chop whiskers) settled his bulk into the strongest chair the farm could provide, then made a signal to begin. What followed may not suit those inclined to stray from the realities of country life. For a boy it was all 'Merrie England and worth a guinea a minute'.

The mill started up: the farmhands tossed the sheaves; half-way down the rats began to bolt. Working in pairs, the old terriers were put in. They made short work of anything that ran to the netting; many rats would not face the open, and burrowed downwards into broken straw. These were picked up by horny hands and dropped into sacks. The last sheaves provided a flush of all ages, and mice as well, that ran in every direction. Now the boys had a go, and I joined the farm lads to whack away at the largest hump-backed scaly veterans –

> Grave old plodders, gay young friskers,
> cocking tails and pricking whiskers.

– before they got to ground in the tile drains which ventilated the rick. The escapers, nose to tail, were picked up – tile and all – and poured into more sacks. This ended Phase One.

The excitement must have lasted half an hour. Then the spectators moved over to a handy paddock and the leather gaiters formed a tight ring, into which the bagged rodents were emptied a few at a time for the entry of young terriers to try their speed and courage.

There may have been side bets on the best dog, but no hollering, horn-blowing or baiting of dumb animals. One shake was enough. It was a trial done for a purpose, to find the best working dogs; the right ones to breed from. Spectators did not escape attention, for spare rats ran up their breeches, and I noticed those not holding terriers kept hands thrust deep in pockets. My bare legs, curiously enough, were avoided, as not presenting suitable climbing material.

The Duke of Beaufort sat it out, quietly and unconcerned in the middle, an old dog firmly gripped and quivering on his knees.

It may not have been good farming. The rat tally on threshing days was huge – not an advertisement for pest control or good husbandry. But able-bodied countrymen had long since been called up by the county regiments.

In 1917 the United States entered the war to general rejoicing.

> Over there, over there
> Send a word: send a prayer
> Over there:
> Say the Yanks are coming
> They're drum, drum, drumming
> Over there.

On the other side of the Atlantic the euphoria which sweeps every country when young men march behind a band had curious side-effects. We heard that dachshunds were chased off the streets of New York, and that sauerkraut became another vegetable, 'the Liberty cabbage'. Homburg hats changed overnight into the modern Fedora.

The good news of Stars and Stripes was offset, however, by Italy's collapse on the Piave and at Caporetto, while Russian disintegration after Tannenberg released German armies from other fronts intended for a final drive in France and Flanders; but it failed.

That year Patrick Shaw Stewart fell. He was, with Raymond Asquith (an older man), among the most brilliant members of that Oxford generation whom I have attempted to describe. He had been another beau of Mama's. In his tunic pocket the Padre found a copy of *The Shropshire Lad* with his own poem pencilled in the margin:

> Stand in the trench Achilles
> Flame capped and shout for me.

Like Julian's 'Into Battle', the verses appeared soon afterwards in *The*

Times. My mother wrote from France they should be learned by heart. Rolanda saw that this was done. I complied reluctantly, being engrossed in the new hero, Lawrence of Arabia. He, more than any other wartime figure, had caught the public imagination. I could shout 'Akaba-Haj' with the best, charging on the garden pony and brandishing a bean pole.

The casualties brought some interesting visitors to Lyegrove. Convalescent officers came to stay on short leave after my mother returned from France in 1918. She was expecting a baby. Some soldiers become legendary; others are soon forgotten. I remember Tom Bridges, the 'Big Dragoon' of Newbolt's 'Song of the Great Retreat', who roused battalions of exhausted infantry (too tired to get on their feet at St Quentin) with a drum and penny whistle taken from a toy shop. His brother officers in the 4th Dragoon Guards were favourites with the children: Butcha Hornby, crippled for life, had the distinction of being the first British officer to kill a German with his own hand. Rival cavalry patrols had met unexpectedly: both charged each other; a German tried to drive his lance into Hornby, who then killed him with his sword. The man who drove him down from London (minus an eye and a hand, with wound stripes running up to his elbow) was Adrian Carton de Wiart, VC, the ex-Adjutant (peacetime) of the Royal Gloucestershire Hussars, a splendid figure who led his men over the top with a light walking-stick and a bag of hand grenades, pulling the pins out with his teeth. Carton was the happy warrior if ever I saw one, and kind to boys as well. I don't know if it was the same car, but he warned me that all combustion engines were hopelessly unreliable, and how, when soldiering overseas, driving at top speed in Cairo, he had frequently been passed by a more dependable camel. Oliver Lyttelton and Lionel Tennyson taught the rudiments of stump cricket.*

These men were natural leaders, and something rare, entirely devoid of self-importance. The contrast in thinking and conversation of civilians wearing what was called 'mufti' astonished me. The most notable individual was fond of birds. I earned my first half-sovereign (already a *pièce de musée*, having been withdrawn from circulation in 1914) by taking Edwin Montagu to inspect the nest of a nuthatch (*Sitta loesia*) in a dead tree on the drive. These birds were the jolliest inmates of the garden: the nuptial plumage of the cock bird made him the smartest native resident. Nuthatches were said locally to roost hanging upside down, like bats in a belfry, but I find this hard to believe. There was also an interesting migrant, unknown in the Highlands: the wry-

---

* Oliver Lyttelton, DSO, MC, Grenadier Guards, became Viscount Chandos. Lionel played for Hampshire and later captained England in test matches.

neck (*Jynx torquilla*). Montagu was a good ornithologist, and in his dour
way showed particular pleasure in helping to establish, from my description, the presence of this snake-like bird, which in south and west
England is called 'the cuckoo's mate'. After abandoning a stormy political career (he fell when Secretary of State for India), Montagu often
came to Beaufort to look at rare local species. On these occasions I acted
as guide, or, to be more truthful, escort into the wilder hills, where
stalkers had located the nests of rarities such as Dotterel, Greenshank,
Slavonian Grebe and Black Scoter Duck. Another early enthusiast who
came with him was Jim Vincent, then Head Keeper on Hickling Broad.
Both men, and Seton Gordon, were pioneers in bird-watching, egg protection and earliest photography of rare species. Edwin, an unhappy
man, would hardly speak throughout a long day's march. But the
sombre eyes fairly blazed with excitement the day we found Redpolls
(*Carduelis flammea*) nesting in the Orrin Valley. Hickling, where later I
stayed to flight ducks, has become a bird sanctuary, gifted to the nation
by Edwin and Lord Desborough. It once provided famous coot shoots,
considered essential to keep numbers down and make room for wildfowl
and winter migrants. The protection of 'mudhens' and seepage of
impurities has now largely spoiled the Broad.

By way of contrast, Henry Asquith, the former Prime Minister and
silent husband of our garrulous great-aunt, filled me with gloom.
'Reflect, young person,' he used to say, 'reflect before you speak.'

I looked forward to these visits from the outside world. There were
lively individuals to set the pace: Boy Capel – killed soon afterwards
in a car crash, Fred Cripps,* Alastair Leveson Gower, Maurice
Baring and Herbert Buckmaster were high-spirited officers who cheered
things up all round. They cajoled Rolanda, neat and pretty (she had
been kissed behind the garden wall by Buck, recently parted from
his wife, Gladys Cooper), to put on a floor show with Magdalen. They
sang 'Greensleeves' and 'Cherry Ripe' to the tinkle of a spinet and
applause for Rolanda from the 'walking wounded'. It was a sad day
when she was called up for national service, and returned to Wales,
leaving a legacy of sunshine. We sketched, worked out of doors, learned
painlessly to sing, dig the garden and press wild flowers that last
summer. I was well up in two subjects: geography and natural history.

---

* Colonel Frederick Cripps, DSO, Buckinghamshire Hussars, became Lord Parmoor and
remains the best of company, well past his ninetieth birthday. At the expense of his killjoy brother
Stafford, the 'Crypto Communist', he penned this verse:

> It was gay in the days of Maid Marion
> Ere we heard of the word Proletarian.
> But our land's in eclipse in the grip of old Cripps
> The teetotal totalitarian.

Rolanda was an imaginative girl, and Greek mythology also flourished in the classroom.

The last winter seemed the worst, not for bitter fighting, tightened belts and shipping losses – but rather the dread arrival of Sophie Buller, my mother's old Teuton governess. 'Zellie' was apparently famous, an intimate friend of 'The Souls'. She had taught the Lister and Tennant children and even given advice to Great-Aunt Margot, before going on to the Wilsons, where Peter, now head of Sotheby's, benefited from her teaching. At the time I wondered how she had escaped internment.

In *Dear Youth*, my Aunt Barbara gives an account of 'Gagool'* as she was dubbed by young Frasers on arrival at Badminton:

When she reached the schoolroom, she found two children waiting for her, a boy and a girl. She had a marked preference for boys, but this one, Shimi, was not the kind she was accustomed to. He had the lust of the hunter on him, he loved to snare and to kill. He slew with bow and arrow, with sling and stone, and with naked hand, ranging wide and solitary, watching the flight of birds, marking them down with keen eyes, as his prey.

Zellie was apprehensive – she spoke to Laura [my mother] of her fears tactfully: 'Do you think Laura, chérie, that he is entirely human? I do not want to upset you, but I saw a funny expression on his face when he was watching the sheep in the park the other day. Do you think it is possible that he is a little, just a tiny bit, of a wolf, or that he might turn into one?'

Nevertheless she loved him, this Mowgli boy, and his gentle sister, whose hair was cut like the Knave of Hearts, leaving a small window for her pretty face.

The appraisal is bad for business! Let's hope that time, the great healer, has mellowed the grizzly portrayal of my character! I deny the bloodlust charges and the offensive weapons referred to did little damage.

Zellie filled the schoolroom bookshelves with the collection of educational works used at Easton Gray and Gisburne. Among them was a red-bound cloth volume entitled *Near Home*, which the previous generation had enjoyed twice a week as a dessert dish in the geography lesson. It is on record that Ronald Knox, excellent friend of Zellie though he was, felt that for Catholic children the contents were not a happy choice. Ronnie opened the volume at a page where the author makes the following statement: 'Barren Iceland is better off than Sicily – and why? – because it is Protestant.' But 'Gagool', quite unabashed, swept the protest aside with, 'Ach, nonsense!' No one knew whether the contemptuous comment was directed against the author or the Catholic faith.

---

* The fearsome witch-doctor and generally hideous hag in *King Solomon's Mines*.

I was not concerned with religious perplexities, but Zellie and I never got on, for like certain old people, she had an unpleasant smell. Better education came from mother reading aloud while expecting my brother Hugh. She it was who taught me something of the depths and beauty of the English language. We started with the animal kingdom: *Black Beauty*, the *Jungle Books* (especial favourites), *Jock of the Bushveldt*, and, of course, Jack London's works, which exemplified the Moon's adventures. I remain uneducated in simple arithmetic: my spelling is Elizabethan, with a hand best described as cuneiform. The disadvantages were, to some extent, offset by independence and the physical stamina of an older lad. Being left so long alone had encouraged me to watch and listen to people, taking careful note of their conversation and appearance – even their smallest mannerisms – which remained photographed on the inner mind.

In January 1919 we returned to Scotland. Magdalen howled when she said goodbye to Nesta, the gardener's daughter, of curious name: there was another child appropriately called 'Lettuce'! The crippled mole-catcher presented me with a tanned stoat-skin, cured in a jam jar and soaked in alum to reduce the stink. A few eggs, some hornets nailed on a cork board, a fox's mask and a book of pressed flowers were treasures I took from Lyegrove. Perhaps a psychologist can read something into this bundle of belongings, all difficult to handle on the journey home. Who knows? Seen in retrospect, my childhood centred entirely on birds and beasts and an outdoor life. The war had lasted so long that sorrow had grown blunt by repetition and personal losses suffered by my parents became melancholy shadows.

We spent a few days in London on the way north. Aunt Barbara recalls another experience:

When the Master of Lovat was transferred to London he brought a flush to Zellie's cheek. Did he not succeed in tickling a fish in the Serpentine – unique exploit! – and having caught this odious *poisson*, prepare to carry it home in triumph to his mother's house? The stinking trophy fell from his sporran in Hyde Park and having discovered his loss, the kilted angler fell on his knees *en pleine rue* to pray to St Antony for its return.

I do not recall the incident. It must have been a very small fish because I had a very small sporran; at best a moribund roach. But I have not forgotten my delight in finding the lake and wide-open expanse of grass, where sheep, herded by a Scots shepherd, grazed unconcerned, surrounded by the rumble of traffic and labyrinths of bricks and mortar. Later I met a boy running wild in the park as I walked from the house in Upper Grosvenor Street to day school in Sloane Square. It was Peter

Scott – son of the Polar explorer, and future artist, Olympic Medallist and wild-life conservator. Though senior in years, Peter had been excused lessons. Wildfowl studies on the Serpentine were to provide their own reward; his mother was a sculptress. We became increasingly good friends.

More West End surprises of a sophisticated kind provided highlights in that first visit to London. It was time for another haircut: father took me to White's Club, where members' sons were handed over to the barber, stood on a chair and left to be collected, after warnings not to be a nuisance. Seated in the entrance hall, we encountered one of the more eccentric inhabitants. 'Wiggy' Welan supported his years on a Malacca cane; I think it had a horse's head handle, for in his day Wiggy had served in light cavalry. On his own pate, surmounting an artificial crop of hair, he wore a top hat, which never came off, either at meals or in the smoking room. The Club was proud of old Wiggy, a popular member who liked to air his antique views on horse-racing, having lost several fortunes on the turf. He had another weakness – vanity. By coincidence, the barber was combing out his spare toupée that very afternoon. He told me an extraordinary story. The old gentleman kept two wigs in reserve, each with a length of hair that varied according to the season. Thus, on a hot summer's day he might run a hand under his top hat, announcing that it was time for a haircut: then send for a page to make an appointment with the barber. On cue, Wiggy disappeared for a suitable interval, changed top-knots and re-emerged, dusting the nape of his neck with a silk handkerchief saying, 'Ah, that feels much better.' The barber said he earned more money ironing the members' silk hats (now a lost art) than by cutting hair.

White's has remained an uninhibited establishment where members say what they think, no holds barred. During the First World War, the club secretary frequently removed lists of London's better-known trench-dodgers pinned on the notice board under the title, 'storm troops for the next attack'.

After the Armistice there were gaps in the members' ranks. Not all the 'newly elect' met with general approval: they never do. Father told a story about one such individual who, trembling with rage, approached Lord Marcus Beresford, the doyen of the establishment, to complain that someone had been exceedingly rude in the card room. 'I need your instant advice: what should I do? It's an outrage and I want an apology.' 'Well,' asked Lord Marcus, 'what exactly did this fellow say to you?' 'He insulted me in front of other members – actually offered me a fiver to resign. Said I had hairy heels! Should I challenge him to a duel?' Lord Marcus thought this over carefully, then answered with a straight face, 'If I were you, I'd wait until the end of the month before leaving and by that time we might raise the offer to a tenner.'

At White's I discovered that my father (who seemed preoccupied at home) became a different person, casting cares aside, widely popular and sharing jokes among his friends with robust good humour. He was greeted by laughter on this occasion over his complaint in the Suggestion Book outside the dining room: 'Today I choked on the veal and ham pie. It was evil smelling, stale and tasted disgusting. I intend to bring the whole thing up in front of the Committee.'

Hall porters are characters: in good clubs the old servants are friends who look after members without a trace of servility. I cannot remember the name of the courteous gentleman who left his desk to talk to me as I stood on the barber's chair. Down the years I am grateful for some words of praise: it was the first time I ever heard of my father's military achievements and the decorations he had won. Strange as it may seem, coming from a soldier's son, I never saw 'The General' in uniform. He either managed to change into civilian clothes for the rare appearances at Lyegrove or got in so late that I was fast asleep.

CHAPTER THREE

AT PEACE

And the men that were boys when I was a boy
Shall sit and drink with me.

BELLOC

March snow lay deep in the Highlands when father took us back to London and a new house to start duties as Chairman of the Forestry Commission. The Spanish 'flu epidemic, which killed off so many people on the Continent, had reached England and was causing havoc in the cities, whose resistance had worn thin from privation in the war. But in London the mood remained triumphant. Rejoicing reached its peak on Saturday 22 March 1919, when in fine sunny weather the Guards Division and Household Cavalry marched or trotted through the streets of the capital on returning from Germany.

The Victory Parade meant an early start for troops taking part. The unlucky ones were concentrated without transportation in assembly areas at far-away Richmond Park. They returned the same evening without breaking ranks or calling a halt on a route which wound round Greater London and took eight hours to complete. Various relations, who had fought through France and trudged across Germany in the army of occupation, said this final fling was the most gruelling experience of their service careers. Jack Encombe (later the Earl of Eldon), who married Magdalen, was the ensign who carried the colours with Hugh Kindersley, in the Second Battalion, Scots Guards. He had been badly gassed in the Second Battle of Ypres and said he remembered nothing after the sound of tramping feet changed when his boots hit the tramlines on Hammersmith Broadway; but the pipes and drums kept him going to the end. Also on parade were the twin sources of the battalion's milk supply: two cows, Bertha and Bella, taken captive during the Battle of the Somme three years before. They marched round Buckingham Palace and down the Mall before slipping back to quarters in Birdcage Walk. Herbert Buckmaster, who rode with the

Blues, claimed that his horse gave up on the last lap as his squadron recrossed the Thames to Richmond.

The family got a grandstand view from a balcony in Devonshire House in Piccadilly – opposite the Ritz, where the Chrysler car showroom stands today. London always takes a personal pride in the pomp and circumstance that goes with scarlet tunics and bearskin caps, 'when the guards are on parade'. This time the regiments wore service dress and marched at ease. There was no pageantry: it was not needed, for the battalions were welcomed with a roar of continuous cheering that carried with it all the surge and thunder of the *Odyssey*. Regimental bands, playing the tunes that won the war, lifted the tumultuous crowds – and the swinging columns with their splendid bearing – up Piccadilly. I had never heard such sound: it left me breathless and seemed to shake the air. The people cheered as much in memory of those who had not come home as for the men who had survived and lived up to the traditions of their regiments. Old soldiers in the crowd, baring heads as the colours went by, saw again the days of their own service: the stand-to at first light; the counter battery fire; fierce shelling; holding on in desperation; the rumour of a break-through somewhere on the crumbling front; then word running like wildfire down the trench: 'The Guards are moving into the line.'

There was a special cheer for the Welsh Guards who had done so well in their first baptism of fire. Likewise the Irish men, whose loyalty after the Easter Rising in Dublin in 1916 drew even more rapturous applause:

> We're not so old in the army list
> But we're not so young at our trade,
> For we had the honour at Fontenoy
> Of meeting the Guards Brigade.

Both Irish and Welsh had won fame in plenty to join the three older formations – Grenadiers, Coldstream and Scots: *Tria juncta in uno*, grenade, star and thistle. There was a saluting base up the street beside what is now the In and Out Club. As the Micks went by, marching to attention, the regimental band struck up 'Saint Patrick's Day'. The jaunty tune delighted me, but, looking round, I was astonished to see tears running down grown-up cheeks on every side. I was in Lady Kenmare's party. They had lost their eldest son in France. *Lacrimae rerum* – a child is lucky not to understand the tragedy of war.

But now I turn to happier memories. Before public school, father used to take me to Brighton for Easter and Whitsun Bank Holiday weekends.

Two other boys, Tony Wilson and Rufus Clarke, with respective parents, joined an all-age bachelor party (sometimes twenty strong) in the Royal York Hotel, owned by Sir Harry Preston, the ex-bantam-weight champion of England, knighted for his services to boxing and charitable works. To a country lad those early outings provided an eye-opener in patterns of behaviour which just survived the war. The personalities in that widely differing group of interesting people, colourful in their own right and some of them famous, left an impression which may well have influenced my outlook on life.

It began after the Victory Parade, when Buck (Herbert Buckmaster) asked mother to help him find the furniture for the new Club which still bears his name. Brighton is full of good antique shops; on this occasion, after a rapid survey, mama declared that better things could be bought in Bath, and on that pretext illogically disappeared, with Rosie Ednam and Venetia Montagu, on a shopping spree to Paris for the spring collections. It was still the age when no well-dressed woman (if she could afford it) bought off the peg, and Coco Chanel was a friend of the family. In fairness, my mother did buy Buck one picture, a conversation piece called 'Prospero, Ferdinand and Miranda', which used to hang on the wall of the card room. It was rechristened 'Charlie Cochran introducing one of the Dolly sisters to Joe Beckett'.*

In Brighton I first saw hard drinking. 'Claret,' wrote Samuel Johnson, 'is the liquor for boys: port is for men; but he who aspires to be a hero must drink brandy.' Doctor Johnson seems to have missed out champagne, always in great demand among the 'Royal Yorkers', who celebrated the holiday in festive mood. Maurice Baring balanced a glass of port on his bald head after meals in the private dining room provided for weekends. Harry Preston, equally short of hair, tried in vain to compete. One evening he turned to George Webster, the cartoonist, and said – with a trace of envy – 'How does he do it? Just look at that great brain relaxing!' The Services, past and present, predominated, but civilians and men of letters contributed greatly to the conversation.

As well as Baring, I remember E. V. Lucas and Max Beerbohm, before he retired to Rapallo. Valentine Castlerosse, who wrote a gossip column for a Sunday paper, represented the more cynical world of journalism. I did not like his remarks, which sometimes raised a laugh: 'Where there's a will there are relatives', 'No heiress is ever jilted', 'A pundit is a self-promoted know-all' were examples of the humour which he contended that in good journalism – a paper flourished by inventing a story; and then endeavoured to attract an element of truth by building round it.

* Joe Beckett was for a long time heavyweight champion of England.

Hilaire Belloc, a formidable figure in black cloak, dark suit and wide-awake hat, drove himself down to dine from Kingsland. He scorned a dinner jacket (then *de rigueur*) and ate huge quantities of *moules marinières* when in season. Small boys were totally ignored by him. Despite tempestuous gaiety, I remained scared of the close-cropped hair and abrupt, clipped speech that went with the hot angry eyes and astringent, even violent opinions. Like Maurice Baring, he was a brilliant conversationalist. One of the soldiers ragged him, asking why he wrote so many books. Belloc drained his glass and answered, 'Because my children are howling to the moon for pearls and caviare.' I made a note of this *bon mot* in my diary, and also learnt some of his poetry. After a mighty meal he was prevailed on, when in the right mood, to sing his own poems – plain-chant style, in a high, reedy voice: 'Tarantella', 'Ha'nacker Mill' and 'The Winged Horse' were in the repertoire. (The last was a poem that later epitomized the soaring spirit of the airborne forces*.) He also sang French marching songs, for Belloc in his day had served as a conscripted *poilu* in the artillery.

Of Baring my mother wrote:

Maurice was a patron of the arts in the most agreeable sense of the word. The work of a famous musician would take him half-way across Europe. He would hurry to Italy to see Duse act a single performance or to France to listen to Sarah Bernhardt: he had a genius for admiration. He seldom missed a classic race and to the delight of his friends on the turf, could repeat the names of all the Derby winners at random or in chronological order from the first race run in 1786. It is impossible to exaggerate M.B.'s generosity to friends and sometimes total strangers: his gifts were enhanced by their suddenness: a hat, a cheque, a first edition of great value, a sewing machine, a love poem or a racing debt would be pressed into the hands of those he loved. My children and grandchildren, before entering the bedroom during his last illness, were warned to admire nothing – lest it should be given to them.

There was a cross-section of civilian sportsmen: 'Molly' Clarke (Rufus's father) and Raymond de Trafford, known as the 'Borstal Boy'; but it was the soldiers who provided all the fun. The younger ones were cavalrymen. Some had been at Lyegrove and most of them had nicknames. 'Fish' Turnor (later killed by a bus in Knightsbridge),

---

* I rode him out of Wantage and I rode him up the hill,
And there I saw the Beacon in the morning standing still ...
And once a-top of Lambourn down toward the hill of Clere,
I saw the Host of Heaven in rank and Michael with his spear,
And Turpin out of Gascony and Charlemagne the Lord,
And Roland of the marches with his hand upon his sword.

'Scatters' Wilson, 'Rattle' Barrat (who played international polo for England), Lionel Tennyson (the poet's grandson and a cricketer who went on to captain the Test side against Australia), Fred Cripps and Tommy Graves (the bookmaker peer) led the rowdy element with Buck, who acted as a kind of mess president and set the pace in the evening after the boys were sent to bed. 'Skipper' Ward and 'Burghie' Westmorland kept the Navy's end up.

Buck, described as the 'last Corinthian', backed himself to win walking, riding and driving races over different distances, in London-to-Land's-End style. Buck had fought in the South African War. He caused quite a stir when, starting from London Bridge, encouraged by crowds of spectators, he won a match for big money against a much younger opponent in a walking race to Brighton.

On the strength of the victory he took the front row of the stalls at the Pavilion Theatre for all his friends to see Fay Compton in Barrie's play *Mary Rose*. The leading lady's performance was said to be very moving. Fay (she was Compton Mackenzie's sister) knew Buck when he was married to Gladys Cooper. Before the curtain went up she asked Tommy Graves to tell her after the show the exact moment when Buck started to weep. Miss Compton, whom I thought very beautiful, got a surprising answer. Buck had drunk a lot of champagne that evening and was hardly himself. 'Well, when did it happen?' Fay asked. Tommy was loath to answer. 'If you really want the truth, old Buck broke down in the bar at the interval. The place was crowded. Before the drinks were ordered he heard the place had run out of brandy.'

The pier was a favourite haunt for boys. Rufus and I spent the day with hand lines and a paternoster (three hooks) dangled through the supporting structure that went down in deep water. We listened respectfully to the experts, who, with spinning rods and plummet weights, cast far out to sea. Rufus caused a sensation by hooking a dogfish, too heavy for the line, which broke as the struggling monster was hauled into the sky. Despite the ranks of experts, nothing bigger than a dab was caught during that weekend. There is a lot of luck in sea fishing.

The great Jimmy Wilde once came to lunch – a long-armed, paper-thin skeleton who for his weight was one of the hardest hitting boxers of all time. In the afternoon he walked at a furious pace round the pier.

For race meetings on White Hawk Hill some of the sporting trainers took a night off to join in the party. George Lambton, Atty Persse and Percy Whitticker told of the men and horses they had known, listing the giants of the turf. Sir George Thursby, who hunted a pack of hounds in the New Forest, was the oldest member of this group. As an amateur he had ridden John of Gaunt into second place in the Epsom Derby – only beaten by a head in a desperate set-to. In the days of his youth, Sir George had sought fame and fortune in the United States, but discovered neither. He taught me the rudiments of poker, and I learnt of the 'dead man's hand' (two pairs, eights and aces), held by Wild Bill Hickock when he was shot from behind in the Deadwood Saloon 'out West'. It sounded more exciting than the 'curse of Scotland' (the nine of diamonds) on which government instructions had been written ordering Campbells to exterminate the MacDonalds in Glencoe. Sir George provided a follow-up tale. I recall the assumed accent and his droll manner word for word:

My boy, never gamble with strangers; here's one reason why: remember this story, it could keep you out of trouble and save dollars. On pay days in New Mexico, which came once a month, we used to fork our ponies and ride into town to spend the thirty bucks. The Irish foreman – Clancy by name – was a remarkable character who retired to his bunk every night with a quart of whisky. By morning the bottle would be dry and not surprisingly he had a vile temper; he was also a heavy gambler and a good one too. I once got in a poker game with him in Bisbee; the rest at the table were strangers. Everything went well until a one-eyed waiter took a hand; gradually he began to accumulate a pile of money (there were no chips in those days). Then Clancy started to lose; the rest of us were holding our own on the run of cards. A hand was dealt out. Bets were made. It was a pretty useful pot; each man chested his cards or kept them face down on the deck, and looking round, tried to figure what the other fellows held. Then out of the silence came a growl from Patrick Clancy. 'There's been some cheating going on,' he said, 'I don't want it to happen again. Is that understood?' Just in case it wasn't, Clancy took his six-shooter out and laid it on the table. 'I ain't mentioning no names,' he said, 'but if his hand slips again I'll shoot the son of a bitch's other eye out.' The waiter picked up his winnings and departed pronto!

Sir George's drawl as well as the Western scene impressed those who had never visited the States; that meant everybody except Rattle.

I rejoiced in the 'Old Flamers': the distinguished generals, some no longer young, whose opinions, backed by frontier experience and native wars, held me spellbound. The Poona sahibs and blimps in hard-boiled shirts, built up by the news media, were not in evidence at Brighton. These were all fighting men a long way removed from the red tabs or petulant 'puffy faces at the base' immortalized by Siegfried Sassoon. The veteran generals included: Crawley de Crespigny, John Ponsonby, Tom Bridges, my father, George Paynter and Albi Cator, both the last two being Scots Guardsmen.* Wiggy Welan and Dick Molyneux

* During the First World War the addition of one staff officer to Brigade Headquarters establishment raised the Commander to the rank of Brigadier General.

represented a still more distant past. Dick, who had been wounded in the cavalry charge at Omdurman, in which Winston Churchill also rode, was groom-in-waiting to King George V; his duties could not have been arduous as he boasted never to have spoken on the telephone nor to have had the slightest intention of doing so. The old soldiers stood up to a lot of ragging, regardless of their rank. In the Foot Guards and most cavalry regiments, men are equal in the officers' mess except for the Commanding Officer, who is, at all times, addressed as 'Sir'.

The Old Flamers had many things in common. They strongly objected to any form of 'showing off' or cheap publicity. Each held to the military precept that those who did well were expected to say least about it. The senior men agreed that Lawrence of Arabia's publicity and acclaim (not unlike that of Wingate and his Chindits) did no justice to Allenby, who (like Slim) never sought the limelight. Jellicoe was greatly preferred to Beatty, who came out badly over the Battle of Jutland – which, to my great surprise, they considered to be a German victory. My father, with reference to Beatty, took a poor view of any serving officer who wore his cap at a rakish angle. He held suspect brown-eyed men with waxed moustaches who ate soup for lunch, sported baggy plus-fours and hunted south of London. They all considered Lloyd George a bounder – not so much for new taxes imposed, or selling peerages and knighthoods to the highest bidders, but for the remark passed during a critical period in the war when he suggested we could be sure of victory if only the Germans had Haig (the British Commander-in-Chief) to lead them!

The generals' tastes were simple and more robust than today. After the trenches they were determined to enjoy life, and this they did with gusto. The morning began with a walk before breakfast (which was eaten in silence) followed by more exercise – there was no sitting around: golf, a sail or a game of croquet. After an enormous lunch there was racing or tennis, and after tea, if the spring weather was good enough, a cricket net on the Hove ground. (My father, who was past fifty and admitted that his waistline had made concessions to the passing years, could still bowl well enough, helped by Rex Benson, to make Lionel Tennyson play a straight bat to the better length ball. Lionel later faced Australia's Gregory and Macdonald at their fastest – once, heroically, at the Oval, when he batted with bandaged fingers and the use of one good hand.) Following dinner, for those not playing cards, came a visit to the Pavilion Theatre, at the end of the pier, where there was usually a play on its way to the West End. The old soldiers played pontoon (vingt-et-un) or poker for penny points, preferring both these games to contract bridge. Few could afford high stakes.

Not so uncle, 'Scatters' Wilson, late 10th Hussars, an inveterate gambler who took 'Molly' Clarke, Valentine and the 'Borstal Boy' regularly to the cleaners, also Wiggy Welan, who was growing old. My uncle freely admitted he lived off his bridge game: with the proceeds he sent three sons to Eton, shot pheasants during the winter and kept horses with Victor Tabor at Epsom. Scatters had served in India and raked in the winnings quoting:

> If he plays, being young and unskilful,
> For shekels of silver and gold,
> Take his money, my son, praising Allah,
> For the kid was ordained to be sold.

Uncle Scats slept in like all the literary men. When Tony called his father there was sometimes a handful of loose change, which had its uses on the pier or the aquarium.

Wiggy Welan was a steward at the races, and sometimes got into trouble. One day, returning from official duties, he tripped outside the Royal York and fell into the gutter. There had been an objection that afternoon to a hot favourite (locally trained), which won easily. The stewards sustained the objection and another horse was given the race. The cabbie (who had probably lost money on the enquiry) addressed the prostrate Wiggy in scornful terms: 'Cor blimey, Methuselah! Call yourself a judge of racing when you can't even see the ruddy kerb?'

The Old Flamers' homespun sayings (not to be found in textbooks) were very much to the point. I profited from their wisdom. The generals were not snobs – quite the opposite – but they had no time for upstarts. 'He's certainly not by Saint Simon' was an innuendo which referred to the impeccable breeding of a famous stallion. Again: 'Only fools and people on the make have a right to be pompous.' How true! Fred Cripps taught me the definition of a gentleman: a person who conceded to the wishes of others before seeking to impose his own. 'Good manners cost nothing' was more sound advice. It is impossible to win an argument dealing with people who hold prejudiced opinions, for they cannot change their minds; that a man is only as big as it takes to lose control of his temper. Ladies' names were never mentioned and risqué stories in mixed company were frowned upon.

Among the generals, Sir John Ponsonby was the favourite. He possessed that magic quality which inspires loyalty and affection. Sir John suffered from an impediment to his speech which had held up promotion; it was with difficulty that he got a commission in the Coldstream Guards after soldiering in Africa. It remained hard to follow what he said. Once a girl he was escorting into dinner asked politely: 'Have you got a cold, Captain Ponsonby?' 'No, why do you ask that?' 'Well, the fact is I have not understood a word you have said to me since we were introduced.' 'Oh dear; then let me explain. If you care to examine the roof of my mouth after dinner, I can show you the roots of my hair.' Immensely kind, his men loved him, treating him like a father, without really knowing what he was talking about. He retained a sense of humour after rising to command a Guards brigade. There was a good story about his *papier mâché* hat, worn instead of a steel helmet. It blew away on a church parade but saved the day in a critical situation. During the retreat from Mons, in confused fighting at Landrecies (where the battalion was cut off by the enemy's advance), John Ponsonby set off with his batman to find a missing company. Emerging from wooded country they both took evasive action in a roadside ditch, when a troop of Uhlans – the advance guard of the pursuit – came riding up the *chaussée*. The day was warm; the Germans dismounted to water their horses in a convenient pond and then proceeded to relieve themselves into the grass-grown ditch where the gallant colonel was crouching with his orderly. Sir John said afterwards but for his celebrated 'battle bowler', preferred to the regulation helmet, the splash of falling dew-drops would have given the show away! At the time his indignation knew no bounds. Getting his scattered rearguard together he addressed them as follows: 'Men of the Coldstream Guards! This afternoon your commanding officer got pump-shipped on by a Boche patrol – an event without precedent in the history of the regiment. I call on you to avenge this outrage.' Those who understood cheered up considerably, and the battalion counter-attacked with fixed bayonets, shouting, 'Up the Lily Whites!'

Carton de Wiart and Tom Bridges, both cavalry men, were more romantic figures, each with many gallantry awards and of commanding appearance. Tom Bridges – a hero in the Great Retreat – refused to be photographed, but Rosa Lewis had a picture of him at the Cavendish. He owned a pet lion cub at the time his leg had to be amputated. Before the chloroform was administered Bridges gave orders that the limb should be fed to his friend.

Sir George Paynter, a fine horseman, twice rode the winner of the Grand Military Gold Cup at Sandown. During the shooting season, his father rented Balblair House, looking out over the Beauly, which is my present home. Sir Albemarle Cator (another soldier with a great war record, who loved foxhunting) died on his horse at the opening meet of the Quorn Hounds. The generals started the Scots Guards indoctrination ball rolling before I reached the age of nine. They were contemporaries of Uncle Hugh's and remembered Great-Uncle Henry, a veteran of the Crimea, who had commanded the regiment. (He was the originator of Lovat tweed, noted for its invisibility on the hill when out shooting.) Colonel Henry had been known as 'Pope' to distinguish him from 'Pagan', a Fraser of Saltoun who commanded a Grenadier battalion. They were instanced as representing a name that went back in time to the regiment's first adjutant.

I had reservations about these blandishments. I was too young to be interested, but despite inattention they taught me something of military tradition and of the things that go with it: how no one who had not worn the king's uniform could know or understand the heartbeat of a good battalion – proud of itself and the regiment whose reputation was in its safe-keeping. Nor could a civilian in peacetime comprehend the discipline and corporate spirit which sustained each individual soldier, or how the whole greatly exceeded the sum of its parts. At the time I was not impressed by what sounded very like chains and slavery. I also suspected that father had put the generals up to it. He never mentioned the subject during conversation, nor did he appear displeased when I suggested I would like to be a vet.

The year 1921 marked the beginning of public school years at Ampleforth. At Easter I had my uses on the farm, working with the lambing shepherds or spring planting in the new plantations. There was a sleeping-car connection to Inverness on the last day of term which left York before midnight, and I never missed it. By that time the Brighton parties were coming to an end. The quiet old watering-place was becoming too easy to reach from London. Long weekends were spent further afield. Buck was busy with his Club and now took members to Le Touquet on Bank Holidays. Harry Preston and Maurice Baring were no longer well; even the old soldiers began to fade away. Some found it cheaper to live abroad. Rattle started the game of polo in the South of France; for Lionel cricket had become a full-time occupation. The two sailors, Burghie and Skipper Ward, got married. With one exception they have all passed over to the other side. Fred Cripps is the last survivor of this generation of sporting Englishmen who held that courage, conversation, conviviality and wit were qualities from which other blessings are descended. All that was best in Brighton went with them. The Royal York has been pulled down. The pier has buckled half-way along its length and the theatre beyond forced to close down. A lunatic recently set fire to that splendid folly, the Royal Pavilion (embellished by Rex Whistler), where Max Beerbohm saw the ghost.* Now skinheads, mods and rockers roar in on motorcycles

* I am glad that the Pavilion still stands in Brighton. Its trim lawns and cheeky minarets have taught me much. As I write this essay I can see them from my window. Last night I sat there

to beat up the esplanade – neon signs and Hamburger Heavens have done their worst, shattering peace and quiet. O Tempora! O Mores! But when the weather is warm, I sometimes still visit the acquarium.

The First World War marked the end of the golden era of low income tax (death duties were unknown) and of the prosperity which had begun in Queen Victoria's reign – and terminated the sense of national confidence and security at home and abroad that had been created by her name. England had used up all her resources and accumulated wealth. The best of the younger generation – the country's future leaders – had fallen in France and Flanders; Lloyd George's reparations and free trade policy – the dumping of cheap goods – proved disastrous to a nation trying vainly to readjust itself. The first socialist government (1924) did no better.

In the Highlands the depression was less acute than elsewhere, but only in the pastoral areas, where industry was unknown. The returning soldiers were treated as heroes and jobs were waiting at Beaufort. The Lovat estates gave massive employment, for my father, who was considered an enlightened landowner, ploughed more money back into the property each year than he took out of it. More immediate needs were derived from other sources. Rents were low and the escalating cost of upkeep considerable. Father, in fact, was land poor in the sense that he owned too much and was in debt. The chief source of revenue came from timber; thanks to an enlightened grandparent, a sixty-year rotation allowed for a hundred acres of mature woodland to be cut down and sold each year, while a hundred acres of new land, planted annually, replaced the felling. The Beauly salmon fishings, both sporting and commercial (nets), rated next in importance with the farm rents. Sport (grouse and deer) was a tricky business at the best of times, for the ground over most of the estates was high and difficult to manage or negotiate. Late snow often wiped out stocks of birds, and the cost in upkeep of private roads, bridges, paths, grouse butts, draining and heather burning, as well as the maintenance of shooting-lodges, was enormous. To cope with requirements full-time employment was provided for squads of estate carpenters, masons, foresters, saw-millers, gardeners and fencers: each group being supplied with its own transport. The keepers, stalkers and fishing ghillies, about thirty-five all told, were doubled with extra help during the summer season. As many more families worked on the Home Farm and in bye land, as hill shepherds, cattle men, etc., or attended to the studs of highland ponies, Clydesdale

---

in a crowd of townspeople while a band played us tunes. Once I fancied I saw the shade of a swaying figure and of a wine red face!' From *A Note on George the Fourth*.

horses and the smithy. The riding stables, carriage horses and garage, with the workshops (painting and plumbing) required further personnel, while the domestics at the castle seldom numbered less than twenty people who included laundry maids, stillroom maids, pantry maids and housemaids, three in the kitchen, a seamstress, the hen wife, the orra man to carry wood and coal, and Willie the Moon, who controlled the lighting system. My father, who enjoyed the measured phrase, used to say, 'A man can keep a good cellar or a good butler, but he cannot keep both.' Nor were there footmen at Beaufort. Mr Vickers (the best loader in Britain at the big shoots) presided, with Guardsman Grant (Uncle Hugh's batman) in the servants' hall. They preferred to be called ex-servicemen; as such they could set their hand to anything.

This memoir lays no claim to social history or changes which took fifty years to make themselves felt in remote areas of the Highlands, and then only with the discovery of North Sea oil. A fair wage for a fair day's work, security, a decent house, mutual respect between master and man, all spelt a happy life in the community. In reply to those whose reactions may recoil from the patriarchal days of my childhood, I can say working-class conditions were a great deal worse in the outside world, particularly in cities and the over-populated industrial areas of the South. They probably remain so – sadly the lure of bright lights makes it increasingly difficult to keep young people on the land; the widespread closure of village schools has hastened that decline. Penal taxation and the population explosion have destroyed the existence I have described. The blame is sometimes laid at the wrong door; if wages (by present standards) were extremely low, it caused no complaint in a self-sufficient community where good relations and helpful neighbours rated higher than politics or trade unions. A son followed his father's calling – sometimes for generations – the 'tied house' did not present the problem that it does today. On the other hand, there was no lack of ambition: some of the best young men went overseas to seek a fortune – but not under duress. The crofter's earning capacity was supplemented by the land which he worked at his door. In the Beauly Valley this meant a fair living bonus way above the standard set by ordinary wages; equally important was pride in the home, also a strong sense of moral rectitude matched by sturdy independence. This has not changed in the hills of the Highlands.

The reader can compare those worthies of the past with the people who today drive up in motor-cars to collect their National Assistance payment (some with no intention of ever taking their jackets off). The good old days or 'the bad old days' are largely a matter of political indoctrination. I believe a lot of grumblers are miserable because they cannot obtain some of the things their parents never had!

So much for the way of life in which I grew up. If it sounds luxurious, there was no room for slackers in the Fraser family, father had a sense of driving purpose and I was put to work in the holidays as soon as I could fetch and carry.

A fair criticism of landed proprietors was they appeared to be unaware Great Britain was hurrying from a rural economy to an urban society, where sweated labour, bad houses, malnutrition and insecurity were commonplace, indeed unavoidable. Some could not cope with the change. The Industrial Revolution took many by surprise. The transition period (during which time the population doubled itself in less than a hundred years) was not a happy one.

Father was first and foremost a great forester, but his knowledge of tropical agriculture was also considerable; the pioneering efforts met with mixed success. Soon after the South African War, work began on an irrigation scheme to fructify the Gezira Plain (that area of arid sand between the Blue and the White Niles, where the rivers join above Khartoum). Encouraged by K. of K. (who was Consul General in Cairo) and helped by Sir Reginald Wingate (the Sirdar), the first 10,000 acres of the Sudan Plantations Syndicate were extended by a further 700,000 acres when the railway reached Medani. This area of desert was transformed, through pumping stations, into fertile soil, of which one-quarter was put annually under cotton crop. Bumper yields of sorghum and millet supplied local needs. Sir Frederick Eckstein and Sir Julius Wernher raised most of the capital for this enterprise. D. P. MacGillivray (later knighted), the son of a tenant farmer in Stratherrick and the kind of Scotsman to whom the Empire owes so much, was appointed General Manager. The year my father died, the cotton crop alone was sold for £2½ million – quite a sum in those days.

The next venture (in Brazil) was a disappointment, though soil and climate offered far greater potential. The Parana Plantations Land Company (founded in 1925) acquired a controlling interest (in the state of Sao Paulo) in a railway line and in the vast area of Mato (jungle) that lay beyond. It soon ran into financial difficulties. Scrub clearance was more costly than expected, and the extension of a further hundred kilometres of track to the Tibagy River (where I helped to build a considerable bridge) required more capital. Also, there was a surprising shortage of settlers once the jungle was cleared. Even a visit from the Prince of Wales and Prince George, whom I escorted up from the Argentine, failed to dispel the fear of bandits. A second bridge, thrown across the Paranapanema, linked the two states, but there was trouble on the other bank, for, during the years 1929–30, a counter-revolution was sweeping the state of São Paulo and the more remote provinces. But worst of all, the higher land was found unsuitable for growing cotton, being subject to frosts.

'There's an awful lot of coffee in Brazil.' At the time there was such a glut of this currently scarce commodity that dry beans were being burnt locally for fuel, and export had come to a standstill. (During the Second World War there were to be difficulties in maintaining rolling stock on the new line, requiring engines which were made in Germany.) The Brazilian Government – never helpful – made it virtually impossible to take money out of the country, and finally expropriated the property after paying a nominal sum for 300,000 hectares of land and railroad – worth a fortune – and already growing high-quality coffee. This was a bad deal all round, for a popular cousin, Ronnie Maxwell, who, after transfer from the Sudan, became manager of the enterprise, died up-country from typhoid fever. His assistants, two Fraser boys from Ballcraggan farm, later established fazendas of their own, and Simon, the eldest brother, was a rich man the last time I saw him in Rio. Conception and realization are difficult to match south of the Equator – especially when dealing with Latins in tropical country.

South America was a failure which came at a bad time. Father did better in Forest Investments – a fine lumber business in British Columbia that logged timber to the Pacific Coast, one of many profitable pioneering ventures controlled by a great Scots Canadian, H. R. MacMillan. Forest Investments helped to pay the taxes arising from my father's death in 1933, but in the interval many of the outlying portions of the Lovat estates were sold to pay off overdrafts. The high ground (only suitable for grouse and deer, and involving neither crofters nor tenant farmers) was the first to go. It included most of the shooting lands in Stratherrick, Corriegarth, Killin and Glendoe, the grouse moor of Ardachy and Inchnacardoch at Fort Augustus, and on the West Coast the deer forest of North Morar. Two large arable farms at sea-level outside Beauly, Barnyards and Tomich, were broken up for the settlement of demobilized soldiers, but the scheme was not a success, the land being too heavy to be easily worked. There was much heart-searching before parting with Struy in the Aird of Lovat, for it lay in the middle of the property and was one of those rare places where the sport provided made it possible for tenants with a strong pair of legs to bag a grouse, a stag, a salmon and even a little hill partridge on the same day. But old Lord Derby's long lease was at an end, and the price offered was unheard of at the time. (It changed hands again a decade later, in the 'Hungry Thirties', for a quarter of the original sum.) During the depression, at the time of father's death, deer forests were changing hands at a pound an acre, while the Forestry Commission was offering no better than ten shillings for ground over 500 feet. Struy went with the rest, reducing the size of the estates by 100,000 acres, and to more manageable proportions. But the sale that I remember best, and one that will be talked about as long as pedigree cattle are bred in Inverness-shire, was the draft of beef shorthorns sold on the home farm in 1920. The pedigree herd, cut off from the New World by the submarine blockade, enjoyed an enviable reputation for the size and substance of its cattle. Beaufort blood lines were in demand as improvers of less growthy stock; they were seldom exhibited, for father did not approve of over-conditioned animals. The bull Broad Hooks Champion had been sold to an Argentine buyer for 1,500 guineas as far back as 1906: a record price and the first in the four-figure bracket until after the war. At the Beaufort sale in 1920 thirty-six head (six bulls and thirty females) realized 27,000 guineas, the ten in-calf two-year-old heifers obtaining the astonishing average of just over 1,200 pounds apiece, and the bulls doing even better. The tradition has grown up that every man associated with the herd was drunk for a week; which may be a slight exaggeration, but it is a good indication of the general rejoicing at the time, for a pound was worth something in those days. The herd celebrated its hundredth anniversary in 1958, but artificial insemination has now destroyed the export trade for pedigree bulls and unfortunately the beef shorthorn is no longer fashionable.

I have outlined some of the diverse activities, successes, failures and responsibilities that go with the running of a Highland estate: something that is called a 'big spread' in North America. It remains a full-time preoccupation and a fascinating one. It is also a challenge, for, without the right qualifications and a capacity for hard work, the modern proprietor will inevitably fall by the wayside. 'The eye of the owner fattens the ox' is a Spanish proverb. Nearer home, I recommend the Yorkshire equivalent: 'There's no muck like t'master's footstep.' The days of the private agent seem numbered except on the very largest estates.

Father's health began to deteriorate after I left public school. Before his death at an Oxford point-to-point, I was put into harness to help with the general management of the property.

CHAPTER FOUR

EARLY TRAVEL

There are some that love the Border Land and some the Lothians wide,
And some would boast the Neuk o'Fife, and some the Banks of Clyde,
And some are fain for Mull and Skye and all the Western Sea
But the road that runs by Atholl will be doing yet for me.

VIOLET JACOB

A train journey has its excitement, particularly for children. I remember the summer excursions best. The Inverness express left King's Cross at 7.20 pm, travelling by way of York and completing the journey of 550-odd miles in only slightly more time than it does today.

In early August many families (the non-residents loaded down with gun-cases, cartridge bags and fishing-rods, and with all varieties of gun dog straining at the leash) converged on No. 1 Platform for the start of the Scottish season, which followed after Cowes and Goodwood. The station was spotless, and so was the stationmaster, in a frock-coat and silk hat. This imposing individual moved like a Grand Vizier among the throng, greeting old friends, introducing others, ready to sort out the difficulties of young and old alike. I watched this performance open-mouthed: a word of cheer in the ear trumpet of some querulous dame, the dispatch of a porter swifter than Malise of the Cross of Fire to get a paper with the latest cricket scores. When pressed, the same official once rescued, with equal efficiency, a litter of white mice at large in a sleeping-car. Like the first swallow, or cricket umpires emerging from the pavilion before a Test match, he seemed the harbinger of good things to come. Some knowledge of sport and the right way of putting it across gave added authority to the lofty figure as he moved majestically among the crowd: 'I hear conveys will be small on the high ground this season because of the late snow,' or the discreet aside which supplied tidings of whose river had been fishing well. 'Rain is needed in the West, and they say His Grace has lost his case with the Water Board.' The stationmaster was clearly a man of parts. Magdalen thought he was the King of England! Not everyone received the VIP treatment or the same seating accommodation. Class distinction was much in evidence and compartments were numbered from one to three.

There were no visible bookstalls; during the day boys with trays hung around their necks paraded the platform, hawking their wares with a mournful street Arab cry of 'Newspapers, magazines, chocolates, cigarettes, green grapes and bananas' – I could put on a very fair imitation in a Scottish accent.

After consulting the guard's van, whistles blew, the green flag waved, and 'Top Hat' sped us on our way up the Highland line. He seemed omnipotent, but the sleeping-car attendant, filling my mother's hot-water bottle, said the man was a snob who lived off gifts of game for half the year and salmon all the summer. I made a mental note to become a snob, whatever that might mean.

Prior to the Rattler's departure (my father's name for all locomotives) the young were sent to bed. Magdalen and I, sharing a sleeper (first class only), stretched head to toe. My last conscious thought was a never failing delight in the various gadgets which could be switched on and off around the sleeping berth. Bright lights, blue lights, side lights, control levers for heating adjustments, an electric fan, a drugget, even a small drawer for loose change and a hook to hang a watch on.

Meanwhile, the grown-ups, who scorned high tea before starting, had disappeared along the train to tuck into a substantial meal: the dining car went as far as Doncaster. They were sometimes pursued by father, who arrived late from White's or his desk at the Forestry Commission, brandishing a small gold shield the size of a postage stamp which provided free rides for a Director of the London and North-Eastern Railway.

At Perth, water and a second engine were added to the train, and by Pitlochry we were awake and very hungry. I learned to read the time off clocks on station platforms, for the train stopped at every lamp-post. At Pitlochry the hour approximated 6.30 am. Here there was a halt which allowed time for early birds to acquire a wicker breakfast basket, handed through the window. The children's turn did not come until Kingussie, further up the line; by then our parents were astir. At Aviemore in summer weather it was traditional for the gentlemen to stretch their legs and give the dogs a run, while we finished off the hampers ordered ahead by the London station authorities.

Mr Vickers, who looked in for leftovers, grumbled that times had changed for the worse now that a florin was charged for fried eggs, sausages and bacon, various jams, scones, oatcakes, and lashings of buttered toast, sugar, tea or coffee.

My uncle, Alastair Fraser, gold mining in South Africa, and a child-hood hero who came home on rare visits and smoked a foul tobacco called Matabele Mixture, took a different view. He maintained that strong men under harsher skies dreamt of Kingussie breakfasts – the fresh butter, bannocks, and Dundee marmalade – as they shivered on the Khyber Pass or sweated in the tropics.

> There are pines in Rothiemurchus like a gipsy's dusky hair.
> There are birch trees on Craigellachie like elfin silver-ware.

At Aviemore, coaches were uncoupled onto the Boat of Garten line to Grantown and the Spey Valley. The connoisseurs admired the timber, which includes vestiges of the old Caledonian Forest, or commented on the snow-fields left in certain corries on Brae Riach and Cairn Gorm, the colour in bold contrast to the dark woods which clad the lower hills.

On the platform was a faded billboard showing the head of an Indian brave in his war bonnet advertising Mazawattee Tepee Tea. The name intrigued and humiliated the children. My father kept a record of the IQ of each member of the family who read out this tongue-twister correctly. Unsightly hoardings were then much in evidence.

> Large friendly names that change not with the year;
> Lung Tonic, Mustard, Liver Pills and Beer.

wrote Siegfried Sassoon on his first glimpse of England from a hospital train carrying wounded back from France. Now advertising has very properly been forbidden by the local authorities.

After the rails left Perthshire (the county boundary is at Drumochter, where that rotund hill, the Sow of Atholl, looks across the pass to the Boar of Badenoch) father was usually busy on a public relations exercise then considered normal practice. With communications by road, train, post, or telephone both limited and precarious, he conducted a baron's court in an empty compartment of the train. Various delegations climbed on board from stations up the line to discuss a variety of subjects that ranged from the Territorial Army to hill farming, forestry, shinty matches, local shows and the Inverness County Council. There were other interviews of a private nature with old Lovat Scouts, emigrating families, and overseas visitors of Highland origin who came to shake hands and have a word with the Chief of the Clan. Despite the early hour, good use was made of drams, with many a toast from the black bottle; friends often chose to continue the journey, talking all the way to Inverness, for my father had the happy gift of

putting strangers at their ease and everybody felt at home. The powers
of a Highland Chief have long since departed, and their territories
dispersed, but father's reputation still raised him *primus inter pares*,
beyond the limits of the Fraser country.

Magdalen was excused duty, but as we grew older I was expected,
in father's absence, to keep my wits about me. Train journeys took on
a new significance; our leisurely progress puffing up and down the
Grampians provided opportunities to combine a farm walk, nature
study, news bulletins and learn the history of the countryside, both past
and present. In daylight hours books were set aside, and I attempted
to quantify the scene. Thus father, working in London or by letter when
abroad, got early news of pastoral activities up north. The message
according to the season; how the hay crop or grain harvest round Perth
compared with fields at home; which grouse moors were heather-
burning; if sheep shearing had begun in Badenoch, or the lambs had
been weaned; the fishing heights on the salmon rivers as we crossed
the Tay, the Spey and the Findhorn; if wintering deer were in good
order, or whether the snow-line had forced them down off their own
ground. All this and more besides was duly written or reported on
reaching London, my father listening attentively to matters which had
a bearing on his own estates.

Sixty years ago, 2LO, the London code sign on the crystal set and
cat's whiskers did not provide weather forecasts. I remember father's
annoyance when I failed to report serious damage to woodlands
on Tayside. It was bad luck; the day had been calm and sunny in the
Beauly Valley and dusk had fallen when the train reached Blair Atholl,
so there were excuses; but father heard the news at White's before I
saw him. It took weeks to assess the damage to plantations ripped
up by a freak storm which broke after leaving home. A year elapsed
before blown trees were finally cleared away by timber merchants,
who make a killing out of gale damage. Fire risk to woodlands along
the line remains a hazard, despite the advent of Diesel engines.

This sharpening-up process improved my powers of observation, for
I learned the names of every property between Perth and Inverness,
many of the farms, and all the high tops along the skyline.

In return for the use of our eyes, we were rewarded with descriptions of Scotland's story. Some of the tales my mother read aloud;
others she embroidered from memory. All the legends and folklore
of the North; Fingal and his tall heroes, the Roman legion that
marched with Severus from York to Inverness, King Brude the Pictish
overlord, Columcille the missionary saint, Malcolm Canmore, the Wolf
of Badenoch and the death and doom prophecies of the Brahan seer.

All the bloody deeds of rival clans, slogan, feud and foray. The Lords
of the Isles, the Massacre of Glencoe, Bonnie Dundee, Montrose, Allan
Breck and stirring Jacobite intrigues, the Curse of Moy, Cluny's Cage,
and, of course, the Prince in the Heather. They fitted into place, for
the actors and events which quicken the grim past had staged their
best performances beside the winding train:

> And ever by the winter hearth
> Old tales I heard of woe and mirth,
> Of lovers' slights, of ladies' charms,
> Of witches' spells, of warriors's arms,
> Of patriot battles won of old
> By Wallace wight and Bruce the Bold;
> Of later fields of feud and fight,
> When pouring from their mountain height
> The Highland clans in headlong sway
> Had swept the scarlet ranks away.

As we crossed the river Nairn, gathering speed through Culloden
station on the downhill run to Inverness, mother liked to quote some
lines from William Howitt's *Visits to Remarkable Places:*

> We drove out to Culloden, and stood on the moor at sunset.
> Here the butcher Cumberland trod out romance.

She enjoyed those unhappy long-lost things and battles long ago, and
I am grateful for inheriting that love of poetry which illuminates the
mind.

Father talked of more topical events. An abstemious man, he maintained that whisky was Scotland's greatest gift to the world, but conceded that four men – Wallace, Bruce, Burns and Walter Scott – had
created a nation, and that Rabbie was perhaps the greatest Scotsman
of them all.

My mother enlightened us about fairies, or the Little People, as they
were known locally, and so did Hallie Sutherland, my close boyhood
friend. I have said how well mother read aloud, but looking back, she
must have excelled in firing our imaginations in the preternatural
realms of enchantment which provide hobgoblins, wizards, witches
and old Nick himself. In her own youth she had learnt the legends of
the Border and Upper Tweed, being a Tennant on her mother's side
and brought up in the glen above Traquair.

Though she admired everything Sir Walter and the Ettrick Shepherd
had to offer (nor must True Thomas, who met the Queen of Faerie
beside the Bogle Burn, or the works of Andrew Lang, be forgotten), her

chief loyalties lay in Carrick. She explained many poems, teaching us
how the rough ore thrown into the melting-pot of Burns's genius came
out as purest gold. Her mind rejoiced in the Land o'Burns. There
Tam o'Shanter and Souter Johnnie became living people; had not
Robin himself foreby in the roaring-ranting days freely declared 'Auld
Glen' (one of the founders of her folk's fortune) to be the best friend
he ever had? Elsewhere her imagination ran riot in the world of
fey.

Her tales of the Little People interested us greatly, for we claimed
with Highland pride that the sensitive creatures, shunning the
materialism and overcrowding of the South, had retreated to more con-
genial surroundings in Caledonia stern and wild. My sister disapproved
of other manifestations, such as witches and warlocks, whose malice
made Magdalen's blood run cold. A river keeper had told us that when
a witch and warlock (the latter being the male of the species) got
together and mated, they seldom if ever bred a child, but if they did, the
result was a hideous monstrosity, too terrible to describe, that usually
died at birth.

There were more agreeable sidelights on the coming and going of
the Little People, who, though good-natured if properly treated, were
easily upset. Doctor Halliday Sutherland (who ruined himself in a
protracted lawsuit with Marie Stopes on the subject of birth-control)
had wonderful stories from his own childhood in Lochaber.* He wrote,
'In the West the Wee folk lived in woods facing South, where they
could enjoy both sunshine and shelter in rough weather. They were
mostly gentle creatures who never hurt you unless you molested them.
But, if you did offend, they could be vindictive. They might take a
baby out of a cot, and put a changeling in its place, or steal hens'
eggs, hide the spade and even scatter a peat stack.'

One crofter had witnessed a fairy funeral crossing the road from
one wood to another. There was a little hearse about two foot in length
drawn by four tiny horses, and followed by a large procession. The
man waited respectfully until the cortège had passed on its way, and
then thought it was all his imagination. But a neighbour, who was
less considerate, drove his horse and cart, carrying sea tangle for the
'tatties', through a similar parade. As he passed over it, shouting in
Gaelic that the road was wide enough for all to pass, there were shrill
cries of rage and malediction, and though the procession faded from
the road, he clearly saw angry Lilliputian figures in the woods on either
side. When he reached home, the wife greeted him with an ominous

* See *Hebridean Journey*.

remark: 'The summer visitors arrive next week and I've been moving furniture. Yon staircase to the guest bedroom is awful narrow. I could nae get the armchair out of the kitchen up there this afternoon.' A week later the 'Fir nan Tighe' – the head of the house – was dead, and the coffin had to be taken through a window because the staircase was not wide enough to let it pass.

Female fairies are jealous, and that is why you never mention a woman's name in conversation when walking at night. This slight on their vanity might be overheard, and once they learned the other person's identity, the fairies could entice her unwittingly away. Again, the Little People dislike boasting, and that is why a Highlander, in his canny fashion, when looking over the harvest, never says, 'What a grand field of oats,' but rather, 'Aye, yon crop's no' so bad.' Otherwise there might be trouble.

Lady fairies are associated with the colour green, but some little men in Skye prefer a reddish apparel; this could mean they dye their garments with crotal, which is made from lichens and the iodines extracted from local seaweed. When fairy women wash their clothes in a forest pool, the green tinge is clear evidence to keen observers.

Hallie was taken to the 'wishing tree' at Fasnakyle, on the Knockfin beat of that forest. In my youth all the local children from Tomich and Guisachan were passed (as infants) through the round hole in the living trunk of the elder tree's smooth wood. It brought certain good luck to those who carried out the rite. Rowan trees, planted in front of a croft house, were often plaited to form an arch to keep away evil spirits. Hallie was intrigued by a local belief that, if one blinked one's eyes in a chance encounter, opportunity for a closer acquaintance with the Wee Folk was lost forever. Fairy rings and figures of eight are common enough in some of the clearings in the woods round Beaufort, but they are not taken seriously by local people, for keepers and crofting families know them to be the work of roe-deer, who play in circles in late summer evenings – some say for courting purposes, but more probably by the does, taking evasive action to wean their fawns.

On vacation from Oxford there was time to make a visit with Hallie to the Rough Bounds and Moidart, where he had lodged as a boy during summer holidays. We stopped at Morar, part of the estate, then crossed from Mallaig to Skye, and so to Dunvegan, the MacLeod stronghold, to view the Fairy Flag. The 'elf spots' can still be seen on what looked like a brown bundle of tussore silk, that is said to date from the Crusades. Every Highlander knows the story of how a Chief's son, born in the remoteness of time, was found wrapped in a strange bundle high in a turret of the castle. The nurse had been ordered by MacLeod himself to bring down the infant to present him to the assembled clansmen. As she descended the stone stair, a great voice filled the hall, chanting a solemn warning that the flag would save the clan on three occasions if waved in circumstances of direst need. History relates that this has happened on two occasions: once to stay a cattle murrain which swept through the livestock on the island, and again, when all hope of survival appeared lost, the MacLeods turned defeat into victory at the Battle of Waternish. This epic contest has been commemorated in MacCrimmon pibrochs, the most famous being 'The Desperate Battle', a tune often played in championship competitions when the best pipers strive for gold medals at the Inverness Gathering. The flag has yet to be waved – officially – for the third and last time. It is feared that a Factor, Buchanan by name, in a drunken escapade in 1799, destroyed the magic powers when he broke the glass case, took out the Relic and streamed the elf spots over the battlements. Retribution soon struck. The young Chief was blown up at sea in HMS *Charlotte*, Much of the estate had to be sold off, and the clan MacLeod in Skye became so reduced in numbers that there were not sufficient men of the name left to row a four-oared boat across Loch Dunvegan. Few Highland stories have happy endings, but this is one of them. When we visited Dunvegan, some fifty years ago, Mrs MacLeod of MacLeod seemed an old lady. Dame Flora, as she was known throughout Gaeldom, recently died in the ninety-sixth year of her age: Clan MacLeod societies girdle the earth and no ancient name has its members more loyally associated.

Tales of enchantment are as enjoyable as great traditions, but even in the realm of fantasy people had to work in order to eat. I provide this example of long ago; it concerns the experiences of one Murdoch, a noted deer-stalker in days of yore, who went out at sunrise in the forest of Gaich to take some venison – for Murdoch was a poacher. Coming over a hillock as the daylight broke, he spied a herd of deer, but not being quite within range, he wormed his way closer to a handy knoll. Arriving there, he crept into a firing position and, looking over and beyond, was astonished to see a number of neat little women, dressed in green, in the act of milking some hinds which were feeding on the rich grass along the burn. These creatures he knew at once to be fairies. One of them had a hank of yarn for hobbling the deer thrown over her shoulder. As he watched, the hind she was milking turned her head, made a grab at the yarn and swallowed it. The irritable little fairy, seizing another band which was tied round the animal's hind legs (to prevent kicking), struck the deer, saying at the same time in Gaelic, 'May a dart from Murdoch's quiver pierce your side ere darkness falls.' The fairies, it seemed, were well aware of the old poacher's skill as a deer-stalker.

In the course of the day, Murdoch went on to kill a fat beast, and in gralloching it found to his amazement the identical rope that he had seen the deer swallow in the morning. The object was preserved for a long time in Badenoch, and, it is said, was shown to William Scrope while he was writing the classic work *Deer Stalking in the Scottish Highlands*, which includes this story. Scrope had the Landseer brothers to help him illustrate the famous pages. The authenticity of the tale is weakened by the fact that hinds are not killed in summertime when they have calves at foot – even by a poacher.

The existence of the water horse, or 'Eachuisge', past or present, is firmly believed in the magic West and the Outer Isles. Being an amphibious creature, he must have obvious connections with the Loch Ness Monster and is described in Skye as resembling a magnificent black steed, but without a mane. In daylight the water horse can assume human form at will, turning into a handsome young man, who on occasion has sought out the local maidens should they be available in that island's solitude. Seton Gordon, who wrote so charmingly, tells how a young woman, herding cattle in a shieling on a summer's day, met a handsome stranger, who made love to her. Afterwards they lay in the heather and the man put his head on her lap. As she caressed his hair, she suddenly noticed grains of white sand in his black curls. Her suspicions were aroused, but she continued to stroke the stranger's head until he fell asleep. Then, rising cautiously, she slipped off the kirtle on which he rested, and fled for home. As she neared her bothy, the girl heard the infuriated Eachuisge neighing shrilly as he galloped in pursuit; she just managed to reach the house and bolt the door before he overtook her.

'Tomnahurich', the hill of the fairies, forms part of the cemetery that serves the Royal Burgh of Inverness; it is not a place to visit at night – least of all to dance reels by the light of the moon among the gravestones to a fairy pipe. In the town everyone knows the story of Farquhar Grant and Tommy Cumming of Strathspey, who, in a drunken frolic, once climbed the hill to fiddle on a summer evening. When they emerged, they were surprised to see that the timber bridge on which they had crossed the River Ness had turned to stone, and the people who met them in the street jeered openly at their strange clothing and appearance: for, in Rip Van Winkle style, the unseemly ploy among the gravestones had put them both to sleep for just two hundred years.

We learnt many stories before the train finally arrived in the Highland capital.

The last lap of eleven miles to Beaufort offered a choice of conveyance. A trot with Sandy Stewart in the two-horse wagonette, which seated the staff and took most of the heavy luggage, or a hurl in the open Panhard, which (for mechanical reasons that I cannot explain) had to negotiate the steeper gradients of Glendoe* and Culnacirk out of Glenurquhart chugging backwards in reverse gear. I still use the same old number plate, ST9, which once belonged to the ninth car to be registered in the county. A third alternative for the run home was the Hotchkiss – a hearse-like vehicle favoured by mama, and only used in stormy weather, for its stuffy depths brought on immediate car sickness.

The men-at-the-wheel (for·'chauffeurs' was hardly the right word) played an important part in our lives. Johnnie Sutherland, old John-the-Coachman's son, and Danny MacDonald, both ex-Lovat Scouts, had recently returned from Mesopotamia, seconded from the regiment (in Salonika) to learn their post-war trade the rough way in armoured cars, grinding across desert sand in Allenby's drive against the Turks.

They were entertaining characters. It was Danny who slanted a remark into history after a chance encounter with the C-in-C on the final Victory Parade. Noticing the conspicuous Scout bonnet as he rode down the ranks, Allenby reined in and asked MacDonald where he had spent the war. Danny replied smartly, 'Gallipoli, Egypt, and your desert campaign, Sir.' The wound stripe on his sleeve elicited another question, 'Where had he become a casualty?' To this friendly enquiry Danny – heedless of the horror in the General's entourage – uncoiled from attention, scratched his head and said, 'Dammit, I dinna rightly ken the lingo, but they were telling me in hospital; it would be somewhere west of the Euphrates on the Rothiemurchus side of Baghdad.'

Both Scouts had taken part in a curious incident. I refer to the train journey in 1914 when father's mounted brigade were sent from Blairgowrie to Huntingdon. This started the rumour that a force of Russians had landed in the north of Scotland, and were on their way South. The fantastic story, which spread like wildfire, had some foundation, for it fitted in remarkably well with the movements of the Highland Brigade. More than a dozen troop trains had passed through Newcastle and York, travelling in succession during the hours of darkness. Many of the men on board were reported to speak a foreign language, wear curious headgear, and be uncommunicative and shy in

* Beina Bhacaidg: the 'Hill of the Hindrance'.

manner. When asked from whence they came by benevolent ladies staffing a canteen on York platform, they could only murmur, 'Roscha' (Ross-shire). Those who have witnessed the hysteria of non-combatants in big cities will not be surprised that witnesses were soon prepared to swear that the strangers had snow on their boots, while others, even better informed, had learned from high officials the exact numbers in the Russian Expeditionary Force. Those who disbelieved such bunkum were suspected of being pro-German.

Tarmacadam had not reached Inverness in 1920, but the route now known as the A9 followed a similar course along the Firth to Beauly. Compared with today's traffic, the highway seemed deserted, but our progress on the home run was seldom without incident. The meeting of two cars was something of an event, for which both parties drew up and exchanged pleasantries, enquiring about each other's health, the state of the crops, and, when the women did the talking, all the latest gossip. Nobody was in a hurry. The horses sometimes played up, but I cannot remember any runaways or actual boltings among the carriage animals. There were few blood tits or high steppers on the road, and no buses or charabancs to panic them. A notable exception, in breeding, was old 'Mark Forrad', an entire bay horse, kept to improve local stock, who had won a premium at the Doncaster Stallion Show. He travelled the county for many years, adding bone, quality and size to the light-legged mares in the neighbourhood; but in winter he worked between the shafts of the governess cart until he grew too big and round to fit between them. The good-natured old fellow, perhaps because of increasing obesity, was accused of leaving fewer foals than his 'travelling lad', known as 'Jock Stud', in their respective wanderings up and down the roads of Inverness-shire.

Alarms and excursions were frequent among the younger nags, but all part of the day's work. Traction engines seemed to prostrate the equine mind. So, too, the startling appearance of the stone breakers, armed with long hammers, and with goggles to protect their eyes, who worked at strategic points along the route, cracking piles of whinstone rock into small pieces of road metal. In such encounters, the youngest or most able-bodied member of the party, ignoring proffered advice, sprang down, with soothing words and a tight grip on ear and bridle, then forced the reluctant steed to pass, like the Levite in the parable, upon the other side.

The Aird of Lovat was viewed with conscious pride, for every prospect pleased the eye. At Phopachy Point, where the highway runs closest to the sea, seals were counted at low tide on the sandbanks. Their presence indicated the movement on the coast of returning salmon which would soon run up the river Beauly. The fertile plain, seen drowsing in the sun, bestowed an air of peace and plenty upon the countryside. It was rich in history too: the Lovat lands had withstood the inroads of the Norse Jarls, Celtic Mormaers and the raids of ravening Islesmen and the Western clans.

When Norman blood settled the region with longsword and chain-mail (intermarrying with the resident Gaels), they brought with them the benefits of education and improved agriculture. William the Lion and later his son Alexander II had crossed the Aird with punitive forces to restore law and order in Caithness during the thirteenth century. It was at Phopachy that the Reverend James Fraser (born 1634), Minister of Wardlaw and Kirkhill, wrote the Polichronicon Manuscript, tracing family history back to the year AD 916. He was a person of lusty prejudices: men of the Fraser name, long since crumbled into dust, return to life in his hands, as their doughty deeds are suitably extolled. It is an enthralling story.

A son of the manse, married to another Fraser and father of twenty-four children, the Reverend James lived in stirring times. He well describes the scene as hostile armies marched and counter-marched across his parish. The figure actively involved was the great Montrose, the royalist general, or King's Lieutenant. Scotland's proudest soldier has remained my personal hero, though the clan suffered severely in his wars. After the betrayal in Assynt, Montrose was led past Phopachy on the long ride to Edinburgh and execution. The Wardlaw account describes high fever and torn clothes; hands lashed to his back and feet tied below the belly of the shelty on which he sat, without a saddle. When the Provost and Bailies of Inverness stepped into the street to offer sympathy for the sorry circumstances, Montrose gave the gracious answer, 'And I regret to find myself the object of your pity.' He died in the high Roman manner. No man has better served the Stuart cause. Montrose is a figure who haunts all those who travel the rough roads of Scottish history.

And now the horses' heads are fairly turned for home. At the Tinkers' Hole, a gypsy encampment on Inchberry Farm, rough-coated lurchers – big savage dogs, half-starved and wholly wild – rush out to snap at heels, urged on by the resident urchins. A bag of stones was kept under the box for pelting the canine assailants, as, with derisive shouts and much cracking of whips, we trundled by at an increased speed.

A final hazard, experienced in September, which slowed the carriage down, was the exodus of many thousands of sheep, all walking in the same direction as they drifted to the markets of the South. For days on end, droving turned the King's highway into a stock route. Farm economics have changed; so has the taste in mutton. With the fall in wool prices during the 1960s (barley and afforestation have both become more profitable than sheep breeding) for there will soon be no shepherds to look after the more remote flocks that remain, it is hard to get people to go back to empty glens. But even today thirty thousand lambs pass through the ring in the big August sales at Lairg and Thurso, the market centres which cover the produce of Sutherland and Caithness, where trees are hard to grow. Then, there were no cattle floats, and the drovers slept with their flocks, as they nibbled their way across country to journey's end in the abattoirs of Perth and Aberdeen. Big dealers rented grass parks and clover aftermath along the route that lay ahead. The moorland heather beside the road was all right for the hardy Blackface, but prudent buyers saw no need to lose carcass weight on the line of advance; the two- and three-year-old Cheviot wethers, then in demand, were big, heavy animals that increased the numbers in the road, eating up the grass like locusts on the march. The cattle went by train; many village butchers owned farms or rented grazing to fatten beef for their requirements. The slaughterhouses, now condemned, were kept in close proximity to the shop.

We are almost home. Gradually the landscape changes; the champagne country has been left behind. Between Bogroy and Balchraggan attention was drawn to a former arm of the sea reclaimed by a great-grandfather. The land still shows a water table of black alluvial soil, criss-crossed by open ditches. Farm boundaries marked by oak trees, and well-stocked fields surrounding the big steadings, taper up to a ridge where the Beauly Valley turns into the hills.

The village lies on the north bank of the river. The waters are tidal, and small boats once plied a useful trade with the Continent in salt herring, timber, and parboiled salmon preserved with vinegar, in exchange for such Low Countries' exports as claret, cloth, lace and brandy. A mile above the tide is the stock ford of Ross. There was no bridge across the Beauly until 1810, and the ancient ford was of great importance; it is now forgotten. The village drew its name – wrongly attributed to Mary Queen of Scots on her visit in 1564 – from the order of Valliscaulian monks, also French-speakers, who built the 'Beau Lieu Priory, in AD 1230. The old Gaelic name *Mannachin* meant 'the place of the monks'.

Up-river, the landscape reverts to Caledonia's sterner mood. The signpost pointing west to Drumnadrochit and Kiltarlity separates the Norman from the Celt. Wheat gives way to oats and lighter grains. The wind-shaken barley merges into tall deciduous trees and aromatic fir woods soon follow the carriage up the valley. Broad-leaved on the lower levels, the timber climbs to foothills, showing a variety of scattered hardwoods, then into forests of spruce, larch and pine that reach long fingers up between crofts and inbye grazings. Between the tree-line and the moor, bird cherry, gean, birch and gorse provide bright colours in the spring and autumn. Away to the west, and twenty miles beyond, like giants at a hunting, lie the high tops of Glenstrathfarrar, their shadows changing with the sun.

So into the Park, along the bank of the salmon river that appears suddenly at the lodge gates. The hurrying waters and deep pools form a natural barrier on the north side of the policies. Past the smithy at Old Downie, the trotting horses are greeted by whinnies from the forge: across two burns where planks rattle on each bridge. A climb, and then the castle – set high above the river – 'a thing to dream of, not to tell'.

The red sandstone tower, focal point in the golden countryside, transmits an air of profound tranquillity. In summertime, its long shadow reaches down to touch the river, which reflects the earth and sky. Perhaps that sense of perfection should be left where it belongs: to childhood memories. But I still have the same warm feelings whenever I see Beaufort and I hope I always shall.

CHAPTER FIVE

LOCAL COLOUR

Come, I'll show you the red deer a-roving.

LIZZIE LINDSAY

'Ahi-Aii-Aii-Yah!' The war whoop of the Ogalala Sioux carries half a mile, and when Halliday Sutherland's rod tip whipped over to the tug of a sea trout, the sound went further.

Hallie, demobilized from the RNVR, had been invited to fish the home beat before returning to medical practice. He was another hero: mother said he drank, but father approved of the cheerful extrovert. The bright autumn morning was the first of many visits from Invergordon. The red-bearded naval surgeon, who had made his mark in tubercular research, was related to neighbours in Easter Ross. His enthusiasm and increasing success as an author did the rest; for he had natural charm. Hallie organized his summer holiday serving as a *locum tenens* armed more often with gun or rod than with stethoscope or Gladstone bag. The enthusiasm was infectious, and he was notably kind to young people. He could also tell a tall story to please his audience – especially about Red Indians.

When the Wild West Show visited Scotland in 1901* Hallie's father was practising in Glasgow; my friend recalled that as a child he had met Sitting Bull, the Sioux Chief, and Buffalo Bill Cody. Hallie fired my imagination with tales of Redskins, intrepid frontiersmen, the great plains, and the way the West was won.

Buffalo Bill had already taken London by storm during the Golden Jubilee celebrations. The Circus caused a sensation. What shows they must have been! Until then the noble savage of Fennimore Cooper, Longfellow and 'Ballantyne the Brave' had moved with solemn dignity among shelves containing the collections of my parents' Victorian childhood. With the changed outlook, the likes of Deerfoot and Hiawatha seemed no better than collaborationists. After the last

* See *A Time to Keep*, by Halliday Sutherland.

of the Mohicans went to their long home (penny dreadfuls were not allowed in the schoolroom) there grew a gap in the Redman-Paleface story. During the same period the Indian had learnt bad habits from the pioneer.

Hallie projected an all-action display of the Sioux on the circus warpath in Glasgow. He re-enacted various scenes: Sitting Bull (the Medicine Man behind his warriors), Crazy Horse, and Red Cloud (Chiefs who had made history on the Little Bighorn, wiping out Custer and the US cavalry). Hallie added ideas of his own: smoke signals, the Pony Express, painted mustangs, a mail-coach beating off the first attack, the wild gallop through a prairie fire, the settlers' last stand, naked braves swarming over the stockade. The final massacre – then mounted troops riding to the rescue; moonlight on the battlefield, and Buffalo Bill removing his stetson in the presence of the dead. 'Too late, too late.' Many a rough hand dashed a tear from a sunburnt cheek! The mime was better than the movies and my friend was better than the originals.

Hallie's blithe spirit erred on actual facts. The tall braves who came to Scotland, finding it cold and wet – with many evil spirits – did not include the man who led the Ghost Dancers in the last fight for freedom in the Black Hills.*

But there was a near massacre in Glasgow: on arrival in Scotland, Colonel Cody publicly requested that none of the Sioux should be supplied with liquor, for fire water was bad medicine and meant trouble. The appeal was overlooked. Unfortunately, one brave more venturesome than the rest of the cast, stepping into Sauchiehall Street, found a bar which served him whisky. When the Chief, for it was Red Cloud, returned to the showground, all who saw him fled. All but one – an intermediary who rashly advanced and advised the red man to sleep it off in his tepee. The suggestion was not well received, and Red Cloud went on the warpath. Swinging his tomahawk and raising the war whoop of his people, the Sioux struck down the agent and made for the next paleface and more scalps. Just as suddenly, the madness left him; the tomahawk was thrown down and Red Cloud, without a word, went meekly to the nearest jail. The interpreter, with a fractured skull, was removed feet first to the Infirmary. Next day the Chief was committed to the High Court on a charge of assault with violence. While his victim remained unconscious, the case could turn into the capital offence of manslaughter. Hallie's father was the prison doctor, and reported a contrite warrior sitting cross-legged on an antelope skin on

---

* Sitting Bull was treacherously murdered, unarmed and in cold blood, after riding back to give himself up, by a trigger-happy VIIth Cavalry 'pony soldier'.

the floor of his cell. Little Hallie was sent round with fruit and other tokens of goodwill. Glasgow's indignation was directed against the publican who had sold the whisky; the interpreter quickly recovered; all was forgiven and the show receipts went up again. But not before a deputation of silent braves, with a guide to lead the way, presented a war bonnet of eagle feathers to the doctor's wigwam in Pollokshaws.

The names of those who performed in Scotland are in the files: Red Cloud, Kicking Bear, Iron Tail, Little No Neck – a Sioux waif saved from the massacre and found beside the bodies of his parents at Wounded Knee – and Young Man Afraid of Horses, who was asked by reporters, 'What's happened to Young Man Afraid of Mother-in-Law?' History relates that the noble savage did not reply to this facetious sally.

Though I recommend the habit, dignity is no longer a civic virtue. The benign effects of circus and civilization left the red man unchanged – a stranger who kept his own council. Off their horses – then as now – the Indians were a disappointment to the staring crowds who flocked to see them, not speaking once in a whole day and communicating only in sign language.*

Then, as now, the white man who destroyed their habitat considered the savage no better than a freak. Time was when, in the struggle for survival, the Sioux brave with equal opportunities was the better scholar of the two.

I write with feeling for a race who only asked to be left alone and was destroyed by civilization and its greed of gain; the names of Sitting Bull, Red Cloud, Crazy Horse and Rain in the Face should not be forgotten. They, too, were Americans and showed the same heroic qualities as their conquerors.

Years later my lot fell to open the Calgary Stampede in Alberta. The buck-jumping, steer-wrestling and roping is a sight worth seeing. Part of the colourful affair is a downtown dress parade of cowboys and rodeo competitors, led by a war party from the nearby Micmac tribe. The same week the VIP visits the tepees. After a puff of the peace pipe and an initiation ceremony for the tycoon, the visitor is dressed up and given an appropriate name. Among the foothills of the Rockies, I can be identified, clad in eagle feathers and porcupine quills, as Chief Eagle Eye. Not very convincing! How a once proud people must hate the rôle now exploited for a tourist attraction. The spark has gone but the impassive reservation Indian remains unpredictable and is sometimes dangerous.

* Cunningham Graham in the short story 'Long Wolf' has left a moving account of one who stayed behind. Cody died in 1917.

I often heard father extol the merits of Colonel Cody: pioneer, scout and Indian fighter. He had been a spectator at the circus in 1887. After the South African War, both he and Baden-Powell endeavoured to indoctrinate those same qualities of determination and 'go it alone' independence recently shown by commando and plainsman. Boer War critics had said with scorn that no British Expeditionary Force was capable of marching out of step and needed a brass band to get them started. Father maintained Buffalo Bill was in part responsible for the success of the boy scout movement.

'The Wild West' brought a breath of fresh air to suburbia. Father described how at Earl's Court the Prince of Wales had ridden on the Dead Wood Coach with three future monarchs. Buffalo Bill drove into the arena at full gallop – an unheard of precedent in those formal times. The Prince, hanging onto his top hat, had remarked after a triumphant exit: 'You held a pretty good hand up there on the box, Cody, with four Kings to play with.' 'Better than that,' the Indian scout replied, 'four Kings, sure, but if you count me as Joker, that makes a Royal Flush.'

I remain an admirer of the Red Indian. The vanished race whose sign language once served the length and breadth of a continent shows a marked affinity with Highlanders of yore. They shared similar attributes: a sense of identity, courage and respect for traditions built round the hills they loved so well. Historically, mountains have always been the home of freedom and independence, while plain-dwellers tend to be subordinated by conquests. I sense a kindred spirit in the Indian's veneration for the Red Pipe Stone Quarry, focal point in a limitless horizon that stretches up to peaks hung suspended between earth and sky. There is the same message in the slogans of the clansmen, who invoked their mountain as they charged in battle. 'Cruachan for ever!' has carried the Campbells through many a bloody fray. 'Stand fast, Craigellachie!' was Clan Grant's call to the rock above them in the hour of need.

At home there were two medicine men in the Beauly district when we returned after the war: Donald Macdonald and Jock Leach, both admirable physicians. In such a far-flung practice they spent a great part of their lives on the road in dogcarts – machines, as they were called in Scotland.

Doctor Leach had a cob which, it was said, could trot all the way to Braulen, a distance of sixteen miles, only breaking its action for a breather on one or two steep gradients, which were presumably negotiated at a walk. After a check-up and a talk to the shooting tenant, a Mr Buchanan (who was in the whisky business), the doctor and his rested pony were ready to trot back, change horses and go out on another case.

Mr Buchanan suffered from duodenal trouble, and it was said that Leach cured him with the old Scottish simple of drinking 'sowens', an infusion of hot water strained off broad beans, boiled over the kitchen stove.

Mr Buchanan survived to remain my father's lessee in Glenstrathfarrar, the estate's best deer forest, outliving the Aberdeen and West Highland terriers – his Black and White companions on the whisky bottle's label, for he distilled that famous brand. Doctor Leach lent an attentive ear to Buchanan's views on high finance and acted on his advice. When he died, his son, young Doctor Jack, inherited a fortune built up from shrewd investments on the Stock Exchange. But poor Jack hit the bottle and died young, crashing an expensive car.

The glen was a wild place in the old days and one of the last strongholds for illicit stills and whisky smugglers. Before the First World War, the mail delivery handled by the blind postman, George Tate, whose horse knew the passing places, ran once a week and only in the shooting season. George called himself 'the carrier' and refused to wear a uniform. Emergency communications between Braulen and Beaufort were sent by pigeon post. The old stone doo'cot can be seen west of the Lodge between the head stalker's house and the ghillie's bothy. The winged messengers sometimes failed to arrive, for they risked sudden death from the falcons which nested in the Carness Rock above Loch Mhuillie.

Doctor Macdonald, who looked after the Lovat family, was an older man, both in appearance and outlook. Always dressed in an Inverness cape and deerstalker cap (with ear-flaps kept in place during fine weather by ribbons tied in a bow on top of his head), the old medico had a penetrating gaze which disconcerted us as children. This effect was accentuated by mutton-chop whiskers, gold pince-nez and a conflicting smell of peppermint and mothballs, as he bent over the bed to tap chests or examine our tongues.

Doctor Macdonald was an excellent botanist, which endeared him to my mother. I remember my first visit to Cape Wrath and the John o' Groats promontory was to help them collect specimens of the Caithness Primula, which, though common locally, remains one of the rarest of British wild flowers.

An adventurous (and more romantic) journey – for a Jacobite – was a visit to the Hebridean Island of Eriskay, where Charles Edward Stuart made his first landfall to raise the clans in 1745. Tradition has it that seeds of the Prince's Flower (*Coilleag a' Prionsa*, in Gaelic), traced from his sojourn in Italy, fell from a royal pocket on the Highland strand. Better known as Sea Convolvulus or *Calystegia Soldanella*, it still pushes up trumpets of pink flowers in the *machar* (dunes) above the beach. This is a pretty legend but not entirely true: sea currents have carried the rarity to other places along the Atlantic coastline.

I remember my surprise on an expedition up Ben Lawers (Perthshire) to find the Drooping Saxifrage (*Saxifraga Cornua*) when the old doctor, after high tea in the clachan of Fortingall, took off his pincenez and said, 'There are two remarkable facts about this village, once the camp site of a Roman legion. Down the road apiece is the oldest tree in Britain. You Shimi can work out the age from your school books. It was under that yew tree's branches before the birth of Christ that Pontius Pilate first saw the light of day. His father commanded a garrison on this very spot.'

Through children's eyes our doctors were firm friends, but this could not be said about dentists in the county, who were not abreast of the times.

One worthy, a retired army officer, had been invalided out of the Veterinary Corps. After giving up a practice to serve King and Country, he was wounded in the line of duty. Father befriended the veteran no longer able to calve a cow. When the ex-vet hung up his shingle I was one of his first customers. The humorists maintained that his professional skill was acquired by strenuous practice with hammer and chisel on a horse's skull. The amalgam in the stoppings put into my head startle the modern practitioner, but the fixtures (they look like iron ore) have stayed in place.

The Maxwell cousins (twelve in number, six boys and six girls) used to parade with others in Beauly for routine inspection by a frock-coated gentleman from Aberdeen, who set up his foot-pedalled instrument of torture in the village hall.

Known as 'Doc. Holliday', after the Texas bad man, he was, in truth, a dour but kindly Scot. One of his best remarks, in the accents of the Granite City, caused family merriment. Dental investigation took the form of a running commentary as he pried around for suspected cavities: 'In a prreliminarry examination, I was about to prronounce decay, but on furrtherr scrrutiny, I detect it is only chocalat.' (Those who visit the land of the mist and the mountain can supply the accent for themselves.)

In the village people who suffered from periodic toothache preferred mass extractions to the risk of further pain, and many a pretty girl lost early good looks: for false teeth were unknown in the remote country districts.

The joy of Hallie's stories was how he embroidered the simplest anecdote with the wit and colour of a lively imagination. Part fact, part fiction, he was seldom uncharitable. My father loved the droll descriptions of the fierce old worthies of the church. How he laughed at the tales of Red William. To his flock the Reverend W. MacGroanach was a great man, far greater than Principal Rainey, the celebrated Moderator of the Free Church of Scotland. 'Rainey!' they'd say, 'Rainey was only three years at college and oor Minister was ten!'

The following story of Red William's visit to Moy was a favourite:*

Once, in the absence of the Minister, Red William travelled up from Glasgow on a Saturday to take the Sunday service. On the platform was a schoolmaster with whom Red William was to stay, and by chance, the Mackintosh, who invited the clergyman to sleep at Moy Hall. Red William accepted.

'And your luggage, Mr MacGroanach?' asked Mackintosh.

'In my overcoat pockets. I only carry a nightshirt and a toothbrush.'

Nor had he need of aught else, for his flaming red hair and beard were left to nature's care.

'An hoo's her Ladyship?' enquired Red William as they set off in the carriage.

'My wife's very well,' said Mackintosh, 'and you'll be seeing her at tea.'

At Moy Hall the great adventure, as told by the Minister, began:

'There were a lot of people in the drawing room and her Ladyship, who gave me China tea with wee French pastries and such like. I thought it a poor sort of meal for a man who had come off a journey. After the repast, if such it were, she said I must be tired and asked the butler to show me to my room. It was a grand room, but none the less I thought these people were awful queer. It was only six o'clock and o'er early to go to bed. But in Rome ye must do as Rome does, so I put on my nightshirt, got into bed, and went to sleep. I was wakened by a great noise. Some instrument of brass was being beaten with a hammer. I rushed out along the corridor and stood on the landing in my nightshirt. There I saw half-naked women hurrying down the stairs. I knew at once what had happened. At the top of my voice I shouted: 'Flee for your lives, good people, the hoose is on fire.'

Neither my mother nor Ronnie Knox, who were converts and respectful of all creeds, thought much of the following tale which convulsed another catholic, Maurice Baring:

Frosted snow covered the countryside and the roofs of the scattered houses in a Highland parish. The night was clear, snow sparkled under the light of

* See *A Time to Keep*, by Halliday Sutherland.

moon and stars and the wind was freezing cold. In the manse, the Minister, an elderly bachelor, was sound asleep in his feather bed and the hour was long past midnight. He was awakened by a loud knocking on the front door, and grumbled as he arose to attend the call of duty. He lit a paraffin lamp, went downstairs and opened the door. Holding the light aloft he recognized his caller. It was Donald, a sanctimonious parishioner, whose worst fault was a love of whisky.

'Well, Donald, what is it?'

'Minister, I canna sleep.'

The icy wind blew between the Minister's legs and ballooned his nightshirt.

'That's no reason for disturbing me at this time of night. Go on to the doctor's house and see what he can do for you.'

'Na, na, Minister, you dinna understand.'

'What don't I understand?'

'The doctor's nae use, it's a spiritual malady.'

'Well, Donald, what is this spiritual trouble?'

'I canna sleep for thinking of the awful schisms of the Kirk of God. There's the Frees, the Wee Frees, the Seceders, the UPs, the Episcopalians and the Roman Catholics. It's awful to think of the schisms in the Kirk of God.'

'Well, Donald,' said the shivering Minister, 'I'm glad, whatever the circumstances, that you've come round to considering the things that matter. Now look here. Be back and see me at four this afternoon, for it's already morning, and we'll talk it over in my study – but see that you return sober.'

'Na, na, Minister. Ye dinna understand.'

'What don't I understand?' shouted the Minister.

'Ye dinna understand that when I'm sober, I don't care a damn about the schisms in the Kirk of God.'

Our parish priest at Eskadale was a great original, as they say in the south of Scotland. The son of a Buckie fisherman, educated at Saint Sulpice and invigorated by periodic visits to the United States, where a pretty niece had emigrated to marry a Detroit millionaire, Aeneas Geddes preached the best sermons heard by my generation in the parish church. The epistles and gospels, read and explained in a clear voice, seemed the very embodiment of our unspoilt countryside. The changing patterns of the pastoral season – shepherds with their flocks, tillers working the land for the early seed bed, the reed shaken by the wind, fishermen washing their nets, the wheat and the cockle – all were there, the green shoots, the voice of the turtle, and, best of all, 'behold the lilies of the field'. These were growing things and living people invested with a simple beauty that a child could understand almost better than the grown-ups in the congregation.

During the shooting season visitors of all denominations came to hear these Eskadale sermons. There were two notable exceptions. One hot

Sunday morning the Paget brothers, who rented Eskadale House for many years, strolled up the bank of the Beauly to look at the salmon pools, exasperated by the inactivity of the Sabbath Day. They could not resist shouting through the open window to the priest climbing into the pulpit: 'Try them with a Jock Scott, Father.'

At Traquhair Kirk, when my mother was a child, the sheep-dogs lay beside their master's pew and the collection was put into a plate laid upon a white cloth on a table set outside on the green. My aunt, Elsie Maxwell, a much older woman, born in 1870, could remember being told as a child that in wet weather the scattered parishioners of Strathglass, who covered great distances on foot to get to Mass, took their shoes and stockings off before crossing the burns and soft ground in order to arrive neat and tidy. The dry footwear went on again in the church stables, where the coach horses were put up during the service.

After the Reformation, the upper Beauly Valley and the glens of the Chisholm country that run into Strathglass kept the old faith and never lacked a priest, though the Presbytery both in Dingwall and Inverness frequently appealed to higher authority to 'extirpate the numerous papists trafficking in the remote mountains'.

Much later, after Culloden, two thousand men, women and children emigrated to Canada from this high, wild country, which runs south and west to Glen Garrie and Glen Quoich. They took their clergy with them. Today fewer than fifty families attend Mass at Eskadale and Cannich. The last-named parish had produced two bishops in the last century, and one of them known as Mr Grant is still remembered as a leading exponent of Highland dancing and a mighty putter of the stone and heavy hammer.

My immediate forebears, who were religious people, built churches at Beaufort, Beauly, Eskadale, Whitebridge in Stratherrick, and Morar, besides presenting the land for the Benedictine Abbey and School at Fort Augustus.

Once a week, Mass was said in the Chapel at Beaufort. Until I went to school, the rest of the day, after some sick calls, was a holiday, spent, according to the season, in shooting or fishing expeditions with Father Geddes (on his motor-cycle combination to remote hill lochs) or ferreting rabbits in the warren with help from a kennel boy.

At ten I graduated beyond Willie the Moon's school of angling with garden fly (worms) and night line. Father Geddes made his casts and dressed his own flies. With few opportunities among the salmon on the lower reaches, which were let to tenants, the priest was a pioneer among the greased line fishermen who learnt to present a fly in acceptable form to a grilse lying in water formerly believed too slack for casting. In the upper tributaries of the Farrar and the Glass he knew each lie and taking place according to the height of the water. If the salmon was there, Father Geddes had him out, and maybe another later in the day that took its place behind the same stone, where less observant people had never thought of fishing. His services to obtain a salmon for some important occasion were always in demand. He kept his secrets to himself, though he gladly taught how to mend a floppy cast and hang a fly, which is the best way to catch more than the next fellow.

An admirer of the Canadian dry fly and greased line technique whose early exponents were having successes with dour summer fish in low water, the holy father confided that the principle was sound enough, though never likely to succeed in Scotland, where salmon lie in fast, shallow water at the heads of pools, wherever they can find a current. He had proved to himself that a moving air bubble rather than the size or colour of the fly tempted a stale fish. (This can be produced by a hairy dressing like a stoat's tail – then unknown.) But the surest lure of all was a strip of the skin of a water shrew, cast square across a pool, with the line pulled back through the rings of the rod as the sun left the water at the end of a hot day. John Buchan's story (in *John Macnab*) of the poached fish in the slack water of the Lang Whang was written after a day on the river with Father Geddes. (The original of John Macnab was Jimmie Dunbar of Pitgavennie, who commanded a squadron of the 2nd Lovat Scouts in the South African War.)

On the mantelpiece in the Chapel House sitting room, where a peat fire burned winter and summer, the priest kept a framed treasure: a quatrain, with a Latin tag, on a scrap of paper. The clever words are attributed to Sir Herbert Maxwell, another companion on the river:

Salmonum Piscator: Piscatori Hominum

You fish for souls: For Salmon I.
On different schemes we each rely,
You angle men to firm believing,
While I succeed by sheer deceiving.

Father Geddes did not shine with a gun, and our surprise was unbounded during a day over very rough country, when he was sent ahead to intercept a wood stag which was causing damage to a larch plantation. Three single reports were faintly heard, with a long pause between each round: I think they were the only shots during the drive, for some woodcock, that had been hoped for, were not at home. Opinions were divided as to whether the same deer had been fired at three times or if a fox or wildcat had also sneaked out through the pass; no other animal would move as far ahead of the advancing beaters.

We were all proved wrong. Father Geddes had accounted for three cock capercailzie *Tetrao urogallus*, or giant wood grouse) – an unheard-of feat. Each bird was as big as a turkey and as ancient as the Caledonian Forest. 'Did they fly well, Father?' 'Yes indeed, they looked no bigger than sparrows,' was the reply. The guns were impressed and not a little envious, for a hat-trick by a bad shot with 'bull' caper in winter (when they are fully feathered and very wide awake) is equivalent to winning the Newmarket autumn double, at about the same odds.

It was after the shooting season that the Urchany shepherd gave the show away. Working with his sheep, high up on the open hill, he drew his telescope and stopped to watch the drive; the caper had got up singly 300 yards ahead of the advancing line of beaters and, black against the westering sun, flew the half-mile length of the young plantation, to alight in turn in a sentinel pine at the west end of the wood. The sun must have set blindingly that afternoon, or they would surely have seen Father Geddes, crouched in ambush, slowly raise his gun to blast them in turn out of the topmost branches. As he said, they were very high birds! His promotion in the Church (he became Canon Geddes) coincided with this red-letter event.

During Prohibition years, Canon Geddes visited his relations in the United States, and his luggage then included two canvas portmanteaux marked 'Glass – Very Fragile'. He received preferential treatment when the liner docked. 'Have you the crystal this time, Father?' was the cry from officials below when, after a summons on the loud-hailer, Canon Geddes appeared smiling over the side. 'Say, fellows, go easy [down the chute] with that crystal.' The Glenlivet was ashore in no time!

Halliday Sutherland met the Canon and was persuaded to listen to one of his famous sermons. The two men afterwards became close friends, in spite of the doctor's leg-pulling:

One day an Irish priest met a member of his congregation who had been absent from Mass for many months. He was not a boy to be bullied, and the priest determined on diplomacy. He made no reference to religious duties, but talked of secular matters, until the lad enquired, 'What's sciatica, Father?'

'Ah! And now yer askin' me. I'll tell ye just what sciatica is. It's an awful disease that ends in madness and a painful death. It comes to men who don't go to Mass, who lie drunk on the roadside at night, and go with dirty women. An' now ye'll be tellin' me – how long have ye had the sciatica?'

'Glory be to God, I've niver had it, but I was readin' in my paper today that the Holy Father has it bad at this moment, and is lying in pain on his bed in Rome.'

The Belladrum Estate, near Beauly – it was said to have been lost in a gambling debt – is surrounded by Fraser property and changed hands for the second time to a rough-and-ready coal and iron master from the Lowlands – a Mr Merry, who was rich enough, among other things, to win the Derby a century ago. Not supposedly a religious man, he still wished to make an impression. His first action was to hand over to the Established Church of Scotland a handsome cheque, remarking as he did so, 'That's the highest premium for insurance against fire I ever heard of.'

The Church of England has got off lightly in this ecclesiastical leg-pull.

Two tombstones in Eskadale churchyard provide a pendant for this chapter. They were erected over the graves of John Carter Allen (1795–1872) and Charles Manning Allen (1797–1880), two brothers, claimants to princely blood who had fought for Napoleon at Waterloo and had subsequently come to London and then to Scotland, where they described themselves as 'the Sobieski Stuarts', announcing that they were the legitimate heirs of the Royal House of Stuart. While in London the brothers learnt Gaelic. On arrival in Scotland, my cousin Christian Hesketh in her monograph on the origins of tartans wrote:

They were received with deference, first by Lord Moray and later by Lord Lovat, who invited them to settle on any part of his estates they might fancy. They chose for their abode a small island in the Beauly river, and at Eilean Aigas spent their time in painting, scholarship and sport.

Imposters though they must have been, they did know a great deal about the Highlands, and through their extensive knowledge of Gaelic were able to acquire much of the information contained in their published works. In 1842, the first of these, the Vestiarum (*Vestiarum Scotocum*), appeared. It was a treatise on tartan, supposedly based on three ancient manuscripts, of which the brothers claimed to be the sole possessors. It need hardly be added that no outsider was ever permitted to see them.

... It is still far from clear why the brothers, who were scholars and capable of better things, should have lent themselves to an idiotic fraud. Possibly their faith in the manuscripts sprang from the same mad, mysterious source as the belief in their own kingly descent!

The Vestiarium was assailed, and rightly assailed, on all sides. Its origins were dubious. Its theories were absurd ... but it must also be remembered that with all its faults, the book remains the sole authority for a number of well-known tartans, and very pretty ones, still in use today ...

For many years the strange pair, one of whom was now married, continued to live on their land-locked island. ... Finally, the house of Eilean Aigas was given up; they moved to Austria, then back to England ... one of the last glimpses that history affords them is of two old gentlemen, handsome and courteous as ever, sitting side by side in the reading room of the British Museum with knives, pens and paper-weights spread out in front of them, all embellished by a small gold crown.

In a memoir, my aunt Elsie Maxwell wrote that the brothers 'certainly had a marked resemblance to [the Royal Stuart] family, and they owned some very personal Stuart family possessions. My grandmother thought highly of them and welcomed them to Beaufort.'

When I married in 1938, my mother moved from Beaufort to Eilean Aigas, where she remained throughout the war, looking after the bedridden Maurice Baring. On her death I gave the island of Eilean Aigas to my brother Hugh, where, as often as parliamentary duties permit, he resides today with his large family.

It is a peaceful spot, this sunny hillock of heather and tall trees, surrounded by the rushing waters of the Beauly. That and the deep-chorded cooing of wood pigeons are the only sounds; with springtime come the primroses and bluebells; by early summer the air is laden with the scent of wild azaleas and, in white drifts below the trees, of lilies of the valley. But pride of place must go to the Eucryphias planted by Compton Mackenzie in the moist peat soil which faces south along the river walk. Rising high above their neighbours, with feet protected by a jungle of rhododendron bushes, they flower, white pyramids of blossom, throughout August and September.

Against this background literature and politics have flourished since the *Vestiarum* was written, for Sir Robert Peel, Compton Mackenzie, Maurice Baring, my mother, brother, and sister-in-law Antonia have all written books or composed speeches there.

Some of the furniture made by the Sobieski brothers remains in the house, together with the big folding doors that were thrown back, in royal style, when the brothers granted an audience. When I was a lad there were old people in the congregation who remembered being told by their parents how the pair were rowed up-river to Mass in an eight-oared barge, with the Royal Standard flying from the stern.

Spare a thought for Highland Dress. In kilted regiments, company commanders on occasion offer a prize, competed for by all ranks, for the smartest Highlander on parade, and the winner from each company is then inspected by the CO to be judged the best-turned-out soldier in the battalion. Long may the custom continue, for the small vanity unconsciously serves to erase the memory of the Hardwicke Act, which prohibited the wearing of the tartan after the Battle of Culloden. The repeal of this Act was moved successfully in the House of Commons by the Marquess of Graham (afterwards the third Duke of Montrose) and seconded by the Hon. Archibald Fraser of Lovat, MP for Inverness. In 1782 the kilt and bagpipes, also considered rebellious and an instrument of war, were restored to Gaeldom and their rightful place in Scotland.

On the occasion of George IV's visit to Edinburgh in 1822 the kilt, tartan and the bagpipe – indeed all things Highland – erupted into sudden prominence. The climacteric of this vulgar affair was a Levée held at Holyrood House when, to the gratification of all present, His Majesty appeared in a kilt of Royal Stuart tartan. The King, incidentally, wore flesh-coloured silk tights that covered his knees and disappeared into his stockings. A view of this bizarre scene comes from a contemporary, Elizabeth Grant, who blames the Monarch's strange attire and the flesh-coloured tights for preserving decency – to David Stewart of Garth, the general who superintended the dress parade: 'Everybody stormed from every point in the country to get to Edinburgh to receive him. Sir Walter Scott and the Town Council overwhelmed themselves with the preparations. The King wore full Highland Regalia . . . someone objected to this dress, particularly in so large a man – "Nay," said Lady Saltoun, noted for her wit, "we should take it very kind of him, since his stay is so short, the more we can see of him the better." ' Sir William Curtis, the Lord Mayor of London, was kilted too and standing near the King: 'Many people mistook them: amongst others John Hamilton Dundas, who kneeled to kiss the fat Alderman's hand. When finding out his mistake he called out "Wrong by Jove!" and rising, moved on undaunted to the larger presence.'* Whatever the outcome of this national pageant, the King, before leaving Scotland – and I write as an interested party – gave an assurance to restore the forfeited Jacobite peerages; a graceful gesture that cost little but meant a lot to the clans concerned.

In the Highlands bowler hats and reach-me-downs sadly outnumber the kilt save on festive occasions or at funerals. *Tempora mutantur* and perhaps barbed-wire fences may have something to do with it. A kilt must hang from the hips and not sag either in front or behind. But it always looks well on the right man – swung boldly as to the manner born. Still and on too much distinction may miscarry.

---

* Byron would have none of it and mocked the whole affair:

> Tell them that youth once gone, returns no more:
> That hired huzzas redeem no land's distresses:
> Tell them Sir William Curtis is a bore.
> Too dull for these, the dullest of excesses.
> The witless Falstaff of a hoary hall
> A fool whose bells have ceased to ring at all.

All the pistols, powder horns, crystal buttons, leather belts and buckles seen at certain junketings spell out 'Clan Mayfair and Victoriana' which Highlanders can do without.

The kilt has retained its popularity abroad. When the King and Queen visited Canada before the war, there were more kilted regiments in that great Dominion than could be found in our own Army list.

A last story of the *nouveau riche* Londoner with a very shaggy sporran, who acquired a fine property in the North. He revelled in all things Highland, and strove to keep up appearances in spite of the Cockney accent. His butler – servant of the late laird – due shortly to retire, disliked the new régime. To impress the guests after a large dinner party, the host called for the butler, rapping the table for silence. 'Angus – bring in the pipers.' After a pause the ancient servitor returned with a silver salver heaped high with the daily papers, which he handed to his discomfited employer.

Tartan, like good wine, mellows with age and the vegetable dyes in the durable cloth look best after a lifetime of hard usage. What is left of my own garment, which I am told is a disgrace, belonged to my father before me; it looked faded when he died nearly fifty years ago. 'The Kilt is my Delight' (reel tune), and I feel only half a man in trousers. It is a serviceable dress – cool in summer and, when wrapped tight, it protects lumbar regions on a winter's day. It provides freedom of movement, but remember a certain discretion is required when sitting in easy chairs.

CHAPTER SIX

AMPLEFORTH

My master whipped me very well: Without that,
Sir! I should have done nothing.

DOCTOR JOHNSON

It is difficult to write amusingly about one's formative years without a certain bias. Indeed, there is a risk of becoming pompous, for growing up is a serious business.

Nothing makes more tedious reading than the hardships, real or imaginary, of manly little fellows pitchforked out of mother care into the stern realities of school. Were children so different long ago? I doubt it. I am sure they cried themselves to sleep – as I know I did, having been packed off to boarding school at an early age after proving incorrigible at home. The upheaval was no worse than expected. Ampleforth was a happy place and I enjoyed the time I spent there. The outlook of others, as much as education, leaves its influence on character and achievement. We are all affected by the strong pull of heredity and environment, but these require the right background to yield any positive advantage. In the gregarious sense, I was luckier than most, with many cousins (Frasers, Stirlings and Maxwells) sharing those early years in the North Riding of Yorkshire.

Ampleforth was *avant-garde* and unconventional. A boy's opinions and tastes were a matter of personal choice, and nonconformists were not necessarily considered mad and seldom boycotted. Wealth and ostentation were frowned on; discipline was an accepted fact of life, creating a built-in confidence in those who gave the best example.

If boys were left to their own resources they were expected to work hard and show public spirit in return. A Catholic school, with monks as masters, need not view the world through stained-glass windows, nor do religious orders all favour cold water or a restricted diet in the pursuit of virtue and the greater glory of God. I look back with affection on industrious and dedicated men. 'Laborare est Orare' is the Rule of Saint Benedict: sound advice, and the school lived up to it, both by precept and example. There was always plenty to do. During the mid-twenties, the senior boys, directed by Father Clement the 'Clerk of the Works', in their free time landscaped a nine-hole golf-course, laid a cinder running track, and built a stop butt for the rifle range.

The teaching staff did not take salaries beyond what was needed for their sustenance, thus helping to balance monastic books. The school fees, in the case of large families and dispossessed nationals, were geared to meet the parents' resources. We were happy with a great headmaster, Father Paul Nevill, who encouraged personal liberty; in the Sixth Form a sense of responsibility was expected. This was readily given.

Ampleforth was small enough for three hundred boys to pull together, imbued with the enthusiasm that came from success in games and studies. Since my time the school numbers have more than doubled. In 1925 the new house system had just begun. There was no time to feel lost or lonely and the reorganization had good effect.

For country boys there was much to offer. Ampleforth had its own pack of beagles, that showed fine sport and could win top awards at the Peterboro' Hound Show. In the summer Jack Welch, the professional huntsman, acted as a spare cricket coach. The Abbey ran its own home farm, and owned twelve hundred acres of land and timber in the valley. There were friends among employees, forthright characters who spoke slowly – choosing their words in the broad vowels of the Vale of Mowbray. Tal Benson, the ex-Grenadier village taxi-driver without a tooth in his head, was a general favourite; a keen sportsman, who knew every meet of 't'otter 'ounds' in the shire, and once drove a school party over to the Grand National at Aintree. Josh Scaife, in the school wagonette, moved at a more leisurely pace behind the grey Percheron gelding 'Fartin Jubiter'; another staunch ally, indeed a rustic trader horn, for he sold poached rabbits for Bill Stirling and myself as far afield as Thirsk. In a cottage below the playing fields, the mole-catcher walked illicit fighting cocks and kept the ferrets (when not in a desk) to net the rabbits along Bolton Bank.

Set games on muddy playing fields did not appeal until I grew to reasonable strength. I ran the school fishing, and also the aviary, breeding various rare pheasants, blue budgerigars and cockateels that sold for enough profit to pay for the seed bill.

A mile of the Holbeck was stocked with Loch Leven trout. The Fishing Club experimented unsuccessfully with Rainbows in the Fairfax ponds. The monks and the school doctor retained the shooting rights, and small boys were in demand as beaters during the autumn term.

There were kind neighbours, many of them Catholics, in the Sin-
nington country, who entertained the boys on holidays. (We had no half-
term breaks.) Sym Feversham was a generous host at Duncombe Park,
where pied flycatchers nested along the banks of the River Wye. Trout
and grayling also abounded in that charming stream. At Arden, the
Mexboroughs offered a lake full of small trout. Richard Sykes provided
Sunday coursing on the Wolds at Sledmere; Gordon Foster, MFH at
Oswaldkirk, the Scropes at Danby and Squire George at Castle
Howard asked older boys to shoot for a half-day at weekends, if not
playing representative games.

The countryside (Vale, Dale and Wold – what good names) remains
delightful. The ruins of the great abbeys – Rievaulx, Fountains and
Bylands – in their matchless setting, breathed an ageless tranquillity,
inviting even the giddiest youth to pause and ponder on the mutability
of human behaviour.

In contrast, there were some rough reminders of more recent action.
At Coxwold a man-trap hung on the wall of the village inn. In the
Wombwell Arms at Wass reposed the silver back pouch of Sir George
Wombwell, pierced by a Cossack lance; it had saved his life when he
was unhorsed in the Charge of the Light Brigade.

Rules were strict, but boys knew where they stood. To be caught
far out-of-bounds spelt loss of privilege for the term; for a second offence
the truant could expect a flogging – six of the best with an ash plant,
which was extremely painful. In upper school, boys could eat in an
hotel (without parents), on the understanding that no one under sixteen
either drank or smoked. Father offered £100 to keep me off tobacco,
and so far I have not yielded to that temptation. Corporal punishment
in hindsight was overdone, but there was no bullying at Ampleforth.
I attribute this to the self-reliance which characterized the school.
There were, however, painful moments.

A member of the Sixth Form, or a monitor, kept order in the Big
Study during revision periods which lasted for three-quarters of an hour
before supper. Any talking behind desks, falling asleep, or fooling
around, risked trial by ordeal. The stentorian bellow to step up to the
presiding rostrum sentenced the culprit (without mercy) to a Force 9
resounding flat-handed swipe across the face. The captive audience
looked on with interest. It was a matter of pride neither to flinch nor
turn the head. In this way I lost a catapult and shooting practice with
lumps of chalk that carried the length of the hall. The bruising blow
really hurt: but the rough treatment was taken in good part; there were
no ill feelings afterwards.

One sad memory of a first summer term – I was rising ten at the time – was when college boys, who were allowed to visit relatives in the Junior House, strolled over after Sunday Mass, and proceeded to eat the radishes in my garden behind the aviary. I wept on that occasion.

Another indignity – a rag which was overdone, but was not a case of bullying – came from missing the train back to school. From Cause-wayhead (a station between Stirling and Bridge of Allan) it was possible to catch the mid-day connection to Edinburgh, then York. At Keir close by, all the Stirling's cousins foregathered at the end of the holidays for a last morning's shoot. The keener sportsmen were early astir, flighting ducks before dawn and risking hopes of eggs and bacon; the school trunks stood piled in the hall. Dogs and keepers waited in the gun room; we were in the field before eight. Uncle Archie caught us up after his breakfast. At eleven-thirty that morning Peter Stirling wounded a partridge in a turnip field. The kennel boy, with dog and Peter, followed the runner in hot pursuit. It was a big field of roots. At the far end, near the bottom of the hedge, they put up another covey. There were distant sounds of firing, with another bird falling into the stubble beyond. By this time the Laird arrived, in a bad humour: he had been held up by the doctor. He honked the horn. It was time to go back and change. The boys closed on the road and the roll was called. One was missing: Peter had disappeared. Eventually he was found, but too late.

There was no time to change into school clothes; the luggage had gone ahead, the sandwich lunch so lovingly prepared was left behind in the general confusion. We were driven to Edinburgh in silence. There were other boys on the York train. Our appearance caused offence. We arrived dishevelled in what was left of plus-four suits, to be greeted derisively by the rest of the school. My braces had been removed; a new pair of green stockings hung round skinny ankles; also missing were natty woollen garters with red pompom adornments. Those garters, my special pride, were hurled from the window by the cads of the Scottish contingent as we crossed the Tweed. The cap went next – after I had scored a black eye in its defence. Bill accounted for two bowler hats in the mêlée. The herd instinct is strong in youth, and boys, like horned cattle, will turn and gore a stranger in their midst. Today there is greater tolerance. My sons tell me they were not embarrassed by their mother's hat on speech days, while new boys have escaped with their lives after being seen with teddy bears and golliwogs on pillows in the dormitory!

I had great affection for that suit of plus-fours, especially the trousers, for thereby hangs a tale.

Another summer – it must have been just before the Twelfth – Bill Stirling came home to fish the river. The water was too low, but we had good pigeon shooting, and stalked roe bucks in the early morning. Bill owned a new gun, a 16-bore. I think it ejected and was certainly made by Purdey, but it only had one barrel. Bill was bigger than myself, and had done more shooting at grouse and pheasants. I was kept to a 20-bore, an economy which rankled, but at least it had two barrels, although the gun did not eject. One-upmanship as shown in the Game Book, credited me with greater success among salmon and deer. We were a little over thirteen years of age.

In competitive mood, during a lull among the pigeons, having first taken the precaution to stuff a game bag into those same pants, I suggested a trial by ordeal to decide the penetrating power of our respective weapons. Bill readily agreed. The scene: a harvest field at Wester Lovat, with the corn newly cut and stooked. The rules were simple. We started one hundred yards apart; each then advanced five yards, halted, turned about; crouched in a honey-pot position, head down with hands clasped over knees; both contestants presenting a similar target. One shot was to be fired in turn at the other's backside.

The first exchanges fell short. At fifty yards the salvos must have started to sting. Past the halfway stage Bill started to rub his bottom, and the advance slowed down sufficiently for some one-sided badinage: 'Surely that new gun of yours throws a very wide pattern?' But Bill still kept a'coming. At thirty yards he sprang into the air with a yell, and I offered to finish him off with the second barrel of my smaller weapon. But now it was my turn to appear hurt. At any moment he might hear the pellets rattling on the defences. At the next stern chaser – his turn at about the length of a cricket pitch – I hopped up and down and flapped my wings before retiring behind a stook, where the game bag came out quick and was shoved under a sheaf of oats. Then we called it a day. Poor old Bill had to sit on one side of his chair at dinner that night. The skin was not broken, but he only saved punctures by leaning with jacket and trousers pulled out away from his situpon. I did not confess until we joined the Scots Guards years later.

The three Stirling brothers all used hammer guns, shooting with great effect until well after the war. Hugh, the fourth son, was to lose his life in the Desert. Peter broke all kinds of duck flight records with his old Purdeys on Lake Ekiad in Egypt, killing three hundred wild fowl, mostly teal and shoveller, in one morning during the war. David was the most dangerous member of the family. As a small boy he had a passion for pulling back hammers and stroking his trigger in the cold half-light, waiting for geese to come over the sea wall before dawn.

Sometimes the pressures were the wrong way round and a premature explosion startled the Estuary. The gun was cocked when it should have been on safe, and loaded when believed empty.

Sport plays an important part in education. All athletic activities were taken seriously. The school only feared Sedbergh on the rugger field. The younger monks in the community, a third of whom saw to our teaching with an equal number of lay masters, could also play good football. (When the newly-formed Guards Armoured Division was training on the Wolds after Dunkirk the Welsh Guards, pre-war army rugger champions, came over to play the Abbey School. Some fleet-footed monks lent weight to the forwards. History records that on this occasion, the bad language and most of the fouls came from the monks' side of the scrum. They went on to win a good game, thereafter making Ampleforth a port of call for many new friends in the Foot Guards. This side included Basil Hume – a boy in the school who was recently appointed Cardinal Archbishop of Westminster.)

The Officers' Training Corps was also something to be proud of, with a Coldstream Drill Sergeant who wore the Mons Star. When I joined the Corps, Gus March-Phillips (later a commando hero) was my Platoon Sergeant. Our Senior Under Officer, who also became a regular soldier, was one of several brothers called de Guingand. Freddy was destined to become a general; he helped to enhance Montgomery's reputation acting as Chief of Staff with the Eighth Army and, later, the 21st Army Group in Western Europe. Another cadet, Rory Chisholm, won fame in the RAF as a pioneer night fighter pilot.

Looking back at the academic aspect, I think earlier education had been at fault, for I spent much time unlearning what had been taught at home. The eleven-plus had not been invented, but passes, credits and distinctions in examinations (Common Entrance, School Certificate and Higher Certificate) were the order of the day. Although I was capable of speaking fluent French, was well grounded in the English classics, some history, geography and Greek mythology, with a love of colour in painting and poetry, these assets were crushed out of existence in the treadmill designed to obtain diplomas in potted knowledge. Having a retentive memory I did not find examinations difficult in preferred subjects, but the time spent on mathematics, physics and chemistry proved entirely useless in later life. I never understood algebra, geometry or equations, and only mastered the most vulgar fraction with great difficulty. Ineptitude in mathematics put an end to Spanish, which was treated as an extra subject, and only available if a boy was holding his own in regular classes. Subsequent travel and cattle interests overseas have taken me to South America on frequent post-war missions; I rue the day when the Blanco y Negro correspondence courses were called off in favour of decimal points.

Classics, like the New Testament, should be studied in translation and not with the use of a crib. I fully accept that Latin and Greek teach a boy to think. Nor did the *Aeneid*, and the 'well-worn thunder of the Odyssey', create the same difficulties as logarithms or elementary calculus. Ovid supplied the touch of colour which was lacking in much of the syllabus. Even his dreariest work gives an impression of fluency that justifies the self-praise: 'Sponte sua carmen numeros veniebat ad aptos/et quod tentabam dicere versus erat' ('My thoughts turned themselves into lines of their own accord and whatever I tried to express changed into verse'). Born in 43 BC, Ovid had the advantage of a head start, and his influence can be traced in Chaucer – even in Marlowe and perhaps Milton. You cannot do better than that. The love of poetry, however, gets you nowhere in examinations. I well remember the headmaster's irritation (and he was a man I much admired) when, after being told to read Macaulay's *History of England*, I confessed a sneaking admiration for the author's powerful verse. For me, I suggested – with one eye on the door – Lars Porsena of Clusium brought history to life. The Head bent his brows, but he listened in silence while I declaimed a sounding passage from the *Armada* as an alternative to 'Horatius':

> With his white hair unbonneted, the stout old sheriff comes;
> Behind him march the halberdiers; before him sound the drums;
> His yeomen round the market cross make clear an ample space;
> For there behoves him to set up the standard of her Grace.
> And haughtily the trumpets peal, and gaily dance the bells,
> As slow upon the labouring wind the royal blazon swells.
> Look how the Lion of the Sea lifts up his ancient crown,
> And underneath his deadly paw treads the gay lilies down.
> So stalked he when he turned to flight, on that famed Picard field,
> Bohemia's plume and Genoa's bow, and Caesar's eagle shield.
> So glared he when at Agincourt in wrath he turned to bay,
> And crushed and torn beneath his claws, the princely hunters lay.

After hearing this in silence, Father Paul wrote a neat line in red ink at the foot of my essay on Whig statesmen: 'Delusions of grandeur: please keep your feet on the ground.'

Macaulay's rolling phrases may seem a trifle stale. I last heard this stanza when Duff Cooper, the Minister of Information, read the *Armada* to the nation on Sunday, 3 September 1939, the day England declared war on Germany.

1 My father, KT, KCMG,
KCVO, CB, DSO, in the
uniform of the First Life
Guards, shortly before he
raised the Lovat Scouts,
1899

2 Aged three with my
mother, 1913

3 Beaufort Castle, high above the River Beauly

4 My grandfather, Lord Ribblesdale, Master of the Queen's Buck Hounds

5 Aged seven at Lyegrove, 1918. The war is over

6 Magdalen, aged six. There was a year between us

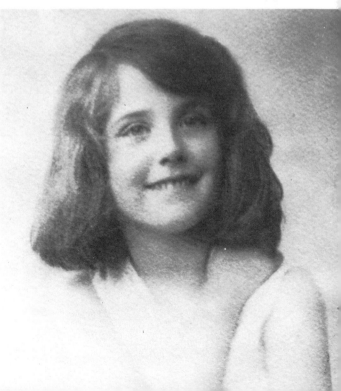

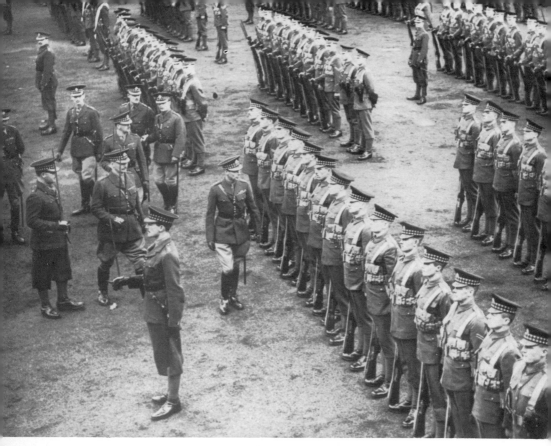

7 'The days of swords and puttees', 1935. Our future King inspects my platoon, Left Flank Company of the Scots Guards. Jim Mellis and Sergeant Burt stand behind me

8 Highland wedding, my marriage to Rosamond Broughton, 1938

For school reports, I can claim the shortest verdict on record, penned by a South African Rhodes scholar who taught science. It read, 'Application – Nil, Progress – Imperceptible'. This was scurvy treatment from a fellow naturalist, after I had shown him the hole in a clay bank under a thorn-bush where kingfishers nested in the beck at Gilling Station.

In his final year, the school-leaver could expect an appraisal in crystallized form. Going through the papers in my father's desk after I left Ampleforth, I found my last report. The headmaster's summary seemed fair enough:

Fraser is intelligent, with natural ability, but early work suffered from lack of interest and a contemptuous disregard for certain subjects. His Masters have complained that he can be idle and inattentive.

In other respects, he is a boy of limitless energy, in spite of early criticism I consider him to be one of the best examples of independent and original thinking in my History Group in the VI Form. If he is lazy as a Scholar, perhaps he has the genius to avoid the things that do not interest him. In this, his last term, he won the English Essay prize over the whole school.* His knowledge of the flora and fauna, also the sporting facilities of the North Riding is remarkable. A popular boy, admired for his native wit and cheerful approach to life, Fraser's achievements in games and studies have made him a natural leader among his contemporaries. I am grateful for his good influence which has been of considerable benefit to the school.

If Fraser has acquired a small part of his personal charm from Ampleforth, his time here has not been wasted. He should do well at Oxford.

V. P. NEVILL,
Headmaster

No doubt this golden handshake was commonplace and freely dispensed all round.

What did school teach me? I reply without hesitation 'a code not to be found in the Comprehensive system: to love God and serve the King. To learn responsibility and be loyal to superiors; to give of one's best and take a beating cheerfully; to feel, but not to show emotion; to lead and not be driven; above all, to show tolerance and consideration for other people, to realize that authority can never be abused, to have good manners and never lose one's temper.' I have found the last precept often difficult to comply with!

Career masters now do useful work in advising school-leavers uncertain of their future. Father had his own ideas. By the time I was

---

* 'The Banner of the Seven Yaks' Tails' was a portentous satire – now mislaid – that told how Genghis Khan, without benefit of education, left a yurt in the Gobi Desert to conquer China and the greater part of the known world.

knee-high, it had been decided that Oxford and the army offered the best opportunities. So went the fiat forth, and I was glad to comply, for father saw things clearly and his own record spoke for itself. The kindest and most sanguine of men, he expected the eldest boy to shoulder early responsibilities. It would have been unthinkable to question his opinion on family matters. Father never spared himself and he expected unqualified success in any long-term plan or serious undertaking.

CHAPTER SEVEN

OXFORD

And laughter, learnt of friends, and gentleness,
In hearts at peace, under an English heaven.

RUPERT BROOKE

That, briefly, was my school background. I spent the summer and
autumn working in South America. Then I went up to Magdalen with
four Ampleforth freshmen in 1929 to take an Honours Degree in
History.

A note from the Dean read, 'Your rooms are on No. 4 staircase in
Cloisters.' Frank Greenway, an admirable scout (a servant who looked
after the undergraduate rooms on two staircases), helped to carry my
luggage from the Porter's Lodge.

I was shown the stone alcove, containing a bath and rusty geyser,
installed for the then Prince of Wales by Doctor Warren, the ex-Presi-
dent. Five pounds a term was charged for this luxury, which saved a
cold walk down stone passages on a winter's morning.

A new life had begun. What were my first impressions? Very favourable of course. Yet, looking back, I fear Oxford was passing through
a dangerously irresponsible period in undergraduate history, culminating before the outbreak of war. The country was in recession; dole
queues of two million unemployed stretched across Britain. In a lecture
at Cambridge, Arthur Balfour had once warned an audience, 'There
are times in the history of societies when there spreads a mood of deep
discouragement; when the reaction to recurring ills grows feeble; when
learning languishes, enterprise slackens and vigour ebbs away.' It happened again to my generation, but the malaise did not rub off on Magdalen. Oxford had slept unawares through the industrial turmoil which
led up to the General Strike. Coal, cotton, steel and shipbuilding collapsed in 1929 and so did Ramsay MacDonald. Two years later, part of
the Home Fleet, a very small part, staged a mutiny at Invergordon – a

sign of the times that caused a sensation in the outside world. I met Winston Churchill for the first occasion when he came to breakfast at Beaufort on his way to deal with the offenders.

In August 1931 the Socialists, having played havoc with the nation's finances during their two years in office, were faced with a crisis of such dimensions that an emergency meeting of the Cabinet decided to resign and leave it to Stanley Baldwin, Leader of the Opposition, to clean up the mess.

The Prime Minister, Ramsay MacDonald, accordingly sought an audience with the King that same evening to tender his resignation, while his colleagues thankfully retired to bed, or so they thought. They woke next day to discover that the King, instead of accepting MacDonald's resignation, had persuaded him to form a coalition with Stanley Baldwin and the Liberal Leader, Sir Herbert Samuel.

Soon afterwards, Ramsay MacDonald (who remained Prime Minister) wisely decided to obtain a fresh mandate from the people by holding a general election. This resulted in a sweeping victory for the national government, who between them won all but sixty of the constituency seats. MacDonald continued to lead the coalition until 1935, when he retired and handed over to Stanley Baldwin.

In 1933 Germany, previously reinstated by the Locarno Treaty, left the League of Nations for the second time. Six years later we were at war.

Oxford University was still dominated by privilege and class structure. Quite unaware of this crumbling edifice, I included myself among the fortunates who joyfully embraced all the college had to offer.

I will not attempt to describe Magdalen: life had not changed there since Compton Mackenzie sketched its beauty in *Sinister Street* with a sureness of touch which cannot be rivalled here. For half a century time stood still for the indolent undergraduate who strolled down Addisons Walk or looked across from 'New Buildings' to the deer park. The quality of this little world has been summed up by Oscar Wilde – another paragon of Magdalen – in a famous epigram: 'If a man is a gentleman – what he knows is quite enough; if he isn't – what he knows is bad for him.' But Magdalen was unquestionably a great institution.

All who went there were expected to take a Final Honours School: if Thirds were numerous, the number of Firsts was exceeded only by those obtained at Balliol. But the success of the college, in its widest sense, was due to the absence of cliques.

At Magdalen the rowing Blue was not a man apart, consorting only with his fellows. Men following totally different pursuits mixed on equal terms and were often the closest friends. This good fellowship was largely due to a tradition practised by members of the Junior Common

Room, in holding a 'wine' after dinner in Hall on Sunday during winter terms. The 'wine' was followed by an 'after Common Room', or smoking concert, given in the JCR* rooms of some hospitable host, who, with one or two others, provided refreshments. Guests were invited and added greatly to the enjoyment; the singing demies added to the success of these concerts. Sometimes there was a conjuror or a hypnotist – so-called, at least. On fine nights the party usually finished in one of the quadrangles in a pig-a-back cavalry encounter with 'jockeys up'.

There was little drunkenness in my day except on big occasions. I seldom visited a pub. The beer was cheaper in JCR and the Buttery. Drugs had not been heard of, and gambling, like drinking, was at worst a periodic visitation.

New College had the only comparable Common Room at Oxford University. New and Magdalen were traditionally close friends: besides close proximity and re-association of former schoolfellows, we shared a pack of beagles and ran the Grind (point-to-point), ridden over a stiff course at Great Tew in the Heythrop country. It was here that my father died of a heart attack in February 1933, after I rode a winner in the Old Members' Race.

Only a handful of men had money, and these were mostly foreigners. Oxford – town and gown – was less than half its present size. Colleges varied in character. Some were noted for rowdy behaviour which often stemmed from their dining societies. Clubs played an important part in life at Oxford. Christ Church provided most of the Bullingdon who dined in a barn; during the summer term they let off steam in a hired pleasure boat, cruising up the Thames. The Bullingdon sported their own finery: swallowtails of midnight blue, with white silk facings which were useful at hunt balls. But it was Loders (a club founded by Lord Cardigan in 1814, and confined to a dozen members of Christ Church) that put away the most drink. At the initiation ceremony after elections, the new member was obliged to drain a bottle of port, quaffed – standing – from a loving cup of singular design. Loders now dine only once a term instead of every Sunday. Other places were less worldly. Jesus College housed a number of serious Welshmen destined for Holy Orders, who never let their hair down. The characteristics of Lincoln, Pembroke and Queen's were unknown. Trinity was famed for its mulled claret. Worcester was, perhaps, the most attractive small college in the University. Brasenose, noted for its concentration of athletes, had two drinking societies, the Phoenix and the Hellfire, that lived up to their names. Brawn rather than brains was the rule among the last-named 'hearties'. Their club had simple rules: 1. Wine to be drunk at all meals;

* Junior Common Room.

2. Members ditto. At Magdalen the short-lived *Multum in Parvo* held high wassail once a term: severe damage to the college furniture in my Presidential rooms ended the festivities – to be banned thereafter by the Dean.

Boojums, at New College, still survives. An Oriel Society was said to provide the most urbane deportment of them all. A dozen members had the privilege of dining in Hall once a term after the regular meal was over. They dressed in tails, and oysters were always on the bill of fare when in season.

The adherents to serious political persuasions were less colourful but better behaved. The Fabian Society was viewed by the Bullingdon as a suspect body of revolutionaries who talked seriously of social justice and how to put the world to rights. The Chatham and the Canning went to the other extreme, and were less well attended. Clubwise the Grid was larger than the rest and represented a good cross-section of the Varsity. Vincent's soldiers on, its coveted members' tie being the most distinguished form of recognition.

The Caledonian sat down to dinner on St Andrew's Day at the Randolph Hotel, two-hundred strong, with guests besides. Kilted reels were danced in the lounge and Mr Gordon, the President of Magdalen, an honest Scot, left his Lodgings to grace the top table for this important evening. The OUDS* I never met.

In Oxford two races lived apart: men and women. It seems astonishing to record how completely a university has changed in its attitude towards the weaker sex. Women were not seen by my father's generation: undergraduette colleges were the daydream of suffragettes who wished to invade the forbidden city.

In the 1890s my uncle, Sir Francis Lindley, wrote:

Then all our energies and thoughts were directed to sport, work and games. If a man was seen talking to a girl in the street, he was quite liable to be sent down unless he had some good explanation.

During Eights Week, a group of young ladies descended on the University to watch the rowing from the College barges. They were escorted by their mothers and elderly relatives. The Commem. Balls were after term but the same rules applied. It was all great fun, but it was no part of Oxford life.

Twenty years later, Harold Macmillan supplied the following:

Thro' Balliol eyes there were no women; ours was an entirely masculine – almost a monastic society. We knew, of course, that there were female students and that their Colleges were situated on the suburban periphery. They never came into our rooms and they played no part in our lives.

* Oxford University Dramatic Society.

They were not, I think, full members of the University and they joined no political societies. Nor were they members of the Union. If they came to lectures, they were escorted by a duenna.

A better communications system, world war and a more permissive society did not open the flood-gates: the change was gradual. By 1930 some of the old prejudices had been eroded. At about this time the Palais de Danse and that pernicious institution, the cocktail party, crept into Oxford. At first, it was a 'town' rather than a 'gown' affair: very respectable and something to be avoided. But mixed drinking and the motor-car clinched the breakthrough at Oxford. Discreet punting picnics appeared on the river – when transport was available, preferably moored at Henley or Maidenhead. Electric canoes (those who could afford to rent them drank champagne) hastened the petticoat revolution. Sisters brought girl friends down to lunch in college; after evening polo, the theatre or a cricket match, mixed dinner parties were not unknown at the George.

These junketings, mainly confined to summer terms, were officially frowned upon. Certain ladies were *persona non grata*: a dark-haired, snake-hipped model of Augustus John's topped the list.* My cousin Tony Wilson admired Tallulah Bankhead. He was in hot water on her account when still at Eton. Tallulah wished to visit Oxford, having already proclaimed she was fascinated by English boys, adding, to startle the fuddy-duddies, she was 'purer than driven slush'. The University authorities immediately put the shutters up; this did not stop Tony driving the actress down in a racing car to deafen our cloistered calm with fish-tail exhausts.

David Tennant, a distant relative, who ran the Gargoyle Night Club in Soho, was married to Hermione Baddeley. She was inveigled to act as chaperone-cum-kinswoman. The trio arrived, wind-blown and none too pleased over a shared seat in the open Bugatti. Jack Wolfson, one of the stricter dons, was in the Porter's Lodge as the actresses tripped in: his gaze was distracted by the Bugatti, parked in the High, across the entrance doorway. Orders were given for its instant removal. Restrained from joining a game of bowls between the Dean and Bursar, Tallulah won the chef's heart in the Buttery (with an autograph) after choosing lunch – served on No. 4 staircase by the admirable Greenway. For a freshman this was an occasion. I remember the meal: fresh asparagus, quails in aspic, and strawberries and cream, washed down with two bottles of hock cup, double brandies and black coffee – with plenty of change left over from a fiver. The cold snack helped to

* She is mentioned in Daphne Fielden's *Duchess of Jermyn Street*.

slow Tallulah down: after climbing Magdalen Tower, looking into chapel and a glance at the deer park, it was time to go back to the theatre.

Nearly all my contemporaries led a celibate existence; few strayed from the straight and narrow, and some were very shy. For those with fast cars, London offered opportunities for assignations. For those less fortunate, and others on more lawful business, a train known as The Flying Fornicator left Paddington at nine-forty to get the supposed womanizers into college before midnight lock-up. A twenty-four-hour pass was not readily granted; today it is no longer required. During term undergraduettes remained secluded: the few who came to lectures or attended 'schools' looked smart and took pride in their appearance. That is no longer true. The sexes look alike and Oxford's Amazons wear trousers and sleep in college.

After Ampleforth I was sent to work in South America. The carefree life under open skies made me restless on return.

In my last year at Magdalen, father, who had been sent down for 'living it up' in the nineties (a sore point, this), came to a painful decision. Britain had followed the United States into recession. Estate finances were in poor shape, and that summer he suffered a heart attack in New Zealand working on colonial resettlement. Hearing from the Dean, on return, that I was to be found more often in Cambridge than the parent university, he delivered a broadside that caught me fairly between wind and water. The interview was brief and to the point. In a word, I was wasting time and getting nowhere. This verdict seemed hard, after three months in the long vac. spent sweating on a coffee-cum-cattle property in Brazil. I had also been forced into a Hellenic tour, listening to lectures delivered by classical scholars. The last straw on this cruise was to catch measles from the Warden of New College's daughter. Now father decided I should take two degrees, read simultaneously, at one fell swoop.

That put an end to the Athanaeum, some good Norfolk shoots, and racing at Newmarket. (To reach Royal Ascot, the Bullingdon, in palmier days, once hired a special train, complete with dining car for the traditional champagne breakfast, returning in style after the Gold Cup.) I had played chukkas in São Paulo and up-country at Tres Barros; now there would be no more polo. Though father was assured I was doing well, Forestry, a pass degree, was added to the Honours History School, then considered an undemanding subject which did not overtax the brain. With a well-arranged timetable, history tutorials included port and biscuits with 'old Luggins' (Mr R.A. Lee) and

'Heretic' Thompson, both eminent historians moulded in the Belloc tradition:

> . . . regal Dons! With hearts of gold and lungs of bronze,
> Who shout and bang and roar and bawl
> The Absolute across the hall,
> Or sail in amply billowing gown.
> Enormous through the Sacred Town.
> Bearing from College to their homes
> Deep cargoes of gigantic tomes.

There were dreamy and absent-minded *savants* as well. Dr Spooner, the Warden of New College who kissed the porter and gave his wife sixpence on the station platform, was before my time. Not so Lord David Cecil who addressed a familiar face after one of his lectures: 'Tell me, was it you or your brother who was killed in the war?'

Reading history was a delightful occupation and I made new friends. A pass degree in Forestry, however, was quite another story. It meant field work and much practical research. There came a further directive: no better digs could be found for third-year men in quest of knowledge and enlightenment than the Old Palace in Rose Place, off St Aldate's. Here Ronald Knox, the Catholic Chaplain, struggled to hold his flock of undergraduates, an assorted collection in which men from Downside, Ampleforth, Stonyhurst and Beaumont predominated.

I had grown up with Ronnie Knox, who spent his leaves at Beaufort. He had tried unavailingly to cram me in school holidays. (Somehow I scraped through Latin and Greek, one being obligatory for the School Certificate; then, thankfully, had dropped both subjects.)

Ronnie, an intensely shy man of deep humility, despite brilliant achievements, dreaded any form of entertainment, yet remained determined in carrying out his pastoral duties. Born in 1888, the youngest child of the Bishop of Manchester, he started his trail of scholarships by winning the Open from Summerfields into Eton, where he became the Captain of the School and winner of the Davies Scholarship and the Open Scholarship to Balliol, the Hervey Verse Prize, and the Latin Essay. At Oxford he gained a First in Greats, the Ireland Scholarship, the Craven Scholarship, the Hertford, the Gaisford, and the Chancellor's award for Latin Verse. He was also President of the Union.

In a family that breathed the air of wisdom, Ronald came last of four brilliant brothers: all mastered the classics at an early age, reading freely from the Greek Testament, memorizing Virgil and writing Latin plays (Ronnie) under the dining-room table, before their father was removed to Birmingham from a country vicarage.

There, Mrs Knox's health had broken down and the children lost the mother they so dearly loved. Ronnie admitted they found it hard to forgive the rigid beliefs of a widowed 'black Protestant' great-aunt who was put in charge of the household. During the last illness she had refused permission for the boys to travel on the Sabbath Day when summoned to their parent's death-bed.

To the Bishop's chagrin, Eddie the eldest son, who wrote scintillating verse under the name 'Evoe' and was later to become the editor of *Punch*, took small interest in religion; Dilly, a classical scholar who helped to break the German cypher code in two world wars, was a difficult character who grew up an agnostic;* while the gifted younger sons, raised by kindly relatives, sought a wider interpretation of Christianity, each in turn becoming an Anglo and Roman Catholic priest. The Bishop was not at all pleased.

At Oxford Ronnie was a withdrawn Chaplain. His conversion, described in 'A Spiritual Aeneid', had not been easy. In lighter vein, the remark 'He who travels in the barque of Saint Peter had better not look too closely into the engine room' and the oft quoted 'Whoa I am only a convert' to the Irish padre, busily pouring out a triple measure of whisky, concealed a deep humility which left him tongue-tied with the extroverts of my generation.

In 1938 Ronnie left Oxford. It was the year that he wrote *Let Dons Delight*, which was considered his most amusing book. By 1942 he had completed the translation of the New Testament. His brothers were surprised at this new venture. 'A bad text,' said Dilly, 'and he doesn't know very much Greek.' 'Or theology,' Wilfred added, 'except what I've taught him.'

Writing of his early sermons, his niece, Penelope Fitzgerald, remembers that, 'Quite casually they introduced an exceedingly brilliant person whose reasoning mind was able to accept the contradictions of Christianity. At the same time they showed that a normal, pipe-smoking, income-taxed Englishman – not a Jesuit, not a mystic, no black cloaks, no sweeping gestures – could become a Catholic priest. As he tried to sort out, step by step, his own difficulties in faith and doctrine, his congregation found – though they did it timidly at first – that it was possible to laugh out loud in church.'

Wilfred, a less gifted preacher, performed a similar rôle as Chaplain to Pembroke College, Cambridge, where he taught history and theology. His piety appealed not only to the intelligentsia but likewise to

---

* Dillwyn Knox, CMG, died of cancer in 1943, still at his post in the cypher department of the Admiralty, in the end too weak to leave his bed. His last words when barely conscious were, 'Is that Ronnie out there boring God in the passage?'

drunks and small children, whom he treated as equals and thinking people. It is told how, while serving on an East End Mission in cassock and black hat, Wilfred was hailed by boozy navvies who shouted from a public house, 'Them ain't the clothes our blessed Lord used to wear.' Wilfred walked unhesitantly across and explained that the remark as a piece of logic was based on four false assumptions, which he then expounded. They parted as friends.

One sensed that as Chaplain, with little small talk, Ronnie's responsibilities had become burdensome; to meet the demands on his literary output, more privacy was essential. Two third-year men were invited to share the Old Palace board and lodging. They were expected to make a contribution to the ex-Beaufort housekeeper's wages (the food was very good in Mrs Lyon's day) and help out the preoccupied Chaplain. Dinner parties for shy freshmen, and literary visitors were bolstered by the sophisticated aides who lived in. Two friends and helpers, Joe Weld*, then Brian Daly,† took duties in turns to oblige (a roster of one dinner party and one tea party each week), with Sunday appearances at the Newman Society, which met in the premises next door during term.

The Lodgers were usually officials. These were entertaining evenings with a full house and many distinguished guests. The Catholic undergraduates, less than two hundred and fifty men in all, were lucky with their speakers in the Society's debates. The year I was President, we broke attendance records with sessions in which Maurice Baring, Chesterton, Belloc and Evelyn Waugh (in that order of popularity) contributed outstanding performances. On such occasions the guest was given dinner at the Grid Iron Club. By asking Ronnie as well, the President and officials made sure of their VIP, who would otherwise have been snatched up by a Dons' High Table. Dinner at the Grid was more fun than the debate, for the literary lion filled the Club with friends, and it was difficult to keep the jollifications to a timetable.

Ronnie sparkled when he dined out, greeted everywhere by a university that could still admire a scholar, author and wit. I think he appreciated the respect he commanded, and the legend of his annual appearance to speak at the Union during Eights Week. Chesterton had the words for him in 'Namesake':

> Mary of Holyrood may smile indeed
> Knowing what grim historic shade it shocks
> To see wit, laughter, and the Popish creed
> Cluster and sparkle in the name of Knox.

* Sir Joseph Weld, MC, of Lulworth, Stonyhurst and Balliol.
† Brian Daly – The Oratory and Brasenose – killed in action in the RAF.

For myself, hoping the success of the Newman would reach father's ears (an overdraft was mounting steadily), the time seemed ripe for a diplomatic approach. I dropped hints that an entertainment allowance would not come amiss. Father countered with a polite enquiry about my efforts as a public speaker. On what subjects had I spoken in the Union, for instance, in serious debate during the previous term? By a lucky chance there had been one question. I have forgotten the motion – some well-worn political theme that had elicited my enquiry 'that although this House knows Lloyd George can stay forever, will "Orwell" [naming the winter favourite for the Two Thousand Guineas] get a mile?' This witticism drew a frosty smile, but father was not amused. He did not live long enough to see my brother Hugh's success at Balliol, where he became a notable President of the Union Debating Society. A year after going down my father was dead, and at twenty-one I was thrown in at the deep end.

Dear Ronnie! Your chaplaincy days and the numbskulls of the thirties must have proved a disappointment after those Balliol men, your own contemporaries, who fell in the First World War – the double firsts, poets and athletes of the lost generation – men from a golden age of which you were almost the sole survivor.

How memories crowd back. The Old Palace went to bed early. We were forbidden to climb in after midnight: R.A.K. hated coming down in a dressing-gown to shouts of 'Ronnie! open up!' from the lane below! We shall not forget the silent breakfasts, *The Times* crossword puzzle propped against the toast-rack, finished before we left the table; the planning of sermons and acrostics; the lazy days punting up the Cherwell (Ronnie's only exercise). That dreadful tea party when, during a pause in the conversation, a gentleman from Ceylon asked, 'Tell me, Father, do you approve of love before marriage?' Ronnie's reply – a shaft of light that pierced the gloom – 'Not if it keeps the clergyman and all the ushers waiting.'

Sadly, the Old Palace was not appreciated. Surprisingly, we passed our various degrees, but any mark made at Oxford must have been left on the college buildings! Brian Daly was a remarkable climber and had an iron nerve. I could hold my own over the roofs and dizzy heights, but he put me to shame after playing a rugger match against the Oratory. It was already dark. The train from Reading to Oxford is a fast one and goes under many bridges; travelling at speed, Brian leant backwards out of the carriage, got a hand grip on the roof, pulled himself up and out, and started to walk along the length of the train. He explained on his return by the same route that the physical strain on fingers was greater in descent because of the swing of the train as his feet groped for the carriage window.

When Ronnie died of cancer, in August 1957, my sister Magdalen* was at his bedside, nursing the padre in Catherine Asquith's house at Mells. Let a Downside Monk, Father Hubert van Zeller, pay tribute to his last appearance at Oxford:

The day fixed for the Romanes Lecture was June 11th. Ronald was so weak that those around him doubted whether he could deliver it. His colour was ghastly, and his speech slow and broken by frequent pauses. We drove over to Oxford in silence. Ronald stayed the night with the Vice Chancellor, J. C. Masterman, at Worcester. He rested all next day, and was taken early to the Sheldonian to be in his place before the audience arrived. He had been given a drug to stimulate him, and a doctor was posted near the rostrum; also a reader to take over the manuscript if he was unable to continue.

The precautions were unnecessary despite my fever of anxiety. The theatre was crowded. Many of the audience knew that he was dying, and that this was to be his last performance in the University he had rejoiced for fifty years. From the opening sentences the lecture seemed to give Ronald strength, and more strength came to him from the people who were listening. He spoke with the old gestures, voice and mannerisms, for more than an hour. When half-way through – to illustrate a point – he recited in full Cory's rendering of a Greek epigram,

> I wept as I remembered how often you and I
> Had tired the sun with talking and sent him down the sky,

most of those present recognized the words as his own farewell to Oxford, and some with whom of old he 'had tired the sun with talking' could not restrain their tears. The applause at the end could be described as an ovation, but the strain must have been very great. When I called him the next morning, he looked desperately ill. I drove him to 10 Downing Street, where Harold Macmillan, a friend of long standing, had invited him to stay for a check-up with Sir Horace Evans, who confirmed he had cancer of the liver. Next morning, they were Trooping the Colour on Horse Guards Parade, and only the Prime Minister's car was allowed to pass through the traffic. Mr Macmillan himself accompanied Ronald to Paddington to speed him down to the West Country. The Prime Minister wished him a comfortable journey. 'It will be a long one,' Ronnie replied.

One last touch to the canvas of this most unusual man. Soon after the specialist's diagnosis, Sir Horace Evans was elevated to the peerage. Ronnie wrote a letter of congratulation, at the same time thanking him for his attention. One of them, he remarked, had left Downing Street

* The Countess of Eldon.
† Based on a passage from *Ronald Knox* by Evelyn Waugh.

that day with a patent of nobility, the other with his death warrant.

In 1958, a group of Ronald's friends, headed by the Prime Minister, opened a subscription for a memorial which has taken the form of a travelling scholarship at Trinity.

CHAPTER EIGHT

OXFORD CAVALRY

THE HIRELING

Then one proud day he plays a leader's part
Across the green grass lands.
Carrying a horseman with a gallant heart
And light and loving hands.

WILL OGILVIE

'Flashed all their sabres bare, Flashed as they turned in air.'

In fact, the officer cadets in the Oxford University Cavalry Squadron did not have the advantage of either sword or lance over pedestrian rivals in the Infantry Company or the Sappers and Field Gunners. These were strangers seldom seen, except when we rode over them on field days. If truth be told, the dashing formation known affectionately as the 'Irregular Whorse' found considerable difficulty in drawing out and returning service rifles to their holster buckets as the squadron galloped into action.

Nevertheless, the waiting list in this Corps Elite remained over-subscribed with a long list of applicants, like the University Air Squadron does today. In a jingoistic age, with an Empire to defend and opportunities overseas, some army candidates opted for three years of education in the wider sense at the University in preference to Sandhurst or Woolwich. Backed with an Honours degree, I joined the regiment of my choice with eighteen months' seniority over younger men emerging from the military academies. These budding juveniles, though proficient in the spit and polish of their trade, were rated more innocent than babes unborn by university rivals, who accepted with equanimity the counter-charge of being no better than armed civilians.

At Oxford, the successful candidate, if accepted for the Squadron, was not committed to become a full-time cavalry soldier when he went down. Territorials and Yeomanry still played an important part in country life and the protection of the Crown. 'Eheu fugaces, Posthume, Posthume, labuntur anni.' Now these traditions are broken; Service families, father to son, have been scattered, and only sorrowing voices from the past lament the changes that mark their disappearance.

The generous attitude to Service conditions was appreciated at Oxford. It was perfectly in order to transfer from a cavalry regiment to the Infantry, and, by keeping the options open, to settle for an army or civilian career.

Likewise, undergraduates who were not thirsting for military service (this applied to Empire builders, future diplomats and men who joined the Colonial Office) enlisted unashamedly for love of horses, free rides at the Government's expense, and the panache of a high-spirited group of men on Christian-name terms with each other, all determined to enjoy themselves. In this we succeeded, and learned a thing or two in the process which did not appear in the training manuals. Getting bucked off remounts, learning to pull tails, soap saddlery, and treating minor ailments do not necessarily make a cavalry officer, but I can think of no better experience in handling horses or creating friendships that have stood the test of time.

> And I said I will list as a Lancer
> Oh, who would not sleep with the brave.

Yeats Brown (*Bengal Lancer*) and T. E. Lawrence were widely read; something of the magic created by their soldiering days imbued us with the same enthusiasm, in marked contrast to an MP who recently summed up life in the army as: 'You do nothing unless ordered to by a superior officer: you salute everything that moves, and paint anything that remains stationary.' A sense of duty can stiffen even the ultimate laxative. Constant vigilance remains the price of freedom. At worst our officer cadets combined *joie de vivre* with clean good looks, physical recklessness and intellectual immaturity in about equal proportions.

Just in case gentlemen rankers are thought irresponsible, it is worth recording that in a group of close friends, four rose to the rank of brigadier general or above, four were killed in action, and two were taken prisoner, while the greatest hell-raiser in the Squadron became a judge. Another was Attorney-General in the Sudan and a freshman in my section, recently retired from the Foreign Office after proving a most successful ambassador.

During the summer term, weekly training took place on Port Meadow, where about a dozen members of the University played bad polo. In the early morning, with no crossing lights and virtually no traffic, twenty minutes' trot from Magdalen Tower was sufficient to reach the

assembly area, and parade as the clock struck seven. But first we had
to find the horses: long-suffering hirelings living in livery stables that
just held out in the back streets of Oxford. Method and procedure lay
entirely with the individual, but the 'lilies of Magdalen' who lived in
college foregathered in the Porter's Lodge – clean, sober and properly
dressed at peep of day. Here we were given the once-over by the Senior
Under Officer, Robert Taylor, a third-year Wykehamist, later to be
killed serving in the 16th/5th Lancers in North Africa.

Before Major Beckwith-Smith (Coldstream Guards) became Commandant, the undergraduates had found creature comfort in irregular
garments ranging from Oxford bags (concealing pyjamas thrust into
gum-boots), with cowboy hats and teddy-bear coats (for cold mornings), to fringed buckskin jackets and chaps, reminiscent of Buffalo Bill.
This bizarre appearance came to an end when half the force was sent
home by the guards officer as improperly dressed, with the pointed
advice that the offenders might care to join the Morris Dancers, then
in vogue, and find themselves a maypole. The message was loud and
clear; now we took pride in boots and breeches.

At 6.15, Bob Taylor consulted his gold half-hunter. As if on cue, came
a clip-clop of grooms and led horses across the Cherwell from the direction of Cowley and the Iffley Road. Some of the posse pulled up in
the Botanical Garden Lane that leads to Christ Church Meadow, while
the rest rode on to University and Queen's. By the time Carfax was
striking 6.30 the motley assembly converged below the clock tower
where the roundabout stands today, to go clattering out of the empty
city, followed at a respectful distance by various grooms and livery men:
the 'Reverend' Tubbs (not in Holy Orders) and Larry from Mac's
garage being chief among them. We rode at ease, a certain levity
permissible before reaching the parade ground, where we formed up
into troops. Our noisy passage, perhaps less spectacular than in Good
King Charles's golden days, failed to rouse the sleeping citizens.
Sophisticated Oxford had seen it all before.

At Port Meadow, then an expanse of soggy turf beside the Thames,
Sergeant-Major Percy Rhimes, 16th/5th Lancers, our rough-rider
PSI (Paid Staff Instructor) helped the Commandant to unbox his
horse. 'Becky' Smith* – a cheerful soul, much loved by all – did not consider it necessary to wear uniform for an early morning ride. This peacetime seconding to run the OTC was reckoned the best job in the regular

---

* Major-General Beckwith-Smith, DSO, commanded the 18th Division and died a prisoner-
of-war in Japan in November 1942. Destined for the Middle East, this Territorial Division – after
spending two months at sea – was redirected to Singapore, where it arrived peacemeal, just before
the surrender. A sad end to a fine soldier.

army, for those who could afford to hunt all winter, fish or race in the summer, and shoot in Scotland during the autumn. The Oxford terms did not add up to eight months in the year and responsibilities, except during summer camp, were negligible. The adjutant also held a plum appointment; he did the paper work and conducted night classes – badly attended – helped by capable staff instructors in other branches of the Service.

For us, Percy Rhimes was the man who really mattered. It has been rightly said that the heart of a good regiment lies in the sergeants' mess: our mentor was a tower of strength in his own right. In face and form, Rhimes possessed the piercing eye and swift-dashing qualities of a stooping hawk. Hard as nails and smart as a new pin, he added a caustic wit which delighted the undergraduates. On or off his horse, the rough-rider was to become a legend in the University. How long the 'Major' served the Squadron I cannot say: two years, possibly a third in his tour of duty as an outstanding instructor. During that period scores of officer cadets passed through his hands: at a formative stage in their careers, and in an historic sense as well, for the League of Nations was coming to an end and ominous clouds loomed on the horizon. We would lose our horses within the decade; not a few of the young men who saddled up in the early thirties would command their regiments before the war was over. Many more were to be killed or captured. All of us learned how to strap on our spurs with the help of Percy Rhimes.

Today is the end of term. There is a big turnout, with almost a hundred graduates on parade; the sun is high, larks sing at heaven's gate, and from a pollard willow, where the mist is burning off along the river, a cuckoo calls. In June the 'Irregular Whorse' will join the Queen's Bays at Tidworth for Summer Training Camp. The horses sense a coming change in their fortunes. They are getting their summer coats, and the blemishes on long-suffering legs, hard-ridden across country three days a fortnight during the hunting season, with two extra afternoons behind the Drag, look less like inverted champagne bottles.

There is time to regroup, light a cigarette and take some weight off backs. At five minutes to seven, the Sergeant-Major calls for a right marker, and Neville Crump, whose father is Master of the South Oxfordshire hounds, leads out a dappled grey hunt servant's horse. 'Get on Parade!' The cadets fall in, dragging their horses into two ranks. 'Stand to your horses!' 'Right dress; eyes front; prepare to mount.' Robert Taylor takes the parade as Senior Under Officer. When he sings out the commands, there is no trace of the painful stammer that at times renders him speechless. We are armed this morning with service rifles. Before mounting, these have to be seized near the muzzle, laid over the horse's neck on the offside, changed from the left hand to the right, then forced down into a leather bucket as the rider swings up and over the high cantle of the cavalry saddle. 'Mount!' There are cries of rage and pain from the rear rank, and a thud-thud on ribs, as kicking and barging sweeps through the Squadron. The Sergeant-Major spurs into action, riding at a gallop down the ranks to restore order.

Today's training is to smarten up squadron drill before joining the cavalry division, but the Commandant is having troubles of his own, with a young horse that has taken exception to the martial scene. The four-year-old plunges at the first command, and then starts a crabwise retreat towards the farthest end of the polo fields. 'Becky' Smith's imprecations grow fainter down the wind, while we follow in review order, at a dignified trot, with tossing heads and jingling curb chains. The Commandant flees before us. As we bear down on the diminishing figure, the rear rank, blamed for every misdemeanour, start to canter, and the pace quickens as some of the strong pullers and weaker cadets are carried to the front. Bob Taylor – in good control, ten lengths to the fore, riding a polo pony with a bang tail – takes the first solid bump and disappears. Then, with a thunder and a crying we are away.

The rest of the morning is spent in section leading, the Commandant having retired hurt, bowled over in the charge, which pulled up in disarray, halted by the big bullfinch at the end of the meadow! When order is restored Rhimes gets us back to faster movements. But the morning's training turns into a shambles. The joys of spring, perhaps the rest and extra oats before camp, or the unaccustomed rifle buckets flapping against flanks and bellies, have got horses into a lather.

While most of the cadets rode well enough, a proportion of the hirelings did not answer to the helm. Here mention must be made of 'the Bolter', one of the less well behaved quadrupeds confined to the rear rank, who had pulled to the front in the first sortie. Now it was his turn to take charge. All he needed was a partner, and one was available.

Among the morning turnout there rode a stranger. Some said a Rhodes Scholar from 'Teddy' Hall. Or perhaps another freshman from the United States – by his twang, a Texan? Others recognized the gangling figure, smoking a black cheroot, as the son of a tea planter from Assam. All present knew the Bolter. 'I like the look of that horse,' said the visitor, surprised that the big chestnut had not found a jockey back in stables. Little did he realize that innocence had committed him to the most notorious animal in Oxford. Swells from the shires, horse-breakers both civil and military, as well as the local vet, had all lost prestige during outings on this animal. Now the gelding had a fresh customer. The groom handed him over with a sigh of relief.

'The Cavalry will withdraw.'

As each troop retired from its fire position, the horse-holders brought up the nags in dashing style, and sections galloped back to join the main body, speed being the essence of the manoeuvre.

'Cavalry have withdrawn.'

But where were they going? Did I say 'galloped' back to rejoin the main body'? No. That was not so simple. As the tall stranger swung into the saddle, reins in his left hand and rifle in his right, the Bolter took off.

Directly in his path lay a sizeable ditch which ran across Port Meadow. The Bolter took it in his stride. Our friend never got his foot into the other pedal: with rifle brandished in one hand (he had no chance to return it to its bucket), he looked like some avenging horseman – a descendant of Ghengis Khan leading a horde of Tartar invaders! Once over the brook, Tamerlane received some advice from the Sergeant-Major: 'Drop your rifle, Sir, and ride with both hands. He'll take some stopping now.' But the intrepid cavalryman scorned to deprive himself of his weapon.

Three hundred yards away, towards Oxford, the meadow was bordered by fencing, ending in a gate; beyond lay a wooden bridge over the railway line. Thank Heaven, the gate was shut. He would soon be in control. But no! Even as he furiously approached, a bent and aged yokel, smock-clad and obliging, sprinted for the exit. His wizened legs, working like pistons, got to the barrier just in time. Mistakenly, he threw it open. The officer thundered past. As the horse's hooves beat a tattoo on the wooden slats, the Paddington express hurled itself below, screeching and whistling on its way to Reading. The Bolter, put off for a moment by this surprise, considered going over the parapet; but the visitor steered him straight, and they galloped on.

Soon the track gave way to houses, then to a busy street in the outskirts of Oxford. John Gilpin's ride to Edmonton was a colourless dawdle by comparison. A patient milk-horse, waiting with its float on early morning rounds, sensed the excitement and the call to arms. Scattering bottles, it raced along with the Bolter, despite inarticulate cries from the abandoned attendant. Now Walton Street looked like a rehearsal for *Ben Hur*, with cavalry supported by chariots.

The publican of the 'Golden Ball' was a large man, broad of back and hard of hearing. As his bulk appeared in the middle of the road to collect the morning paper, the Bolter braked violently, then faltered; his legs flew from under him, and down he came on the tarmac. The undergraduate, regaining control, hauled him off the floor and, turning the chastened steed, trotted back to good natured applause as the Squadron, the day's work done, came into view heading back to Oxford.

Percy Rhimes rode with us to Merton Lane. No smile had crossed his whiplash countenance until we reached the outer suburbs, when suddenly he slapped his Weedon remount on the shoulder and burst out laughing. 'Well, strike me pink! I thought I'd seen just about everything in my years in the Service, but this morning's performance beats all. You ought to put on a show every afternoon at Olympia, but please, gentlemen, don't let's have a repeat performance at Tidworth, least of all when you're wearing the King's uniform. I must remind you that I have friends in the sergeants' mess, and furthermore I'm supposed to have taught you young toffs how to ride. Now gentlemen, the horses are still too warm to go into stables, and even if you're late for your studies, the animals come first. Please dismount and walk them back from here.'

So we re-entered the City of Spires, where, for that group of friends, the sun rose over Magdalen and set behind Trinity – when all the world was young and life was just beginning.

But for our rough-rider, the future could hardly have looked more discouraging. The mechanization of cavalry regiments was on the way, and the thought of an army without horses, with all its consequences, had grown into a nightmare that never left him. The sands were running out.

I have described how the officer cadets rode up the High, the best of them to serve with the Dragoons, Lancers and Hussars, whose predecessors had clattered about the University in much the same way in Carolean times. Regimental mottoes had not changed in the interval, some had a romantic ring: 'Our Arms are our Defence, Their Arms are Recompense' and 'Love and Ride Away'.

As we dismounted at Mac's garage, a humorist enlarged on this last theme – the regiment concerned was soon to get armoured cars: he suggested the motto 'Screw and Bolt' might, in future, be considered more appropriate.

How about some tribute to ourselves? Surely the 'Irregular Whorse' – that body of faceless men – were entitled to recognition? Why remain anonymous down the years? It was David Silvertop,* later to be killed leading his regiment in circumstances of great gallantry in Holland, who hit on the right idea. Over a breakfast of pork sausages at the Eastgate (a meal available until twelve noon for undergraduates

* Colonel Silvertop, DSO, Ampleforth, Magdalen, 14th/20th Lancers.

who overslept), we agreed that the motto 'When we charge, we over-charge' be adopted by the Cavalry Squadron of the Officers' Training Corps. This was duly minuted by Robin Grant Watson,* Secretary of the OE Club, and dated 27 May 1930. The entry can still be seen at the back of a ledger which dealt with the Club Accounts.

In the horse lines the Squadron kicked and squealed. And the cadets learned the hard way that bestriding the animal was less of a problem than its care and maintenance. A horse, in fact, was dangerous at both ends and could be uncomfortable in the middle.

W. O'B. Lindsay† rode a black mare who resented his ministrations with a curry comb, for Sheila could kick the stars out when she was in the mood. W.O'B., who had captained Harrow and played cricket for the University and rugger on the Greyhounds wing, was a big, joyful extrovert who took everything in his stride – everything, that is, except Sheila, whom he approached during 'stables' with trepidation, soothing the mare in a language that was all his own. These blandishments took the form of a descending scale of Oriental superlatives that echoed down the lines.

The Inspecting Officer of the day, from the Greys or the Bays, listened in amazement to this sort of exhortation: 'Oh Sheila, thy servant approacheth to comb thy silken mane. Pay heed Queen of Sheba, I beg leave to bathe both quivering nostrils – yea! sponge those glorious eyes – more beautiful than moonbeams that fall on sleeping lilies! Stay, my Princess, why hump your back when I caress thy belly! Black Pearl beyond price, let thy slave now put a shine on your big backside! Stand still, oh Daughter of Delights, you nearly cow-kicked me that time. Hold up, Sheila, wanton without shame, while I wash your hairy heels in this bucket of water. Whoa there, spawn of Eblis, that knocked a hole in Government property! Here somebody, lend a hand. Catch hold of the whore and hang on to her head while this sponge goes up the strumpet's fundamental orifice!' On occasions Sheila came over backwards.

Under canvas – Tidworth 1929–30 and Aldershot 1931 – we lived as troopers, sleeping with our saddles six to a tent beside the horse lines, keeping our own mess (only men without cars ever dined there) and generally having a splendid time. We had a guest night before returning to distant homes. This was a rowdy affair that produced more casualties than those suffered during the two weeks' training. Life was a gruelling

* Robin Grant Watson, Eton, Magdalen, Scots Guards. Drowned on active service in the Mediterranean.
† Sir William O'Brien Lindsay, KBE, Harrow and Balliol, late Chief Justice in the Sudan.

twenty-four hours round the clock, but we gave of our best. The Cavalry Division were unfailingly kind, and Commanding Officers with an eye on the future young entry saw to it that the Squadron got baths in barracks, dined with their future regiments and were asked to all the parties, with polo or cricket on Saturday afternoons. In return, we were expected to prove competent and capable of hard work in a rugged training programme designed to eliminate the weaklings.

Reveille sounded at daybreak. The horses were groomed, watered and fed by 6.30, after which we had an hour and a half to shave, change into uniform, snatch breakfast, and parade with a clean horse as a well-turned-out trooper, with polished brasses and buttons. 'Boots and saddles' was sounded by five minutes to eight, before which tents were brailed in hot weather, or hauled up again after overnight ragging, and swept inside and out, with blankets folded for kit inspection. During roll call, cadets stood at attention on the grass, while the regular Officer of the Day, in a retinue which included Troop Leaders, and the Sergeant-Major, armed with pencil and crime sheet, searched the lines for untidiness or deficiencies.

After field training, we were back in camp by mid-day for the hour taken up by intensive grooming before water and feed. The adjutant then inspected the shining coats, every silken head now straining and whinnying towards the forage shed, where a man from each section had hurried to collect oats and chop. 'Feed!' The nosebags went on, and after the seven-hour stint we staggered to the mess for a beer.

The cadets were hard-driven. There was an hour off for lunch; in the afternoon there were lectures and practical demonstrations, sand-table exercises, TEWT (Tactical Exercises without Troops), and intelligence tests, during which it was difficult to keep awake. Sometimes a vet talked of minor ailments, or demonstrated, with skulls and skeleton joints. Sections with a farrier-sergeant took turns in the smithy for cold-shoeing. By four o'clock, the officer cadets returned to the horse lines for evening stables and the last water and feed. Then we were free to change and go out, or fall exhausted on a palliasse stuffed with straw.

The in-lying picket (two men detailed from each troop) reluctantly kept watch during darkness in the horse lines, to see that no animal was cast in the night, got kicked or pulled heel ropes. This fatigue was imposed for misdemeanour, the unfortunates still parading next morning after their all-night vigil. For a consideration, grooms put on the blankets on cold wet evenings, while picket men dined happily in the mess.

The ostlers, who remained responsible for the general safety of the

hirelings, lived a gypsy life of their own, ensconced in a wigwam of hay
bales and tarpaulins. Hardy campaigners, they enjoyed their fortnight
in camp, for the tasks were varied and lucrative: cleaning boots, or
shining bits and bridles for 'some of the young gentlemen who had gone
to Salisbury for the evening', being examples.

Sometimes we rode out for long field days, with a haversack ration
and four pounds of oats in a nosebag on the saddle. They were the
best of all – the 'cops and robbers' manoeuvres against the regular army,
riding in shirt-sleeves over Salisbury Plain, a game in which we were
the 'baddies' with the Oxford horses, always sent farthest: mount and
man were considered more expendable.

In camp the hard grind soon separated the men from the boys: personalities emerged, and this applied to horses as well as riders. Some
fell by the wayside on both counts. I owed my survival to experience
gained in South America, but above all to 'Steady Johnny', an ugly
hunter with a fiddle head and long upper lip, who carried me uncomplainingly across country in and out of uniform, through hard seasons with the Bicester and the Old Berks. There were frequent falls,
but Johnny, a bold jumper, never refused, although asked some very
foolish questions. Johnny was all horse. I once rode the old fellow into
third place in the OTC Race. He lacked the pace but I never got to
the end of his stamina. May the turf rest lightly on his bones.

During the last days in camp, the third-year men, and anyone else
with a good enough mount, took the Certificate A proficiency test in
equitation. This exam was of first importance to Percy Rhimes, who
no doubt had side bets on the outcome with his cronies in barracks.
The Sergeant-Major rankled from a sorry display the previous summer,
when half the candidates had fallen off or refused.

The jumping course was a stiff one; a variety of obstacles, including
banks and drop fences, among which 'the Pit' appeared most forbidding. Because cadets were physically strong and determined to succeed,
certain instructors took a sadistic pleasure in contriving falls, without
good reason. Going down the small jumping-lanes with arms and stirrups crossed is fine for grip and balance, but to sit facing a troop horse's
tail over the same course can destroy confidence, and is worse for the
mount than for the man. There were two crushings from bad falls, and
a horse was crippled one wet morning. The officer responsible was not
seen again; I heard he got a wigging, not for the damage, but in failing
to provide a demonstration – expected by the class – of what to do
and how to do it!

Rhimes, the rough-rider, was a resourceful master of his trade. Some
of the hirelings who went hunting, or followed the drag, were no mean

performers: Rhimes knew the second-in-command of the Queen's Bays by reputation. Our examiner was a thruster of the old school who did not like to hang about. Certain arrangements made overnight saw to it that only Major MacNaughton's show-jumper, the apple of his master's eye, would pull out sound for the coming test.

So it was that on a cold, grey morning thirty cadets seeking certificates assembled in a huddle to await their fate. Beckwith-Smith had gone to Ascot; Bob Taylor had the rest of the Squadron out training, and there were no official spectators. The juxtaposition of each horse and rider had been planned, according to ability, by the PSI, who paraded on foot, with a long whip under his arm.

At five minutes to the hour, MacNaughton, followed by a galloper, tittupped out to join the candidates on his bay charger – all muscle and twitching veins. Their approach quickened perceptibly on sight of the motley crew – some dismounting to tighten girths, others, like condemned men, puffing a last cigarette.

Roars of sound preceded the field officer, who pulled up, in imminent danger of a collision with Shan Hackett,* the nearest undergraduate, who sat mounted on 'Tom Thumb', a diminutive cob and the best jumper in the Squadron. 'God damn your eyes, what the hell are you ball-less wonders all up to, sitting round a dog's dinner?' 'We are waiting for you to give us a lead, Sir,' piped up Shan, saluting smartly from where he sat, his head about level with the other's knee.

That was enough. Instead of falling in the parade or taking a point of vantage to check individual performances, the major, with an oath, caught his startled charger by the head, wheeled him and set off down the lane, shouting, 'Catch me if you can, you bloody young bastards.'

The ride was over a 250-yard distance, with perhaps a dozen fences. In theory, the competitors went down it singly, at intervals of ten lengths at a fair hunting pace. This was not allowed to happen. As Major MacNaughton approached the first obstacle, nicely balanced and measured to a stride, Tom Thumb fizzed past him in the air; for three more fences the diminutive cob led the field officer, then dropped away beaten. Pat Tweedie at once challenged on the other side; by now the major was showing signs of apoplexy. Then the bay charger, making light of the bank and the slither into the pit, showed his class and went sailing away like a bird on the wing, free from the close attention of the ill-disciplined pursuit. The last two jumps were built of railway sleepers, solid and unbreakable, to be cantered into with precision and taken slowly off collected hocks. This time science was of no avail. Neville Crump, later to train winners of three Grand Nationals,

* Later General Sir John Hackett, GCB, DSO and Bar, MC.

had been 'follerin' on' aboard the hunt-servant's horse. The reduced
pace provided his opportunity; galloping into the timber as though riding a finish in the Maryland Hunt Cup, he flew the last, with a derisive
'Whip up, Major!' and stayed on to win the unauthorized scramble.

Labouring far to the rear, the 'awkward squad' took advantage of
these distractions to jump the course in pairs and trios, with doubtful
candidates sandwiched between bolder horses. The result: four refusals
and three falls – mostly in the pit, where Rhimes was standing ready
with plaited kangaroo-hide whip. 'Jump up again, Sir; he can't
look round in front; you can get the cinders out of that ear later.' Smack,
smack, and more certificates were recorded.

To his credit, our Inspecting Officer looked in for a glass of port after
stables and then stayed on for lunch in the mess, having accepted Mark
Twain's view that a difference of opinion is what makes horse races.

Guest nights in full dress uniform were very splendid affairs. The
lavish hospitality was equalled by the variety of regimental invita-
tions. An undergraduate could find himself seated between a Master
of Hounds, a polo-playing Rajah or a Cabinet Minister. Tailcoats were
de rigueur for civilians; cadets with no mess kit got by in a short jacket
and black tie. Preliminaries before the Loyal Toast were on the formal
side, after which the musicians withdrew and crusted port began to
circulate. I do not remember guests of honour; speeches were out of
line, but good stories provided better entertainment. It was a splendid
evening, and old soldiers got the chance to let their hair down and recall
their prowess.

Fred Cripps, a noted raconteur, drove over with Crocker Bulteel
(who ran Ascot Racecourse) to dine with the Queen's Bays. Both had
served with distinction in the Buckinghamshire Hussars. Fred described
contemporaries in his Yeomanry Brigade before the First World War.

Among the trusty lieutenants of the Oxfordshire Yeomanry sharing
Fred's camp were Winston Churchill and F. E. Smith, later to become
Lord Birkenhead. The latter's brilliance and mordant wit were apt to
get him into trouble; nor did he take 'Yeodog' soldiering too seriously.
His superior general, a fire-eater of the old school, was fully aware of
this ignorance. He decided to sort out young Smith at the first opportunity.

Going round stables, he came to F.E.'s troop and questioned the daily
corn ration for each horse according to Standing Orders. 'How much
were they given?' Smith called up his troop sergeant and asked the
same question. He received the reply, 'Ten pounds of oats, Sir'. The
general turned to F.E.: 'Mr Smith, I asked you because I wanted to
know if you understood Regulations and could provide the right

answer.' 'Quite so,' said F.E., 'and I asked my sergeant because I wanted to know if he knew the answer.' The general was not amused! To keep his end up, after checking the appropriate passage on stable management, F.E. approached the Brigadier the next day and enquired: 'Can you tell me, Sir, the correct allowance of water per diem for the horses in my troop?' 'Certainly, Mr Smith, the answer is ten gallons.' 'Well,' said F.E., 'I personally never let my horses drink the water the men have used for washing.' A hasty search in the book revealed the ten-gallon ration laid down was inclusive to horse and man! Again the general was not amused.

On guest nights the regimental strings played 'The Roast Beef of Old England'. After camp cooking, the meal was magnificent, but Fred Cripps said soldiers ate too much and all Englishmen dug their graves with their teeth. I asked if the green cigars he smoked from morning till night might not affect his taste-buds. Fred believed it heightened his perception, and over-eating only damaged the palate. He instanced the light lunch James de Rothschild had recently given him in Paris, consisting of a sliver of turtle fat with a glass of Château d'Yquem 1870; the next course *foie gras* with a vintage Burgundy, then truffles cooked in champagne and served in a folded napkin; to end this snack Fred settled for a Doyenne de Comice pear, black coffee, brandy and a well-matured cigar.

I thought of these things, tripping over tent-ropes, in a stagger back through the lines. Old Fred could keep his turtle fat. Troopers preferred a more solid breakfast, and dawn was already in the sky.

Camp was fun; we were fit and the evenings were free. Horses and water meadows have little in common, yet I associate leisure during high summer with broken nails, curry-combs and mayflies on clear chalk streams; the contrast with the dusty road a never-failing source of pleasure and surprise; the long shadows and swallows skimming over limpid pools; river banks overgrown with almost tropical vegetation – comfrey, figwort, flags, moon daisies, agrimony and giant docks. (Lord Grey of Falloden said fishing was a form of madness, and happily, for those who have been bitten, there is no cure.) Tidworth was within reach of streams in the Newbury-Lambourn area; there was just enough time to catch the evening rise – fishing with the National Hunt trainers Bay Powell and Ken Goode.

Ken, ten years my senior, had served as an officer in the Indian Army on the North-West Frontier. Later he rode five times in the Grand National; it was on one of these occasions (in 1927) that I first met him when a schoolboy in Lord Stalbridge's box. His best performance

(possibly a record) was when he rode home on three winners, owned
and trained, in one day at Torquay. During his wedding, which I
attended in 1930, he shocked the congregation by smoking a cigar
throughout the ceremony. Ken deserves special mention for the
panache which appealed to young bloods of the day – notably his zest
for tackling any challenge in a reckless style not seen in a more conventional age.

During my time at university, Ken rode his last races in a dazed
condition. He sustained crashing falls every season and finally injuries
to his skull, which necessitated a delicate brain operation; but it took
more than a kick in the eye to tie such a good man down.

In camp he showed me two interesting bills – one from his Club, the
Junior Naval and Military, the other from Rowland Ward, the taxidermist: both premises were then in Piccadilly. The items listed were pecu-
liar. The Gentleman Rider had spent an eventful evening in the West
End. After racing at Gatwick, where he ploughed through an open
ditch, sustaining delayed concussion by the fall, Ken went up to London
and after one drink blacked out and remembered nothing. Visions of
India's jungles were busy in his clouded mind and he found himself
reincarnated as a big-game hunter. Crouching low, he started to stalk
stuffed trophies of the chase adorning the Club walls, armed with a
billiard cue: the members were not pleased; Ken was saved from expul-
sion by a medical certificate. Later that week he received a bill which
read like the Law of Moses – an eye for an eye and a tooth for a tooth:

To remodelling lion's nose	£15.00.00
To replacing 3 tiger's teeth	£ 3.10.00
To fitting buffalo with new glass eye, etc.	£ 3.10.00

When war broke out he joined the Grenadiers – a surprising choice
for, much as I admire the 1st Regiment of Foot Guards, they are hardly
noted for their sense of humour or ability to handle eccentric subalterns
approaching middle age. Ken had an interesting war because he made
it so, with a variety of odd jobs serving independently in North Africa
and Italy. His last appointment was that of Railway Transport Officer
(RTO) at Euston Station. He had little time for higher authority, for
this was the kind of thing he had to put up with:

One day, on opening my mail, I found the following letter from a Brigadier
at the War Office: 'I should be most grateful for your advice on the following
"Travel project":

1 I wish to send my wife and three small children from London to Ravenglass
on a visit at the end of August, leaving by the 11.55 am train.

2 The journey is not essential from the point of view of my family, but it is highly desirable as a visit to ailing parents. In other words, I don't want to make the journey, but it ought to be done if it is at all reasonable. Can you possibly give a guide, from past experience and future anticipation, as follows:

(a) Will it be necessary to queue, and if so how long before, to make certain of four seats?

(b) Is the train likely to be so crowded that:
  (1) it will be impossible to get to the WC along the corridor?
  (2) compartments will be so full as to mean virtually no movement?

(c) What would be the best time of the week to travel?

(d) Would it be better by the night train?

Finally, I should be very grateful for a frank opinion as to whether such a journey would be bearable, or ten hours of hell.'

At first I took this astonishing letter to be a leg-pull. But, on checking with the War Office, I found that there was a Brigadier of that name in the room number quoted, and feeling in a frivolous mood, I sent the following answer: 'In reply to your letter of the 5th July in which you ask for a frank opinion of your "travel project".

While feeling flattered at being endowed with supernatural powers, I consider that your letter would have been more profitably addressed to Nathaniel Gubbins or the Astronomer Royal. I notice that you fail to ask for a weather forecast for the day in question, but as you state no date and are uncertain whether you will travel at all, this omission will doubtless be made good in the near future.

After a consultation with the Stationmaster and examining the entrails of a bat, the following points emerge in regard to your queries:

(a) As return tickets are available for one month, and can be bought at any time of day or night, the question of queuing need not arise.

(b) (1) If you choose a compartment next to the WC you will not have to walk along the corridor. An alternative suggestion is that the children – if of immature age – should bring a receptacle with them, and the contents could be ejected from the window, preferably when the train is not standing in a station.
  (2) You do not state what sort of movement you have in view.

(c) Weekends, Bank Holidays, and other days of public festivity, are best avoided.

(d) Night travel is deprecated unless sleepers are obtained. Unfortunately there are no sleepers available to Ravenglass.

In conclusion, I feel that a trip to Brighton would prove more beneficial to all concerned, and I am sure that the RTO at Victoria Station would be only too pleased to furnish you with full particulars.'

Needless to say this set things buzzing. I might have poked a stick into a swarm of bees. The Major-General commanding London District asked my Commanding Officer for an explanation, and I was on the mat. But nothing could alter the fact that a frank opinion had been asked for, and I escaped with a caution. The civilian Stationmaster at Euston enjoyed it so much that he begged me for a copy of the correspondence to include in the Christmas number of the *Railway Magazine*. But I had to refuse him, as I felt that this would be going too far.

Ken is still with us, living in Malta. May the sun continue to shine on a man who remains young and has not ceased to be modest. What joy it is to fish a chalk stream! From Aldershot, the River Arle and the Candover Brook, tributaries of the Itchen, supplied numbers of small trout caught from my uncle Frank Lindley's house above Winchester. He also provided quantities of green food for the mess from the watercress beds; the big fish lived farther afield. Ken rented a small backwater of the Kennet which ran through the garden of the parsonage in Chilton Foliat. Open-ended, it afforded good accommodation for the various monsters reared in Sir Ernest Wills's hatchery to supply the needs of the Elizabethan manor of Littlecote. The story ran that the trout were fed on chopped liver and hard-boiled eggs until they grew too large to be contained in tanks. Then, the size of porpoises, resplendent in girth but easily deceived, they were turned out to cruise around, seeking sustenance in strange surroundings. Those that entered the backwater seldom returned; yet the stock of two-to-three-pounders remained constant, for in chalk streams, if there is a favoured lie, bigger fish on the beat will immediately replace the one that has been removed, where feeding conditions are more acceptable.

Sad to relate, the best trout caught in camp were not landed in accordance with the rules of the Stockbridge Club. The excuse? Shall we say, to feed the mess. I admit that poaching big ones has undoubted charm. A beauty met its fate in a culvert under the road high up the Lambourn River near Eastbury, in six inches of water. The three-pounder had grown too large to turn round. Ken waited at one end, and I crawled in to bolt him among tin cans and broken bottles. The other brown trout, bigger by a pound and a half, was brought to book with the help of Jack Eldon, my future brother-in-law, who came over from Longwood. This was a lanky black bottom feeder, which lived in the tail race of the local flour-mill outside Alresford. It was taken on a wet fly baited with grains of wheat. The capture involved the removal of certain floorboards, hooking the fish on a short line, dropping the rod (to which a ball of string was attached) and then passing the hooked trout, tackle and all, down through the sluice. This sleight of hand met with the approval of the late ambassador to Tokyo, who stood guard, slashing thistles on the towpath, while he kept watch for the village bobby!

The camp fishing was good but the company even better. After hilarious late suppers, it was difficult to struggle back from hospitable homes – especially Ken's, where other trainers (Donald Snow, Dick Warden, Ivor Anthony and his brother Jack) were frequent visitors. Jack had a great singing voice, and, before the accident that destroyed his leg stepping off a hack in a Maryland stableyard, was one of the greatest cross-country jockeys at Aintree.

There were other diversions on the Plain. The outing in Salisbury which I remember best was a birthday celebration; in the course of the evening I got put down for the long count and found a life-long friend. Troop-Sergeant Derek Mills-Roberts,* two years my senior, was an enthusiast who enjoyed soldiering as much as he loved Oxford. Intended for the family business – a solicitor's office in Liverpool – the war was to come at the right time for him. He would take to the new life like the proverbial duck to water. Ability apart, some men are born lucky. Derek, laid up with pneumonia, was to be one of the few Irish Guards officers who survived the Stuka bombing attack on their troopship before the battalion could disembark in Norway. He will play an important part in this story, emerging as one of the outstanding commando soldiers in the war. Neither short nor tall and quick on his feet, the 'Mills Bomb' of those days was jet-propelled by consuming energy and a determination which belies the benevolent father-figure, retired and now living in North Wales. A favourite with Percy Rhimes, he modelled his blistering language on the PSI's and then embellished it, as he bucketed round the troop with a genial ferocity not approached in the rest of the Squadron. He was not a man to be trifled with, drunk or sober! I learned this the hard way outside the 'Haunch of Venison' in Salisbury, after a pass at the barmaid, who seemed to grow prettier with every glass of 'Gin and It'. At the end of a boozy evening, the birthday party broke up and straggled back to the car park. As the girl slipped out of a side door to keep our date, a burly figure, barely recognized in civilian clothes, appeared to bar the way. Appraising Derek through a haze of Martinis, I underestimated the stranger from another troop. The barmaid fled in terror down the alley. Derek had spotted the girl the night before, and, after a few ineffectual swings, I found myself on the pavement with a bump the size of a small orange growing rapidly on the back of my head where it hit the kerb.

The Tidworth party had grown tired of waiting, and it was Derek who got me back to camp. That year he won the Cavalry Cup for the best officer cadet, and I carried off the shooting prize.

---

* Brigadier D. Mills-Roberts, CBE, DSO, MC, Lincoln College, Irish Guards and Commandos.

A night on the tiles, but farther afield, proved the downfall of another member of the Squadron.

The Baron – we never caught his real name; it sounded like the Vicomte de Broncoute – was of Belgian extraction. A stout, bespectacled individual of Teutonic mien, he was immediately dubbed 'Baron Von Braunshnaut'. The nobleman, who had few friends, was anathema to Sergeant Rhimes and a goldmine to the grooms. He had an arrogant conceit. That he was well equipped was reluctantly accepted; that his horse was a superior type of Hanoverian with impeccable manners and certain dressage ability, could not be denied; but in camp such niceties were considered superfluous. The retainer with a cockade in his hat, left behind at Oxford, was replaced by the bookie's runner. The Baron understood the finer points of *haute école* (having won prizes in *concours hippiques* on the Continent), but his style of riding did not appeal to Percy Rhimes, whose scathing comments seared like fire:

'You must sit down in the saddle, Sir, if you want to join those Death's Head Hussars. Don't lean forward, Sir. Quit and cross your stirrups, Sir. Now trot and sit straight. Fold your arms, Sir; grip with your knees, head up, heels down, toes out! You will sit correctly if it takes all day to teach you, Sir. Are you married, Sir? I'm afraid you will be disappointed, Sir, if you sit on them, Sir! Remember our nursery rhyme for beginners, Sir!

> With your head and your heart held high
> Your heels and your hands kept low;
> With legs gripped tight to your horse's sides
> And your elbows close to your own.'

The Baron did not like such pleasantries – still less having a strip torn off him by an NCO. When he showed signs of blowing his top with the rough-rider, Neville Crump, recently promoted to Under Officer and a good horseman, took charge. He disliked the Baron on the grounds he had kicked his father's hounds and had never apologized. He fairly mobbed him up. 'Do you know Sevenoaks, Braunshnaut?' Sullenly the Baron nodded in assent. 'Then stuff five of them up your arse. Did that hurt, trooper?' A shocked silence. 'Well, in that case, stuff the other two up! That should straighten you in the saddle.'

It was an episode in the Long Valley which proved the Baron's undoing. This is what happened, as described by an eye-witness:

The Oxford University Cavalry Squadron was about to mount, but the call 'Boots and Saddles' found Von Braunshnaut unprepared. Tubbs, waiting with his horse, enquired anxiously for 'the foreign gentleman', for names were quite beyond him.

Just at the off, the gallant horseman, who had been in London overnight, appeared and mounted in hot haste, but failed to notice that his girths needed final adjustment. Officer cadets in theory saddled their own chargers, but the rich and lazy chaps often paid the grooms to do it for them. Gathering up the reins he cantered off, despite shouts of warning from the stableman.

Once formed up, the Oxford Cavalry proceeded at a trot to join the big parade in the Long Valley, where General Blakiston Houston was to review the troops. The regiments rode past: at a trot, canter and finally at a gallop. In this last manoeuvre the Oxford Cavalry followed the Life Guards; behind them thundered two lancer regiments. It was a brave show and the horses enjoyed it as much as the men who rode them. At middle pace the band played us by to 'The Irish Washer Woman'; then disaster struck!

The Baron was not far in front of me and I saw his saddle begin to turn. He made a gallant effort to stay on board and Oliver Woods* and Bailey, a powerful New Zealander from Magdalen, edged forward to catch him, but to no avail. His saddle slipped and the Baron bit the dust! Two rocking, galloping regiments rode over him. Above the thunder of the charge came the cool voice of Percy Rhimes, our peerless instructor: 'Keep your head under your saddle, Sir. The Seventh Hussars are close behind.'

At the end of the parade the Reverend Tubbs approached me in the horse-lines; he looked worried, for the Baron was a good customer. 'Where has he got to, Sir?' he asked hoarsely. 'What's left of him is down in the Long Valley,' came the callous reply. 'You'd better take a cab and pick up the bits.'

The unfortunate Continental finally limped in bruised and minus his hat, and immediately had his name taken for being improperly dressed. 'Where's your Service Dress cap?' enquired Neville. The Baron was incensed. 'Up your arse!' came the swift riposte! Braunshnaut was learning fast; even so Crump got the better of him.

Charged next morning at Adjutant's Orders, with loss of army equipment, the Baron's parting shot told against him. Now he faced two offences under various Army Acts:

1. Returning improperly dressed from the Major-General's Inspection.
2 Failing to report the loss of Government Property – one Service Dress cap. Neville supplied the details: 'On being questioned where he had put this article of clothing, the officer cadet replied "Up your arse, schweinhund" or words to that effect. I immediately took his name for making a statement which he knew to be false, and placed him on the Charge Sheet, Sir.' The Baron was ordered to make good the deficiency.

Good athletes abounded in the Squadron. At Aldershot, where the training was less strenuous than on the Plain, there were cricket matches in which we beat the regiments in Southern Command. Peter Oldfield and W.O'B., who both played for the University, were not always available, but Charlie Marks, a gifted all-round athlete from Downside, and Peter Scott, the demon bowler from Winchester, skittled out

---

* Major Oliver Woods, MC, New College; a future assistant editor of *The Times*.

all opposition. Poor Peter was killed in the Western Desert, serving in the City of London Yeomanry.

I would mention one final memory, of another sporting occasion, before we broke up for the last time.

It was at Tidworth that Oxford challenged our hosts, the subalterns of the Division, to a Gymkhana tournament, watched by the Inspector-General of Cavalry; we beat them easily. In a spirited afternoon, the Squadron won every event except the jumping, flooring the regulars, wrestling on horseback, in musical chairs, and a mounted pillow fight – with all competitors riding bareback. In the VC contest (where a wounded comrade, left helpless on a stricken field, is picked up, laid across the withers of the rescuer's horse, and ridden like hell for home), Derek concussed W.O'B. Lindsay, dropping him on his head off the black mare Sheila as they galloped for the finishing line. The NCOs won the tug-of-war. For the finale, Sergeant Instructor P. Rhimes, 16th/5th Lancers, gave a stirring display of tent-pegging, his lance head flashing in the evening sun as, to a ripple of applause, he rode out of the arena – and out of the lives of the third-year men going down from Oxford University. There I shall take leave of a great non-commissioned officer.

The *arme blanche* has gone for ever. People still talk about the 'cavalry spirit' without knowing exactly what it means. That indefinable quality is not easily recognized except in barracks or on active service; then it can lift a regiment and inspire ordinary men. But I will venture a definition of its basic principles: fight hard, play hard, and take what's coming – laughter or tears.

As civilians, we learnt the rudiments of this philosophy in the Oxford Cavalry Squadron. So did the Inns of Court Regiment, who turned out good officers along the same lines, trained by their regular adjutant Bertie Bingley, a splendid 'Cherry Picker'*. I will end this pre-war story with one of his better remarks – to a disappearing Commandant, whose horse had tucked its head and taken control: 'If you're really going to leave us, for Christ's sake say goodbye!'

POSTSCRIPT FROM NEVILLE CRUMP
24th May 1974

Dear Shimi,

At last I have found time to write to you and try and produce something for your autobiography...

The most amusing incident I can remember was the 1930 camp at Tidworth on which Dennis† also remarks. A good few of us went to the Queen's Bays

---

* A 'Cherry Picker' denotes all ranks of the 11th Hussars.

† Sir Dennis Wright, GCMG, later ambassador to Ethiopia and Iran.

sergeants' mess and, not unnaturally, got very drunk. On the way back to camp we passed the tented camp of the Sandhurst cadets and somehow or other managed to get into one of the tents. Unfortunately, being extremely inebriated, we did not realize that there was a body in a camp bed which we turned upside down and stuck a sword through it – this was hanging on the tent pole. I looked up and saw a red hat lying on a peg and said, 'Christ – this must be the Brigadier' and we all ran as fast as we could back to our lines. The next morning there was an identity parade; however, the Brigadier failed to recognize any of us which was lucky and we escaped with a hell of a ticking off. I remember you were at this particular camp and I am sure if you were, it was you who turned the bed upside down – or possibly stuck the sword between the Brigadier's legs.

...I do remember that you were as wild as hell and behaved even worse than I did. Mind you, I was an officer and you were only a flipping sergeant!!

I do hope we meet up again one of these days.

Yours ever,

NEVILLE

In Chapter Two I described how the Old Flamers in Brighton failed, despite prodding, to enthuse any serious thoughts of an army career. It was a man of peace who made up my mind for me. He came from an interesting family. Thomas Robert Dewar, born in 1864, elevated to the peerage in 1919, put whisky on the map. A salesman who believed in advertising and backed his own judgement in business, Tommy Dewar was above all a patriotic Scotsman. Dewar House in the Haymarket became the gathering place for friends from north of the Border and travelling Scots who hailed from overseas. His brother John, later to become Lord Forteviot, was an equally shrewd businessman and MP for Inverness. The family home lay on the Highland line. Until the late Victorian era their father's shop in Perth High Street sold whisky at 3s. 6d. a bottle. Old John traded locally with the 'cretur' supplied from a rented distillery on the Braes of Tullymet. The boys were more ambitious. In 1896 they built their first distillery at Aberfeldy. The English whisky market was not established until the end of the century, and for ten years the brothers built bonded warehouses and vats in Perth to meet the new demand. Then a bottling plant was set up at Dewar's Wharf near Waterloo Bridge. Tommy never looked back; he found time to become London's youngest sheriff and later MP for St George in the East, though his drive to find new markets took him on extensive overseas tours.

It is not as a salesman that Tommy will be remembered by friends of all ages, but for his wit and original mind. He later became a patron of the arts. I gave full approval to the paintings in Dewar House which included Landseer's *Monarch of the Glen*; Raeburn's portrait of *The Macnab*, bought at the end of the war, was probably the best picture; but there were many others – *The Spirit of his Forefathers* (good propaganda) and *The Thin Red Line* (which showed Highlanders drawn up in battle array at Balaclava) appealed to soldiers. Chantry's bust of Sir Walter Scott, and the tavern table on which Burns wrote some of his greatest verse, were added attractions in this port of call.* Here I learnt to take real pride in my native land.

Perth has always been a recruiting centre for Scots Guardsmen. Tommy remembered the tall lads who took the king's shilling when he was a boy: he said they were the finest soldiers in the army and he never missed the Birthday Parade when Scots Guards trooped the colour. He knew his military history. Tommy told how the regiment had won the first Victoria Cross to be awarded at the Battle of Alma in the Crimean campaign. In fact, they won two: one to Ensign Lindsay, who carried the colours; the other to a sergeant in the escort. That started a train of thought.

Among Scots who foregathered at Dewar House was Harry Lauder, beloved throughout the Empire, whom Churchill rightly described on a visit to Edinburgh as our last minstrel. Lauder too enjoyed the Trooping of the Colour and shared Tommy Dewar's admiration for the Scots Guards, inspired, I believe, by his only son (in the Argyll and Sutherland Highlanders) killed in action, who had fought in the Battle of the Somme. Two of the country's most persuasive showmen – each in his different field – caught my imagination in a way that Major-General Sir Albemarle Cator, GOC London district, had failed to do.

I did not witness the ceremony on Horse Guards parade ground until I went up to Oxford, by which time the vet's image had receded and thoughts had turned to horses, perhaps light cavalry? These dreams dispersed as mechanization gradually took over the rôle of the horseman. Rufus Clarke, already at Sandhurst, was on his way to the Grenadier Guards. Tony Wilson had joined the RAF. My father had the final word.

In 1929, Victor Mackenzie's Battalion (ISG) returned from China. They held a high reputation, enhanced by a difficult period in Shanghai which brought officers and men to a peak of efficiency which won admiration during a preliminary attachment. Father did the rest. In

---

* In less good taste was an electrical device which stood against the old Shot Tower on the Wharf depicting an illuminated Scot of giant stature, with kilt and beard, taking off his dram. The whole effect was ingenious, for the figure every so often raised its glass to thirsty lips while kilt and beard shook realistically on a winter's night.

October 1932 I reported with Bill Stirling to Colonel Balfour of Bal-birnie at Chelsea Barracks. Bill jumped out first and I paid the taxi! Two soldier servants awaited our presence on the steps of the mess. Bill picked one of them, who later set Keir on fire. I was left with Guardsman Mellis, a gold nugget who stayed with me for nearly twenty years.

Life in the Scots Guards was pleasant, but not particularly exciting. Much time was spent in London, admired through the railings (I hopcd) by nurserymaids or the passing Chelsea Pensioner. In 1935 the battalion was sent to Egypt; my sojourn there was a worthwhile experience, and I became a more serious soldier. This was short-lived for within a year we were back in England. The old life had grown stale and there were other distractions. This brings me to the events which form the subject of Part Two.

PART TWO

ACTIVE SERVICE

CHAPTER NINE

THE BRAVE MUSIC OF A DISTANT DRUM

Learn sleep upon the ground:
March in your armour thro' watery fens:
Sustain the scorching heat and freezing cold:
Hunger and thirst right adjuncts of the War:
And after this to scale a castle wall:
Besiege a fort to undermine a town:
And make whole cities caper in the air.

CHRISTOPHER MARLOWE

On a recent visit to America, Mrs Thatcher described Great Britain as an eleventh-hour nation. This curious definition, despite the apologetic ring, was intended as a compliment. The New World raised its eyebrows. Back home old fogeys registered disapproval. To raise a ghost the introduction to this chapter has been revamped. Expedience has outdated heroics, but I dare to echo Napoleon's dictum that bravery is never out of fashion.

Dished up in the United States recovering from war wounds, the stories that follow were well received on a lecture tour. World opinion may have changed, but self-respect should not come amiss in a generation whose patriotism appears likely to be eroded by uncertainty. These things have happened before, yet I beg leave to reproach the current frame of mind.

Fundamentally we are the same people, yet our leaders (there are no statesmen today) no longer grasp the helm, preferring the rôle of those celluloid ducks who suffer rude disturbance in the waves of a child's bath, either to capsize or be buoyantly blown off course. An age more concerned with material values and personal advancement will inevitably breed malcontents who belittle the old country, which Mrs Thatcher was supposedly defending. There was truth in what she said. The good lady could also apologize for those who denigrate the crown and parliamentary procedure or mock Britain's institutions and everyday achievements.

There is scant evidence to rejoice the hearts of those who seek a modern Utopia, but lock-outs, strikes and violence on the picket lines are not the answer either.

To the audience who cares to listen, we are finished as a nation. It is a rewarding pastime for Marxist extremists and a small disruptive element among hardliners and the behaviour of certain shop stewards in militant trade unions. Doctrinaire socialism, which yields increasingly to the dictates of the Left, admits such talk is dangerous.

Yet detraction and a good belly-ache are characteristic of the Anglo-Saxon temperament.

Peculiarly Anglo Saxon is the perverse habit of losing every battle until it comes to the last round. 'Plus ça change plus c'est la même chose.' England always pulled through – which accounted for our complacency. Tribulations occurred in the days of the Empire, when maps of the Earth's surface were largely marked in red, when there were greater responsibilities, when Kipling wrote, 'What should they know of England who only England know?' We could rise above mistakes, and we were seldom out of practice! The military record is worth a backward glance. Take a rough hundred years. We muddled the Crimean War, the Indian Mutiny, Egypt, the North-West Frontier and the Sudan. Humiliating defeats by Boer farmers are still talked about by veterans who survived those unpleasant experiences. There was little comfort in the dark days of 1914. In 1940, the Old Contemptibles, like Solomon eagles, prophesying woe, asserted Britain could no longer breed the kind of men who perished on the Somme or died at Passchendaele. Hitler believed von Ribbentrop when told the English lion had lost both teeth and claws. Maisky informed Stalin that a new generation speaking at the Oxford Union had accepted defeat in the Fight for King and Country debate. Another ambassador, Joe Kennedy, who should have known better, likewise misread the young men of the day. Harry Lauder's patriotic songs were booed off the stage. We nearly lost the war – through lack of faith in our own ability.

After abandoning Czechoslovakia, Eddie Knox expressed the general humiliation in 'Hymn to Dictators', which appeared in the 1938 Christmas number of *Punch*:

> O well béloved leaders
> And potentates sublime,
> We come to you as pleaders
> Because it's Christmas time.
> Illustrious banditti
> Contemptuous of our codes,

> Look down today in pity
> On democratic toads.

So much for recent history: peradventure the old days were lousy too. No doubt the heroes of Agincourt, Blenheim or Waterloo echoed the same opinion to young men who had not proved themselves.

What of the future? I remain optimistic: much depends on the next generation. But it will be a sorry business if Parliamentary democracy becomes so effete as to allow other forces to take charge. The new road has many turnings. The potent and emotive myth of national sovereignty is not all sweetness and light. There are times when it is better to close the ranks than seek independence and divide the nation. Is Great Britain falling apart?

Today the kingdom is on trial: lack of strong leadership is the root cause of half the trouble. Belatedly our rulers must learn that if political loyalty is no bad thing, people still come before party and the country comes before both. Apathy is the immediate anxiety; the danger will increase unless the public learns to think for itself. The ability of a handful of extremists to destroy any government has not been fully understood. The reader can decide where we stand. Our problem is not insanity so much as feeble-mindedness, a refusal to think things out at all.

This is my introduction to Active Service. I have dwelt deliberately on the point that nothing in peace or war exasperates a young man more than to be told things are not what they used to be. That is how we felt in 1939.

This was the background to the commando story – the build-up deserves closer study.

Raised in 1940 during the period of crisis which followed Dunkirk, and disbanded in 1946, the men in the green berets played a brief part upon a bloody stage. It is important to establish identity in order to understand the rôle of these army volunteers; then put their value into proper perspective.

The word 'commando' means different things to different people: from knights in shining armour to any kind of thug or café gangster; from shock troops to undisciplined glamour boys, ill-disguised with cloak and dagger. Exaggeration and petty jealousy have added to the confusion. Readers must draw their own conclusions, but it is important to dispel certain misapprehensions.

It would be presumptuous to supply documents to describe such intensely individual soldiers, recruited from every regiment in the army. Each man started a new life from scratch. There was no common pattern of behaviour. Some were good, other middling; the inadequate were ruthlessly eliminated. Various books have been written about those years of active service. There is an official history, and lively accounts of all the big raids and subsequent campaigns have been published by close personal friends – Derek Mills-Roberts, Peter Young and John Durnford-Slater – all men of ability and sound good sense. I have made full use of their work.

This book is different. War memoirs can be spoilt by being written in the first person. The more senior the scribe, the more tedious become the self-justifications and veiled 'rightitis' that go with a brass hat and red flannel. I will confine myself to glimpses of various individuals, to their reactions during battle, and to some of the training experience which merged into the comic, often tragic, situations remembered as bright threads that ran through the 'warp and the woof' of Special Service. Better writers have already supplied factual details of the various operations.

A summary of commando service written by Hilary Saunders, the official recorder on Mountbatten's staff, reads thus:

The long series of amphibious assaults during those five years were carried out in places as far apart as Stamsund in the Arctic Circle to Akyab off the coast of Burma. Assault craft carried into battle armed and determined men over the waters of the North Sea, the Channel, the Atlantic, the Mediterranean and the Indian Ocean: the cliffs of Dieppe, the white sands of Madagascar, the glutinous mud of Myebon felt the impress of their feet; and the firwoods of Norway, the apple orchards of Normandy, the mangrove swamps of Arakan were witness of their prowess. Within their ranks were men from every quarter of the world and from many allied and even enemy nations. They came from all classes; from all occupations. Among them were cracksmen and peers, poachers and bank clerks, bookmakers and university graduates. They fulfilled the precept of Hobbes 'that force and fraud are in War the two cardinal virtues' and in doing so aroused such a passion of hate and fear in the heart of their enemies that first von Rundstedt and then Hitler (1942) ordered their slaughter when captured down to the last man.*

All ranks were young and strong and the spirit of adventure ran hot in their veins. But we must get down to details, for there is a seamy side to this story and there were growing pains. It was far from plain sailing, with certain difficulties to overcome before being accepted as a corps élite.

The commando soldier enjoyed one particular advantage. He was a

---

* I quote advisedly: with David Stirling (lucky to survive spells in prison, ending in Colditz) I shared the doubtful compliment of 100,000 marks reward offered for our proscribed heads, dead or alive!

hand-picked volunteer. There is an old saying that one volunteer is worth ten pressed men, and it is true. In this way only were the men unusual – and with this bonus was to come a special pride. Applicants wishing to join the new force did not consider themselves as out of the ordinary when, bright-eyed and bushy-tailed, they signed on for special duties.

Reading back through magazines and press cuttings, I feel that the 'tough guy' image beloved by reporters seems to have been overdone. Wrongly, the news media made much of the carefree, casual, happy-go-lucky image of commando service. Nothing could be more misleading. There is no short cut to proficiency in war: when dealing with a highly trained, hard-hitting and confident opponent toughness is useless unless backed up by technical skill, a certain mental resilience and iron discipline. The sort of people who practise robbery with violence are seldom in evidence unless the odds are overwhelmingly in their favour.

The severity of commando training, rather than its diversification and originality, have also been taken out of context and exaggerated. All ranks took pride in physical fitness, knew how to swim, handle explosives and sail boats, and expected to live dangerously. None were supermen: until they had received specialist training few showed exceptional skills in the arts of war. Only a handful had seen active service or been under fire in France or Norway. On both occasions they had been on the losing side. So why the enthusiasm to join the new force?

There were many reasons, stemming from the fall of France. Good men objected, in the vernacular, to 'being buggered about'. Perhaps the majority hurried to Sir Roger Keyes' private army in a spirit of adventure. Some joined because they loved the sounding sea, blue water and small boats; others had a personal score to settle with the enemy. Some were idealists; others (occasionally misfits from their own regiments) were seeking a change. The psychopath or drop-out trying to prove himself was not unknown. Some of the 'smarter Alecs' may have lacked civic virtues, but in war there is more than one standard of worth. Not all accepted the Sermon on the Mount. Conversely, I must record that prosaic ex-citizens were sometimes a liability. They were destined to make better targets, destroyed by the solid qualities that embraced an orderly existence. By and large these volunteers were ordinary people of determined character and independent mind who expected the war to go on for a long time; they had no wish to spend months, and possibly years, on barrack squares, retraining and re-equipping after the defeats in Europe. To them, commando service offered new and exciting possibilities. In the reality of war such men found a new freedom from a humdrum existence of convention and routine. It created a classless society in which respect was won by better things than money or position. There is one aspect of total commitment which concerns a volunteer soldier. And that is personal performance: good, bad but never indifferent. Commando officers strove to excel; many of them succeeded. In No. 4 Commando I asked all ranks to remember we came from an island race, reminding them that Britain's best performances had been afloat. We sought for ships, lived on the coast and took to the new life like ducks to water.

> I am a man upon the land
> I am a selkie in the sea.

Slow to anger, the mood of the nation changed after Dunkirk. The Blitz added to that wrath. The public record had been shocking, and many men who had revolted against the thinking of the thirties went back into battle with a new spirit of dedicated ferocity. They rejected the materialism of former times, the policies of appeasement and peace at any price, the League of Nations, disarmament and the shame of Munich. Then, as now, the double-standards of gutless politicians with the clammy hand, the mealy-mouth and dubious breath had soured allegiance to authority in the land they loved. All resented being told that the Germans were invincible. Churchill's fighting speech, made at a time when we stood alone, pulled men together:

Though large tracts of Europe and many famous states have fallen or may fall into the grip of the Gestapo and all the odious apparatus of Nazi rule, we shall not flag or fail. We shall go on to the end. We shall fight in France: we shall fight on the seas and oceans: we shall fight with growing confidence and growing strength: we shall defend our island whatever the cost may be.

We shall fight on the beaches: we shall fight on the landing grounds: we shall fight in the fields and in the streets: we shall fight in the hills: we shall never surrender.

And even if, which I do not for a moment believe, this island or part of it were subjugated and starving, then our Empire beyond the seas, armed and guarded by the British Fleet, would carry on the struggle.

Men who had lost their weapons and their pride in France answered the clarion call: it came to a country bowing to accept defeat. The Prime Minister got the answer he deserved and the country rose behind him. This was the spirit from which the commando volunteer drew inspiration, shape and meaning. From a level start, leaders emerged and confidence grew with every raid. Men responded to a sense of purpose and determination under uncompromising leadership in a cause that required no explanations. Dunkirk* had been a disaster – not a salvage operation, as we prefer to remember it. Defeats (reverses might be a kinder word) were currently repeating themselves in the Western Desert. Sterner methods were required: some new formation must take the fight to the enemy.

That stout soldier, Peter Young, now a military historian and the President of the Sealed Knot Association, wrote in *Storm from the Sea*:

Commandos approached new tasks in a critical spirit. No tactic was necessarily sound because the book said so. The men who rammed the lock gates full steam ahead in the destroyer at Saint Nazaire, or the Brigade that fought its way off the beaches through six miles of German defences on D-Day, performed feats of arms more daring (and with greater success) than the cavalry charge at Balaclava. But the old attitude of 'theirs not to reason why' had long since gone into mothballs. A junior officer once jested 'a subaltern is always wrong with this type of soldier until he is proved right'. The commando fighting man expected to be clearly briefed – 'to know what he was on' – and to be led by a bloke that inspired confidence. That and discipline were the keys to success in raids or night attacks in a hundred fights against the odds.

Intelligent soldiers learned the object of a raid down to the last trouser button. If things went wrong, if leaders fell, or if boats were lost at sea, all ranks used their sound training to improvise and carry on. New battle tactics were no longer the 'load – present – fire' affair of Wellington's day – as outmoded as the static slaughter of Haig's Western Front.† Happy the commander working independently who has courageous motivated men to carry out his attack.

If compliments were scarce there was a wonderful relationship between all ranks in Special Service. We knew that the lives of all depended on the skill and competence of each individual. There were few punishments because in a good commando they were not necessary. If an officer, NCO or private soldier was considered unsuitable he was returned to his unit without compunction. Off parade the relationship between officers and men was informal and easy. During training, discipline was strict and only the highest standard of behaviour was rated good enough. Men worked in pairs, choosing their best friend for the duration of their service. This David and Jonathan 'me and my pal' comradeship lifted ordinary men to many acts of heroism.

* British and allied troops landed in England from Dunkirk between 27 May and 4 June 1940 totalled 338,226 British and 26,175 French.

† The Maginot Line mentality of fixed defences repeated the same mistake in 1939.

Tony Smith, my intelligence officer and a shrewd observer, wrote with truth:

Much has been said on the subject of commando training. I am afraid it did no credit to the skills and intelligence of the methods employed by original minds – a perfect example of private enterprise and individuality beating state control.

Training varied according to the type of operation in view, but round the clock work with the navy formed an important part of any programme. It was a hard and fast rule that commando soldiers trained before a raid with the actual boat crew and landing craft that was to carry them across the water. The value of this dictate was to be proved time and time again.

At first the sailor had an inbred contempt for his opposite number, and co-operation never came easily, but the barrier of prejudice could be broken down if all ranks got together. So it was before the Dieppe raid. Each side came to know the difficulties, needs and limitations of the other. Admiration and respect sprang up in place of hostility and misunderstanding.* I am convinced that this closer association had an incalculable effect on our successful operations.

There was nothing miraculous about commando training. It was of a high standard, of course, but what really counted was the type of man who was trained. Certain attributes – intelligence and determination – were expected, and both were there. Individually the commando was exceptional; collectively he and his fellows were superb. Discipline was rated extraordinarily high. It was probably stricter than in most regiments on a wartime footing, but very different from what the public were led to believe. Discipline was built on morale – good discipline always is – and the morale in a good commando was consistently high, such as can only be obtained in a volunteer force. I cannot stress the importance of the volunteer spirit too strongly. It was the foundation of the whole success of the commandos, and ran like a steel backbone through every phase and aspect of our story.

Compared with the conscript army, we were individualists. I personally encouraged independent and unorthodox training methods. In a bad commando, that never rose to pre-eminence, conflicting views (which go with independence) brought attendant difficulties in their wake. With a weak commanding officer these escalated, sometimes alarmingly. The best types had to be led – not driven. In the very beginning the appointed colonel chose his second-in-command, adjutant and six troop leaders, who, in turn, selected their own section officers, who then found their fighting men. No officer was considered worth his salt unless he ran his own show, and the men of each troop (sixty-five other ranks and two subalterns, commanded by a captain) reflected the character of the man who led them.

* See my account of the Lofoten raid in Chapter Twelve.

I have said there were growing pains. Commanding officers lacking authority found themselves ignored or in contention with self-confident subordinates whose personalities were stronger than their own. This state of affairs was exacerbated by the scattered locations of the troops, who lived in civilian billets, perhaps one troop to a village, which made collective training an impossibility. In the event, only three out of twelve original commanding officers stood up to the pressure of their appointments or qualified for promotion. If they had the guts they lacked the guile. Extroverts, in most cases of a flamboyant kind, they had character and a common heroic quality, but they simply could not go the pace at the weights. Senior colonels either lacked self-discipline or failed to bridge the generation gap between the two world wars, which often led to a breakdown.

Certain qualities were needed for the desired type of commanding officer, the first essential was control over one's self. This, together with swift grasp of a situation and coolness in a tight corner, can make a proper soldier. I have a belief, borne of experience, which sounds hard, but I have found is true, that the man who lets you down in a difficult situation once will invariably do so again. Never give doubtful characters a second chance on active service. A bad CO is also inclined to have favourites – and that is the worst mistake of all.

Looking back, I hold that good fortune, as well as skill, plays an important part in the opportunist's ultimate success. But, by the law of averages, a brave man takes one chance too many. 'Is he a lucky person?' Napoleon was right to favour rapid promotion: it serves as a tonic as well as a challenge to good officers. But there were always heartbreaks. I will give examples.

Brian Reynolds (an Irishman in the Welsh Guards), Gus March-Phillips, Geoff Appleyard and Colin Ogden Smith left No. 12 Commando to take to the ocean wave. Some of Gus's achievements are still on the Official Secrets List. Men of the purest grit, they were all to be killed, drowned or shot out of hand when captured by the Germans in small-scale raids on the enemy coast. First they operated in the Atlantic, sailing the Brixham trawler *Maid Honor* from Poole Harbour to Freetown in West Africa with a six-man commando crew. There they scored various *coups de main* against the Vichy French.

Steering a course back to England, Gus then turned the party into a formation known as the Small Scale Raiding Force, which operated against the Cherbourg peninsula and among the Channel Islands in a motor torpedo boat affectionately known as 'The Little Pisser'. They were joined by more good officers (from Nos. 7 and 12 Commandos), among them Graham Hayes (who had sailed before the mast in a

Swedish grain ship). There were also John Gwynne, Howard of Penrith (who survived, but lost an arm), Philip Pinkney, Peter Kemp, and two splendid Allied officers: Anders Lassen (a dashing Dane who was to win a posthumous V C, added to his Military Cross and bar), and Hans Brinkgreve (a Dutchman, later captured and shot by the SS). Gus was drowned when his Goatly boat was swamped off the Breton coast. Colin, on the same raid, escaped by swimming through heavy seas, and got back to shore several miles from the original landing place. He was handed over by the Vichy French, placed in solitary confinement for nine months, and then taken out and shot on 13 July 1943.

When the Channel Coast got too hot for this intrepid band, they took to parachutes and ranged farther afield. Appleyard was shot down over Sicily, and Philip, against the wall, by the Carabinieri in North Italy. Peter Kemp survived on a transfer to an equally hazardous mission to join Billy Maclean and his Albanian partisans in the Balkans. What was left of the raiding force then merged with the SAS in North Africa.

Brian Reynolds, who continued operating in northern waters throughout the war, almost beat the odds. Then his luck changed. Entering Antwerp harbour the day after Victory in Europe was declared, to celebrate the termination of hostilities, Brian was blown up by a floating mine – killed after three years of brave deeds in foul weather and dark nights, working from small boats in the Skagerrak under constant sea and air attack. His missions included the ferrying of ball bearings (vital to the war effort) from neutral Sweden. Four out of the five motor torpedo boats in his flotilla were sunk by enemy action.

The Germans – stout opponents themselves upon occasion – have a phrase for heroes: 'Always the tallest poppies [*Mohnblumen*] are taken.' These were proper people, and I hope that some day a book will be written about them.

Dame Fortune dealt a better hand to 'Fairy' Veasey, one of the first volunteers, who achieved early success on the Lofoten raid and was to do well again at Dieppe before a transfer to the Mediterranean. He was a very popular officer, much loved by his troop.

> Though not amongst the most select
> In matters of the intellect,
> He had, at times, no small success
> At least for having common sense.

Fairy also had a sense of humour. He pulled my leg successfully whenever we were short of rations (he would have eaten most of them) by telling how he held the pork pie championship of all England (having devoured thirty-three at a sitting) and had to keep his strength up. He was also captain of the Brigade rugger team. It was said that three men could not hold him down, unless they tackled him by his black handlebar mustachios, of which he was extremely proud. But Fairy could not read a map or set a compass course. 'That hill should not be there' was a favourite expression on some dreary map-reading exercise; but his men covered up for the captain on these occasions. Veasey was torpedoed with Donald Hopson* in convoy off Derna. His ship went down in 4½ minutes. He was picked up by an Australian corvette, clad in green beret and pyjamas, shouting, 'What have you got for breakfast?'.

The day came when this officer was overcome by sleep (or some say too much pasta) after days of fierce fighting among the vineyards and sharp stones of Sicily. Cut off from the coast, and waiting for the moon to rise, he was surprised and captured by the Germans and, with another subaltern, taken to the mainland and put on a train under heavy guard, bound for a prison camp in Germany. It was a point of honour that no unwounded commando soldier accepted captivity and the possibility of a firing squad.

The next day, as they approached the north of Italy, Fairy awoke from prolonged sleep, got the better of the guards and, with those in his carriage, jumped off the train. They had no map, and opinions were divided as to the direction. Those who were more technically correct marched towards the Swiss frontier; eventually bumped into the enemy and were recaptured.

Fairy, on the other hand, cunningly walked back down the line, went to ground and settled down to sleep again in a road culvert. After further wanderings and changing into the black (Fascist) shirt of a friendly Italian, he was astonished to find himself one day in Switzerland, where he promptly married a pretty girl. Neither could speak the other's language – or so the story goes; they lived happily ever after. Fairy spent the last year of the war sending postcards to his friends in No. 4 Commando, praising the comforts of the Engadine.

Mental resilience was equally important: I will give an example. Freddy Chapman, Seaforth Highlanders, polar explorer and schoolmaster at Gordonstoun, was my second-in-command on the Field Craft Wing at the Inverailort Irregular Warfare Training Centre. Later he was promoted and sent to Australia to start a similar school 'down

* Sir Donald Hopson, KCMG, DSO, MC, post-war diplomat in Peking and Caracas; later ambassador to Argentina.

under'. Then he was posted to Singapore. After finding himself cut off behind the Japanese lines in Malaya, he remained there, fighting a guerrilla action for more than two long years. A determined soldier, mentally as well as in the physical sense, Freddy was no ordinary person, and his book *The Jungle is Neutral* should be read by all those interested in military history. The hazards of that tropical experience, and the will-power required to cope with the country and the climate, are well summed up in an introductory paragraph:

Twice wounded – captured in turn by the Japanese and then Chinese bandits – he escaped from both. On one occasion he was dangerously ill for two months on end, including a period of unconsciousness for seventeen days lying in a grass hut during the monsoon rains. Suffering at various times from blackwater fever, pneumonia and tick typhus as alternatives or additions to almost chronic malaria, his strength was so weakened that it took him twelve days of hard marching to cross through ten miles of dense jungle.

On another occasion he moved for six days barefoot and without food to attack a Japanese supply depot.

During the same period, and under the same conditions, his companions – six British soldiers – all died, not from any specific disease, but because they lacked the right mental attitude to adjust themselves to the conditions.

Survival of the fittest is a somewhat meaningless phrase in peacetime. It is brought home by the realities of war.

The stars who emerged from commando soldiering were not confined to the officer class or to famous regiments. Take Lance-Sergeant Peter King of the Royal Dental Corps. A tall, fine-looking man, drafted on mobilization into the wrong branch of the Service, he quickly became bored with the inactivity of his professional duties. Stationed near Dover, he decided to raise a private army of his own, consisting of two other men. They 'borrowed' a boat in the harbour and set out on a raid across the Channel. Wind and weather were against them, the engine broke down and they drifted for two days before being ignominiously towed home by a naval patrol. King was court-martialled and reduced to the ranks. He immediately applied to join No. 4 Commando. The man's physique and will to win overcame any doubts raised by his previous conduct. Within six months of exchanging the dentist's drill for a service rifle, King got his stripes back, emerging with top marks from courses set for NCOs. In the battle of Normandy his powers of leadership won a gallantry award and immediate promotion in the field. Captain King survived the war but found the United Kingdom too confined for his immense enthusiasm. He emigrated to New Zealand. On the outbreak of the Korean War he immediately re-enlisted in a local unit, this time as a gunner. Once again the ex-Dental Corps soldier proved his worth, rising to the rank of colonel and retiring with a DSO. There were others like him.

Robert Bridges has summed up collective courage: 'Dear youth who lightly in the hour of fury have put on England's glory as a common coat.' That the men were brave goes without question; that they showed such initiative and self-reliance surprised other branches of the Service. The spirit in a commando was quite remarkable. Though all knew before a raid that the going might be rough – that everyone was in for a bad time and casualties were likely to be heavy – nobody appeared to have a care in the world and men spent their last free time before action playing games and behaving like schoolboys on a half-term holiday.

But I do not think that any volunteer among us would withhold the highest accolade from those Norwegians, Belgians, Dutchmen, Frenchmen, Poles and Jewish and Sudeten Germans who served in No. 10 Inter-Allied Commando. Each nation has different characteristics: all experienced immeasurable hardships to reach England and renew the fight for freedom in a foreign land. They had seen the destruction of their countries and their homes; the fate of those left behind moved each exile with a spirit of reckless heroism that did not count the cost. If they differed in temperament they were united in common resolve to give everything they had or perish in the attempt. The cry 'En avant les morts!' had rallied an army at Verdun after the loss of half a million men.

I have a photograph of just such a Frenchman – a true son of that lost generation. Possibly wounded, probably starved, worn out after his escape over the Pyrenees through Spain to Portugal, he stands proudly on parade – the spirit of *France libre*. Human suffering – perhaps deprivation – are there on his face for all to see.

Many troop leaders refused promotion rather than be parted from the men they had persuaded to join their commando. And rightly so, for in those early days the new formations were recruited with a feudal touch that flashed like fire when put to the test by a good officer. It would be invidious to cite examples, but such a man was Philip Pinkney, who raised his henchmen from the Berkshire Yeomanry.* When his troop went raiding in the Channel Islands Philip is said to have visited the Dame of Sark, entering her bedroom through the french windows. Recognizing the blackened face of an Allied intruder, the good lady sat up in bed with the cry, 'Thank God to see a decent-

---

* Part of Philip's fortune was set aside to help any member of his troop who might need financial assistance after the war. Some thirty men were later given various sums of money under this bequest.

sized man at last!' William Douglas Home based the motif of a successful play on the episode. The Prime Minister, speaking to the Commons
in more serious vein about this operation, caught the public's imagination with the vivid phrase: 'There comes from the sea a hand of steel
which plucks the German sentries from their posts.'

For three eventful years Philip's men stayed with him in various
rôles in different countries, until the time of his death. Parachuted
(with his back in plaster) into Bologna, he was captured and then shot
by the Carabinieri. In a letter to the family his sergeant wrote, 'I could
go on and on, but nothing I can say will properly express my admiration for your uncle. No one I ever met talked ill of him. He was a man,
a gentleman, and a great officer – a very rare combination.'

Could any soldier wish for a finer epitaph? But these things happened
after we had found our feet. The first year of the war was one of confusion and chaos. 'Fact', says Somerset Maugham in his preface to
Ashenden, 'is a poor story-teller. It starts at haphazard generally long
before the beginning: then rambles on inconsequently and tails off
leaving loose ends hanging about without a conclusion.' The ensuing
chapters will show Maugham was right!

CHAPTER TEN

EARLY DAYS AND BAD BEGINNINGS

War is too important to be left to Generals.

CLEMENCEAU

'Not Angles but angels.' Was it Saint Augustine who brought the house down with this ancient pun on straying to our shores in AD 597? Or Pope Gregory the Great, confronted by those fair, apple-cheeked young prisoners, blue-eyed innocents chained in the slave markets of Rome?

Peter Kemp, like Rupert Brooke (though he was no poet), was blessed with the same golden looks when I first met him up at Cambridge, in carefree days spent either riding at Cottenham or at one of those wild parties in the Athenaeum that followed the race meeting. Peter came from a Service family. His elder brother Neil, of whom he was immensely proud, was already serving in the Fleet Air Arm; I only knew him by reputation: he was another of the same sort. Neil lost his life early in the war when the aircraft-carrier *Illustrious* was dive-bombed by Stukas in the Sicilian Channel. He had won the DSC flying a lumbering 'Swordfish' and been previously sunk on *Courageous*, then picked up by a destroyer after spending several hours in the water. In that last sortie, during the sea battle of Taranto, he pressed home a low-level attack on the Italian fleet, successfully torpedoing one of the new 'Littorio' class battleships.

Before these events, the younger Kemp had exchanged his university for the firing line. I would describe him as a soldier of fortune rather than a mercenary: a happy warrior and a most entertaining one; whenever our paths crossed we were destined to have fun together. Peter's gentle charm belied a love of adventure. This personal daring sent him hotfoot to fight for Franco against the Communists in the Spanish Civil War. There he saw such scenes as Goya has portrayed, learnt to make and throw an early brand of 'Molotov cocktail', was three times wounded, once severely, and decorated with the Cruz de Guerra and the Cruz Roja de Mérito Militar. All this happened in the mid-thirties, while the ageing author was teaching recruits to form fours or mount guards on barrack squares. That life began to pall. I left the army and got married.

Then quickly came the war. I joined the Lovat Scouts, a Yeomanry regiment raised by my father for the Boer War, quarrelled with the colonel and was sent to cool my heels in the Cavalry School at Weedon. It was the coldest year I can remember. England went into deep freeze, prepared to put up with all manner of discomforts, during a phoney pause waiting for something positive to happen. Chamberlain dropped leaflets over Poland to encourage an ally cut to pieces and already *in extremis*. The civilian population remained puzzled but cheerful, in spite of air-raid sirens and underground shelter drill. The inconvenience of unheated trains, ration cards, queues, gas masks and blackout curtains was accepted uncomplainingly.

It was the same story in the Maginot Line where boredom became the greatest enemy. While the censor nodded, the Ministry of Information sought for news. Before Dunkirk, war correspondents were allowed to speculate freely on the outcome. Other media tried to paper over cracks in the 'war footing' in a nation hopelessly unprepared for dire emergency.*

Every morning the new CIGS, General Ironside, bounded up the steps of the War Office, three at a time. At weekends in private houses, standing in front of the fire, he pontifically held forth on how to defeat Germany: a harangue that evoked a scurrilous lampoon in which the words 'Ironside: backside: fireside' figured prominently. Once 'inside' his sanctum, it was rumoured that he sat for a portrait painter. Likewise his French counterpart, Gamelin, who had been described as a master of strategy, but remained incognito, a sedentary genius who never left Paris. Both generals got fulsome treatment: they were in need of a boost. A civilian population tends to trust its soldiers too little in peacetime and sometimes too much in war. Ironside, it appeared, was a linguist who spoke nine languages, but as Jakie Astor pointedly remarked, did not make sense in any one of them.

The top brass had to learn their trade, but it was different in the air force and navy, who were active, operational and fully alert. In a conscript army (despite upheaval), service conditions have at least one compensation: they throw casual acquaintances together in strange places. The challenge of potential danger and shared hardship brings

---

* So far as the air force was concerned, the straws clutched at Munich allowed a year's breathing space that was to save England in 1940. Churchill's Commons reply to the Chamberlain 'Peace in our Time' speech had proved prophetic. 'England has been offered a choice between war and shame. She has chosen shame and will get war.'

out the best in the human race and strangers make friends, for no man forgets another beside whom he has fought in action. We learnt to grin and bear it. Like rock climbers, we were tied to the same rope.

But I am outrunning events. Churchill was swept into office, and Chamberlain, the forgotten man, died a few months later.

Nothing had been heard of young Kemp since he returned from Spain; then a mutual acquaintance, Archie Lyall, author, traveller and gourmet with a Rabelaisian sense of humour, reported (with astonishment) that despite his wounds, Peter was trying to join the army and having difficulties in doing so. Archie was an older man. He too had been in Iberia, where he served in a less dashing rôle. The outbreak of wider hostilities found him at a disadvantage. 'I nurse a personal resentment against Hitler for what is nothing less than intolerable interference with my private life,' said Archie, whose bitterness grew with the passing days. 'Up to the age of thirty-five a man can drink, work and copulate. After that he must make up his mind which of the three he will have to do without. I had just decided – indeed made all my arrangements – to give up work, when this ruddy Hun started the war.'

Then came the pleasant surprise: Peter turned up at Weedon, still smiling, but a wiser man, with a mangled jaw and crippled hand, who knew what fighting was all about.

Briefly our respective courses overlapped. The Equitation School provided the last cavalry training for Yeomanry officers in the United Kingdom. The cynic will say that it was a waste of time, but I felt highly honoured to attend. Enthusiasm was much in evidence. We tried our hardest. I was lucky to draw 'Easter Egg', a bay mare with black points, for my charger. She provided an armchair ride and never turned her head – only objecting to the Ha-Ha (with a nine-foot drop into the field below) which we had to jump once in the final test. The barracks, as well as the instructors (whose standards of excellence could well compare with those of such establishments of international fame as Rambouillet and Tordelquinto), were moulded after a grand design, and everything bore the stamp of a former era. Weedon was built in the reign of George II. Some time before the First World War the vast buildings were considered to have outlived their usefulness and were condemned as unfit for human habitation – even for men in uniform. Thereafter successive governments, whatever their political persuasion, had failed to reach a decision on rehabilitation.

In retrospect there was something rather splendid, however uncomfortable, about this austere relic of the past. We learnt to accept tradition. In summer or in peacetime the course would have been pleasant. In some respects it was like Oxford, ten years before. Though the training was antiquated to the point of being antediluvian, the officers on the course took their duties seriously. They were keen soldiers, but generally better horsemen. We were twenty years behind the clock and few emerged to make a name for themselves after they gave up their horses. Boy Butler, a 13/18th Hussar, who taught the class and did it well, was killed in the Desert soon after he learnt to handle a tank. Jack Hamilton-Russell, of The Royals, was another instructor who was soon to be killed, with his brother Tony, in the new armoured rôle. He was a fine horseman and a winner of the Kadir Cup.*

After service in the Carlist Militia, and gaining a commission in the Spanish Foreign Legion, Peter Kemp must have viewed our puny efforts with no small contempt; but he never said so. But there were interesting fellows on his wing. Among them flourished the pioneers of the cloak-and-dagger groups that were later to emerge under such mysterious classifications as SOE and MI5.† Peter has given a lively description of Weedon in his book, *No Colours, No Crest*. After the recent craven performance by the Labour Government who have avoided facing up to the wholesale massacre of Polish officers by Russians in the Katyn Forest, his text makes interesting reading. Other aspects of the wartime scene are described with becoming modesty:

The most distinguished among us was Professor Pendlebury of Pembroke College, Cambridge, an archaeologist whose researches in Crete and Egypt were well known before the war. A quiet man of great charm and sense of humour, he proved his adaptability by becoming a most efficient soldier. He had a house in Crete, where he was sent, after leaving Weedon, to make certain preparations against enemy invasion. His murder at the hands of Gestapo or *Sicherheitsdienst*, as he lay incapacitated by wounds in a peasant's hut a few days after the German occupation, was a tragic blow to learning and a wanton crime.

The most ebullient, by contrast, was Harold Perkins, a former master mariner who had abandoned the sea before the war to graduate in engineering at the University of Prague and become the proprietor of a textile factory in Poland: on the outbreak of hostilities he was attached to Colonel Gubbins' Military Mission, making his escape with Gubbins and Wilkinson through Hungary and Rumania. His work when he left Weedon involved the planning of subversive operations in Poland.

Perkins' friend, Mike Pickles, had been through a bad time during his escape from Poland, travelling from Silesia across country to the Baltic States

---

* The all-India pigsticking contest held at Meerut.

† Definitions of Military Intelligence were obscure: MI5, in fact, stood for 'Internal Counter-espionage'. MI6 and MI9 were more in the cloak-and-dagger line. SOE indicated Special Operations Executive.

dressed as a peasant and riding in haycarts. He and his Polish companions were overtaken by the advancing Germans, but passed through the lines when the Germans halted, only to meet the Russians. Their worst troubles came from the latter, who were looking for Polish officers and shooting all above the rank of Lieutenant: the party were frequently stopped and the hands of each one examined to see if the palms showed a proper proletarian hornyness: luckily, by that time they did.

Whatever the limitations of horsed cavalry in modern warfare, the training filled three of the pleasantest months of my career in the British Army.

The two instructors of whom we saw most were Sergeant-Major Sennett and Sergeant Rourke: under the supervision of Captain Hamilton-Russell, the Equitation Officer. They taught us a great deal in a short time, using a balanced blend of patient encouragement and ferocious bullying, the latter reinforced by a superb flow of vituperation and an armoury of blasphemous taunts. We found it hard to believe that the two genial fellows with whom we had been exchanging pints of beer and dirty stories in the 'Globe' the previous evening, could become next morning the pair of furious and sadistic Tartars who insulted us in the riding school.

For dismounted drill we came under a delightful Highlander from the Scots Greys, Sergeant Stewart. These parades could never be wholly dull because of his practice of marching our squad across the square, as far away from himself as possible, and then shouting an order which, in the distance, sounded like pure Gaelic: each of us would then place his own interpretation on the command, and the squad would break up in disorder. An angry bellow would come from the stocky figure at the other end of the square. 'Na! A' didna say "Wha Whooh!" a' said "Whah whu-u-h!"'

Our theoretical training might have been of use to cavalrymen in or before 1918. We heard lectures on tactics from an earnest young officer in the Inns of Court Regiment, who took his chief illustration from an incident in the First Battle of Mons: lectures on animal management from the Veterinary Officer, who was inaudible; lectures on Military Law, which were incomprehensible, and lectures on the light machine gun (Hotchkiss 1913 model) which often jammed.

The R.S.M. who controlled our conduct and discipline, was an austere Yorkshireman known as 'Gentleman Joe' Taylor: formerly chief Equitation Instructor at Sandhurst, he was the personification of smartness and efficiency. Not a single detail escaped his eagle eye – a cliché which might have been coined expressly for him – not the smallest speck of dust in a barrack-room or the faintest blemish on a button, boot or belt. He had that dour North Country humour which gleams brightest when administering a reprimand or dealing out a punishment.

His strongest weapon was his power to stop our weekend leave, and he used it mercilessly to punish slackness. Weekend leave lasted from Saturday mid-day, until, theoretically, Sunday midnight. In fact, the last train that would get us back in time from London left soon after eight o'clock: if we missed that – and we often did – we had to take a tube to the north of London, and thumb a lift from one of the lorries that travelled up the North Road that passed through Weedon: it was easy to slip into the barracks unobserved in time for a quick brush-up and a shave and a mug of tea before the first parade on Monday morning. But the first parade was always Riding School, when Sennett and Rourke would be at their fiercest, knowing that they had a number of sleepless and hung-over gentlemen at their mercy.

Peter had an interesting war: after that short spell as a horse soldier he was sent on a secret but abortive mission to Spain, a precaution intended to anticipate enemy interference in the peninsula. He always maintained – and I believe him – that although Communist propaganda has branded Franco as a fascist, the Caudillo's victory saved the country from a far worse fate.

Later Peter was parachuted into Albania and ended the war as a prisoner in a Russian jail. Here privation affected his lungs, but you cannot keep a good man down. When the fighting came to an end in Europe, Kemp, David Smiley* and Rowland Winn† were dropped by parachute into north-east Siam on an SOE mission to train Siamese guerrillas in the overthrow of the Japanese forces still in occupation. After Hiroshima, Peter moved to the Mekong River to help France fight Communist risings in what are now Laos and Cambodia. In 1946 he was deservedly given the best job in the south-east Asian archipelago, being appointed Governor of the Islands of Bali and Lambok.

Our friend, Archie, Lyall, went to Belgrade as a press attaché. He was genuinely worried about Peter Kemp's excursions. Archie took a poor view of subversive activities – partly because of SOE's failure to get established in the Balkans before the German invasion in 1941; possibly because he was jealous of this new branch of the Service. When, on one of our rare visits to London, we gathered for a party at the Cavendish Hotel, Archie was heard to remark: 'I have recently designed a coat-of-arms for SOE. This is it: Surmounted by an un-exploded bomb; a cloak and rubber dagger, casually left in a bar sinister. The arms are supported by two double agents. The motto: *nihil quod tetigit non* (made a balls of it).'

At Weedon the married men in the senior class had family problems. There was no accommodation locally, and, except for the 'Globe' (which did not cater seriously for residents), there was no pub for miles around. It was a bad time for young wives with babes in arms. Our eldest son had been born the week England declared war on Germany. Posting overseas could happen any day, and a brigade of the Cavalry

---

* Colonel David Smiley, MVO, OBE, MC, Royal Horse Guards.
† Captain The Hon. Rowland Winn, MC, 8th Hussars; now Lord Saint Oswald.

Division was already being equipped for a move to Palestine: a move which in fact occurred before Christmas.*

Rosie saved up petrol coupons until late November. Then, abandoning the nursery, she drove down from Beaufort to stay with the Cadogans in a search for lodgings. Bill Cadogan, Master of the Grafton, was serving on Salisbury Plain with the Wiltshire Yeomanry, but Primrose had a friend who had a caravan. This was just what we needed. It was towed with difficulty by tractor, for the rubber had perished off the tyres, to the stackyard of a sporting farmer.

Then disaster struck. The big freeze tightened its grip. The gift horse leaked above and below. At night, when we persuaded the gas cooker to work, snow thawed through the roof. It was difficult to determine if the water cylinder was already rusted through, or had burst when the frost began. Each night we filled the tank. In the morning there was no water. That meant a 6 am start, blundering with buckets in the dark to a distant tap which, in turn, froze solid with the falling temperature. Weedon was a draughty place, and after the stacks were threshed there was no shelter. I have felt warmer sleeping in an igloo on a snowfield.

But we survived. It was worse for Rosie, in damp blankets on an unyielding mattress (it seemed to be stuffed with skulls) and with a long, empty day before her, for I was gone by 7 o'clock to seek breakfast in the mess. We had been married for a year and she was still learning how to cook. Some of the gastronomic experiments were not a success. The rejects found their way through the window and into the stackyard, where they frightened the birds. On one dreadful occasion, all shiny-booted and spurred for parade, I knelt heavily on a jam tart while attempting to locate leaks below decks.

The frost put an end to hunting – a Saturday privilege, very properly considered a test of courage and initiative required by a good officer when seeking the quickest way across country. So nobody jumped the locked railway gates that season – a cavalry tradition in the grand manner – when the Pytchley hounds took the right line from Weedon gorse.†

Two memories of Weedon are of small account. One: the boiling weekly mustard bath at Hillsborough Hall, provided by 'Bear' Hillingdon, a much-loved MFH. There was no chance of scrubbing

---

* The division was to spend two years – some regiments spent longer – in 'The Holy-something-Land', being converted into tanks and armoured car formations – a frustrating process during which eight thousand horses were discarded. I hate to think what happened to them.

† Philip Dunne (the Blues), later a troop leader in No. 8 Commando, performed this notable feat just before the war. History does not relate if the oncoming train was passenger or goods.

down in the caravan, but that strong mustard treatment removed aches and pains, turned one lobster red and restored normal circulation; the glass continued to fall.

Luxuries such as butter, cheese or cream had long since disappeared, but one bargain slipped through my fingers. On a shopping expedition to buy rations in Northampton, we made a startling discovery in a dingy alley. The weary Argonauts, sighting the Golden Fleece burning in Colchis' sacred grove, could not have been more enchanted. The shop sold rods, cartridges and fishing tackle. It was a small window in which to hang the skins of two snow leopards, appropriately in full winter coat, whose long, furry tails trailed down to the floorboards. Tinkling the bell, we hurried in. The ancient proprietor knew something about fox masks, but he was not a taxidermist, and the great pelts blocked the view of his cork floats and other merchandise. An officer, home from Afghanistan, with the uncured skins and in need of cash, had asked for their immediate disposal. The price was £15 apiece. It was Saturday. We had no money and the bank was closed. We asked for time and raised the ante. Rosie came back on Monday, but the leopard skins had gone. They are the most beautiful and now the rarest of all furs; the animals should not have been killed in the first place – but what a lost opportunity to brighten our lowly dwelling!

Finally the course came to an end, and I was posted back to the Lovat Scouts in Scotland.

By 1940 the size of the army had increased, but efficiency was disproportionate to the number of new formations and limited by general unavailability of equipment, arms, ammunition – and even training areas.

There were obvious differences between the regulars, the conscripts, and the Yeomanry and Territorials. This was understandable. In many cases the last class of volunteer – men with the experience of annual camps and whose fathers had served before them – often provided top soldiers. This certainly applied to the Lovat Scouts. But if great potential existed in the ranks and among the subalterns, I have reservations about commanding officers. At this stage in the war the majority of the colonels I can remember were in their middle forties, with seconds-in-command of much the same vintage. Many had achieved their rank by seniority rather than selection. Quite untried, they climbed the ladder of promotion without having been tested on any exercise above company level. To be passed over was unthinkable. Such officers were sometimes unfit and obliged to lean heavily on regular adjutants, who were responsible for running courses, administration, night classes, summer training, and dealing with correspondence. Along with staff sergeant instructors, these regular soldiers now had to return to their own regiments – leaving a vacuum behind them.

What was the alternative? I think we should have kept the pros for a year. Our colonel in the Scouts was a banker, and no doubt excelled in that sedentary occupation. Bankers are men who offer an umbrella when the sun is shining, and ask for it back as soon as it starts to rain: I discovered that others in the same profession, when decked in uniform, retained a meticulous grasp of non-essentials which took up training time. The colonel made a study of orderly room bumph, consulted field manuals and took copious notes in little books when visiting the squadrons. The conclusions were then written down like a stay of execution order – in silence.

Seldom out of his office, he became known as 'Uncle Gloom', and I have never met a more apathetic soldier. It was clear we were not going to get on.

A/Squadron Leader was also a banker – a pernickety bachelor who enjoyed life's creature comforts, but whose usefulness did not last through the first winter. Prior to his departure, mornings were spent on kit inspections, refolding blankets and trying to decide the right way to coil a pull-through.

There were other foibles scarcely credible. On mobilization, as many officers as wished to stay, were put up at Beaufort Castle. Because of the cold, the heating was turned on to help the temporary inmates. This took a ton of coal a week, shovelled into the antiquated boiler system to make the scattered radiators smell faintly of warm paint. The long passages might have been described as chilly, and banker No. 2 asked Rosie to supply him with a 'po' to save having to walk to the bathroom across the way. There were two others available on the landing and Rosie was indignant: the story went the rounds. It did not go down well with the hierarchy. Worse was to follow.

I had already got into trouble before Weedon for putting the Squadron Sergeant-Major under close arrest: he had appeared on parade so drunk that he had fallen off his horse before a night march across the Black Isle. The SSM, who was a nice fellow, failed his medical examination soon afterwards and gladly returned to his deer forest in Wester Ross. But he happened to be a pet of the colonel's, and I received a rocket for high-handed behaviour.

Soon I offended again. In the old days when I was a regular, in order to keep fit and get extra leave, I had fought in the Army Fencing Championships (sabre) at Olympia and had also represented the Scots Guards in the spring bayonet fighting team with fair success, twice winning London District medals. The Lovat Scouts had not yet fired their rifles, and I suggested building an assault course in the park, stuffing dummy sacks with straw and running some sort of cadre for NCOs in the use of the bayonet. There were no fencing masks, gloves or padded clothing.

The proposals were viewed with misgivings, but the men were keen, for they did not have enough to do in the afternoons. I appointed myself chief instructor and we got to work. The first group passed out well and enjoyed the exercise, but in the second party, a lance-corporal with slower reflexes failed to parry a long point with the padded training stick and got scratched across the cheek. Consternation. The story was later magnified to a rumour that there had been fighting between officer and men: friends rang up to ask how badly I had been injured in the bayonet attack! Had there been a mutiny? The colonel stopped the course. This was bad psychology, but he did not approve of infantry training tactics. Soon afterwards I was sent to Weedon.

On my return, I found that the regiment had spread out of the Park. The Ross-shire and Sutherland men of C/Squadron had removed themselves to Strathpeffer. My brother Hugh went with them. Regimental Headquarters was still based in Beauly village. A/Squadron now became my personal responsibility. All Inverness-shire born and bred, the men had left the Home Farm for Muir of Ord on the county boundary. The officers settled into the Tarradale Arms; the food may not have come up to the standard of the Ritz, but Donald Maclean, a generous host (he feared the possibility of invasion), saw to it during the three months' spell of hard weather that a bottle of malt whisky was presented every night to the orderly officer. 'For indeed I'd sooner see the Lovat Scouts get the best of my cellar, than have the Gairmans seize her.' What a good man! The Muir – a wide expanse of grass and heather, once the scene of cattle trysts in the old droving days – provided ample space for equitation. While at Weedon, Sandy Fraser rode a bad horse for me 'Tailcoat', who had won some races and took a hold. When the Squadron wheeled at the gallop Tailcoat took off – Sandy stopped him on most mornings by charging into a haystack at the far end of the Muir.

Helped by Richard Fleming, Andrew MacDonald (Blairour), Sandy Fraser and Scrap Balfour Paul, I found myself in command of as fine a bunch of hardy fellows to be found anywhere in Scotland. Another friend, Dick Allhusen (son of my father's old brigade major), was another good officer. He served with A/Squadron before he became our adjutant; later he went to the Staff College. The Black Isle is a useful training area, horsed or on foot, and the men appreciated strenuous exercises. Among the troopers there was a high proportion of stalkers, keepers and ghillies. Soldiering came naturally to the hillmen, accustomed to telescope and rifle. Their quick intelligence, keen powers of observation, strong legs and knowledge of the use of ground were turned to good account. Only Piper Beaton, from South Uist, who spoke little English, disapproved of this kind of activity, for he had to play on dismounted route marches. One day he appeared in the office to complain of the pace, which caused a malady with curious symptoms: 'For indeed it is very bad for the thumbs of my feet, and my water was after boiling as well.' In B/Squadron, whose personnel came from the Western Isles, orders were explained in Gaelic.

It was perhaps my encouragement of the all-round potential, and an aggressive approach to training, that precipitated the inevitable collision with Uncle Gloom. I did nothing to prevent it. Only a thread divides discipline from insubordination between officers of different ranks, and my behaviour was no doubt intolerable.

In March the Scouts moved south to the Dukeries, joining what was left of the Cavalry Division. A/Squadron found billets in Tuxford. Training areas were limited, and we still had not seen a rifle range. Half the time was wasted in the horse lines. The stalemate in Europe was coming to an end. Now it would be difficult to keep up with events; but this had escaped the attention of Regimental Headquarters.

> Alas regardless of their doom, the little victims play,
> No sense have they of ills to come, no care beyond today.

One frosty morning, on noticing some scruffy men, from HQ, with turned-up coat collars standing at a street corner, I suggested impolitely that circulation could be restored and appearances improved by some early-morning physical training.

It never rains but it pours. A man died of heart-failure on the first day of this unaccustomed activity. Like the bayonet fighting incident, the healthy exercise was stopped by the colonel, who ordered a court of enquiry.

It was the end of the hunting season and Richard Fleming put Rosie up on his charger 'Brown Mollie' to ride in the Ladies' Race at the Divisional point-to-point. Brown Mollie was an honest mare, and Rosie got round sedately at hunting pace. As the course was within riding distance of Tuxford, to show the flag, I sent over a matched troop of dun ponies with flowing manes and tails – the pick of the squadron, with the smartest men to ride them. The colonel's permission had not been asked, and Uncle Gloom was incensed. The turnout won approval in that first (and, as it proved, last) public appearance of Lovat Scouts as mounted Yeomanry. The ponies that day were a joy to behold: hard in condition and all fire and feather. Soon afterwards – indeed, within days – the War Office informed the regiment that we were to lose our spurs.

At the races I challenged Bobby Campbell Preston's squadron of the Scottish Horse – the Scouts' old rivals – to a night attack in Welbeck Park (dismounted). A splendid skirmish resulted – highly enjoyed by all and a foretaste of more active service. Dick Allhusen reminds me that before moving off orders were given for all ranks to remove bonnet badges and false teeth should they have any. Some heads were cracked and certain rifles and hats changed hands in skirmishes during the hours of darkness. These items had to be returned. There was another enquiry.

In Europe, the Finns were fighting off Russian invaders. The Lovat Scouts, whose future remained undecided, might well have been useful to Field-Marshal Mannerheim. Then, on 9 April (with little warning, save a preliminary talk with Vidkun Quisling), Germany invaded Norway and Denmark. Britain hastily promised to send an expeditionary force to our Norwegian allies. As a preamble, Richard's elder brother, the author and explorer Peter Fleming, who was employed in Whitehall, suggested that the Scouts were an obvious choice for any expedition fighting in conditions and on terrain similar to the Scottish Highlands.

Uncle Gloom went down to London. The question was put to him. He had reservations. Infantry training was not his line and the subject had never been studied in available text-books. The rôle suggested came as a surprise. He returned fully justified, but only in his own opinion.

General Carton de Wiart, VC, then commanding 61st Division in Oxford, was sent overseas on 13 April with a scratch force collected at short notice. If memory serves, they included a county regiment (the Bedfordshire and Hertfordshire) whose soldiers had hardly seen a hill in their lives. With him also went the French Chasseurs Alpins (without the bindings for their skis), and Peter Fleming and the polar explorer, Martin Lindsay, acting as ADCs. The general's orders were to take Trondheim in a flank attack, supported by the Royal Navy. The same day, on a second visit to Narvik fjord, HMS *Warspite*, flying the flag of Vice-Admiral Whitworth, sank seven enemy destroyers. About the same time the cruiser *Blücher* was sent to the bottom outside Oslo and the cruiser *Koenigsberg* was sunk in the Skagerrak by the Fleet Air Arm. The battle-cruiser *Scharnhorst* was severely damaged by *Renown*. We lost the cruiser *Effingham* and the aircraft-carrier *Glorious*.

Matters were not so good on land; the campaign was to prove a failure. As the general told me later: 'The British troops were issued with fur coats and special boots and socks to compete against the cold. But if they wore all the damned things they were unable to move about and looked like paralysed bears. As far as guns, planes and transport were concerned, I had no trouble at all, for no such things were available.'

Meanwhile, the news of the London interview filtered back to Tuxford. The colonel called the usual weekly conference and squadron leaders reported in a mood of angry incredulity. I remember the scene very clearly. The conference began with the usual typed agenda. The first item concerned the issue of blue patrol undress uniform to NCOs, and whether regimental funds could meet this cost. (There had been no mention of Norway, but it was clear the CO knew that we had heard about the decisions taken in London; the atmosphere was electric.) When my turn came to give an opinion on the issue of patrol uniform, I gave the unequivocal reply: 'I happen to be a poor bloody soldier, not a haberdasher!'

That put the fat in the fire. The meeting broke up. I was offered, by letter, the choice of a transfer (in other words, the sack) or facing a court on a charge of insubordination. Gloom had looked up his manual of military law.

I had burnt my boats with a vengeance. There was no turning back, and I had to swallow the disagreeable knowledge of having lost my temper and made a fool of myself. The step was irrevocable. Virtually a mutineer, the good name of the regiment had priority in a situation already out of hand, for there were others who thought the same way as myself. Had I been younger by a whisker, or without the backing and experience of Scots Guards' connections (which is a very close trades union indeed), this book would not have been written. An adverse report to higher authority meant being dumped into 'the Officers' pool', a dubious collection of 'odds and sods', all difficult to place and some with the same problems as myself. It was a miserable way to leave a squadron that was starting to do so well. I think my father would have taken the same course of action, and deep down I retained a sense of purpose that remained unsatisfied.

Next day I reported to the orderly room, sufficiently restored to tell the colonel I was prepared to apply for a transfer. I asked to leave after the squadron's horses had been shipped to the remount depot. They were loaded the next week; and some great mares were lost for ever to the Highland breed.

The farewells were saddened by unexpected news. The regiment was to be sent to the Faroe Islands to carry out garrison duties. There the Scouts remained for two bleak years – a tale of wasted ability in a task not made easier by the nagging reminder of an opportunity lost. They were ordered to learn foot soldiering – a new trade – and were then forgotten in the harsh realities of a war game whose rules do not accept excuses.

It was a bad start and it meant a slow recovery. The offending colonel disappeared soon after the winter weather started to blow North Atlantic gales over Faroe's rocky shores. Ken McCorquodale (the second-in-command) was passed over for promotion, and a regular Cameron Highlander took charge. Out of loyalty, all the officers stuck it for a year. Some then felt that they had done enough looking out to sea, and sought escape to more active service. Richard Fleming, Dick Allhusen, Grant of Rothiemurchus, Fraser of Reelig, my brother Hugh, Johnnie MacDonald of Tote, Tony Wills and Bill Whitebread, all friends or neighbours, were among those who broke away. Joe Lawrence, an officer of boundless energy, and two good stalkers, Sergeants George Fraser and Frank Sutherland, got posted to join me in No. 4 Commando. Others tried to follow but were refused permission.

There I must leave the Scouts. It was a bitter parting; as in all good Yeomanry regiments, there were close family ties at every level. The officers and other ranks were largely the sons of men who had served in France and Gallipoli, and in A/Squadron the sires of three subalterns had fought in the Boer War and had helped to raise the original regiment. Sent to Canada on a snow and mountain warfare course, the Scouts eventually went to Italy to campaign successfully in the Apennines. During the final hostilities the regiment finished its active service mopping up trouble in Greece and Salonika.

I must pay one final tribute. Two years of exile in the Faroes worked wonders with the band. The Western Isles pipers (ten of them came from South Uist), steeped in knowledge of the bagpipe (its music, virtuosity and fingering, in which they excelled), had all started army life as individual players. Islanders hold a poor opinion of noisy accompaniment and the thump of kettle-drums. Conventional march tunes (which most bands play) they considered fit only for children and beginners. They did not find it easy to counter-march or keep in step. Dick Allhusen took the musicians in hand. Donald Riddell – a capable crofter on the estate, who in his spare time makes cabinets and violins – was promoted to pipe-major and sent on a drill course, followed by a visit to the Army School of Piping on the Castle Rock in Edinburgh.

The piping could not be improved, but the drill and turnout were transformed. The reputation of the band went before them to Vancouver. While the regiment was training in the Rockies the City Fathers, by general request, sent an invitation to Jasper Park: 'Please come and play at City Hall!' The performance was an unqualified success; the vast dominion of Canada is surely Scotland writ large across a continent. That day every Highlander on the Pacific Coast was at the station, with his family and a bottle of whisky, to welcome the visitors. Most pipers have 'hollow legs', but all records were broken on this occasion. If the pipes and drums marched away in good order to 'The Black Bear' – a compliment to their hosts – another tune, 'The Big Spree', was certainly more appropriate.

I go to Canada twice a year, and the jigs played that day in British Columbia are still talked about by Canadian friends. As their fine boat song would have it, which evokes ancient ties:

> From the lone shieling of the misty island
> Mountains divide us, and a waste of seas –
> Yet still the blood is strong, the heart is Highland,
> And we in dreams behold the Hebrides!

My cousin Sandy Fraser (Moniack), who won a Military Cross in North Italy, brought the regiment home. And there we must leave them.

In Europe the scene had changed. The 'Bore War' was over. Norway fell. The Wehrmacht had given Scandinavia a taste of the blitzkrieg attack that had swept Hitler's panzer divisions across the plains of Poland. Indeed, the campaign ended before it had properly begun. The Germans' winter fighting technique was masterly. Battalions of Jaeger ski troops, dressed in white overalls and carrying light automatic weapons, swooped like ghosts down steep hillsides to surprise and outflank the floundering BEF, who had difficulty getting off the coast roads and negotiating deep snow. Dive bombers, working round the clock in the long hours of daylight, gave close support, strafing any movement with machine-gun fire and blitzing harbours, docks and shipping in the fjords. Carton de Wiart was back in the first week of May with a mauled and badly let down expedition. He returned to Oxford and his 61st Division, having made some very rude remarks to the War Office, who incautiously broke the news on the BBC that Paget's army had pulled out farther south and successfully re-embarked; Carton had been left behind, waiting to be rescued, while the Luftwaffe carried out target practice.

The occupation of Scandinavia, with its long coastline and good harbour facilities, greatly increased the German U-boat menace. It also helped to expedite the flow of vital Swedish iron ore from Lulea. But the moral stigma of the German Reich's unprovoked attack on two neutral countries shocked the world.

Meanwhile I packed myself off to London. After visiting 'Butcher' Bill Balfour, who offered employment in a training battalion of the Scots Guards, I went on to see the 'Old Flamer', as de Wiart was affectionately known by his divisional staff housed in the Woodstock Road.

The general (whom I had first met, as a child, at Lyegrove) had been out for a morning ride to review the troops of a nearby brigade. I was summoned to his bathroom, where the interview took place, de Wiart sitting naked and curiously scarred in his tub – quite unselfconscious, without eye-patch or the glove worn over a missing hand.* On his upper lip, a moustache cup added a small concession to the formalities of the occasion.

I told my story. Unlike the emperor who lost his clothes, de Wiart in the nude was still a general: terse and to the point, with that inner quality that commands respect, but with understanding also, despite the blistering temper.

It was at Lancut, the home of Alfred Potocki, an affluent neighbour to my friend's house in the Pripet Marshes, that our paths had crossed again at a winter boar shoot. It had been my lucky day and I was on the mark as the big pigs – black and bristly against the snowy background – bolted across a narrow ride between tall trees. The snow must have slowed them down as each animal, in turn, headed slightly uphill in a climb which flattered the performance; I managed to roll three porkers down the slope into the general's ambush in the butt below, the last pair being a right and left fired from father's old .350 double-barrelled rifle which made shooting easier at a running target.

Before he stepped out of the bath and asked to be rubbed down, Carton de Wiart had offered me the job of sniping instructor to 61st Division.

'There are two snags,' he said. 'One I can sort out myself – unless they heave me out of the War House – namely, to persuade the chairborne wallahs to add another officer to my divisional establishment. It is at full strength already. Meet me at the bar in White's for a drink

---

* The general had been sent down from Oxford. Later de Wiart was to receive an Honorary Degree from Balliol. The citation read: 'This is the famous Balliol man who was torn from his studies to serve against the Boers in South Africa: who was there twice wounded, and who now, after fighting in more campaigns than others have even heard of and receiving nine further wounds, has been elected an Honorary Fellow of his old College.'

tomorrow and I will have one of the answers. The second drawback may not suit your book. There is a flap on in Northern Ireland, and invasion scares, two-a-penny, are escalating around the country. If you want to take a crack at the Krauts a defensive rôle may not appeal to you. It looks like more garrison duty. I quite understand your feelings – I have the same thoughts myself.' What a grand soldier, and how rightly he gauged my sinking spirits.

The roulette wheel spun again, and this time I was on the winner. The 'Parson's Rest' thronged with friends in uniform. The contemporary scene has been described in Waugh's trilogy 'Men at Arms'. Here Basil Seal and 'Uncle' sought by various means to hatch plots for nefarious advancement. Some of Waugh's characters can be identified: Seal was Basil Dufferin (or possibly Auberon Herbert, a charming romantic who later enlisted in a Polish unit at Inverailort). The redoubtable Brigadier Ritchie-Hook, breathing fire and slaughter on the way to war, was no other than the VC General I had come to meet.

Predictably, the first man I ran into was Peter Kemp – knocking them back with a former Company Commander in the Scots Guards, Bryan Mayfield. Military horse-trading at the bar was a common practice. Peter's and Bryan's future employment also hung in the balance. Peter told me the news. Following Weedon he had been at a loose end; after wasting time on a fruitless mission to Gibraltar, still hell-bent on adventure, he had tried to get himself into the volunteer battalion (5th Scots Guards) enlisted to fight in Finland against Russian aggression. Nothing came of this, but it showed the courage of the man.

Peter could not ski down a nursery slope – nor could some of the others – but all were enthusiasts, and experienced in snow conditions, notably the explorers and mountaineers Jim Gavin and Freddy Chapman, Martin Lindsay, cousins David Stirling and Andrew Maxwell, and old friends, Philip Pinkney and John Royle. They had gone to Chamonix for special training, but by March the war in Finland fizzled out; they returned home and were disbanded. The invasion of Norway came too late for the battalion to reassemble.

The broken Norwegian coastline offered obvious opportunities to saboteurs, thereby hindering the Germans' rapid advance from the south. Their efforts gave encouragement to a nation bewildered by the sudden onslaught. Though help was forthcoming, the British forces proved incapable of mounting any effective opposition. The campaign tactics of the BEF showed the inadequacy of infantry training, based on defence and an antiquated conception of trench warfare. Generals

Paget and de Wiart – commanding two little armies, working independently and without consultation – had found themselves caught up in
a monumental confusion of contradictory orders, bad planning, staff
ineptitude, and lack of co-ordination between all three Services.

The fiasco on land was relieved by the magnificent gallantry of the
Poles at Narvik – a great feat of arms that would be repeated four years
later at Monte Cassino by the same nation. There were also bold naval
actions: along the coast twenty-six German transports were sent to the
bottom.

Norwegian soldiers and resisting partisans (a new name) must be
remembered. They had expected better and more positive assistance,
though the remnant of the 5th Scots Guards had tried their best. An
abortive attempt was made by submarine to effect a landing in the
Sogne Fjord to bring in a load of arms and munition. With the help
of guides, this party of officers would make a forced cross-country
march on skis to destroy railway bridges on the Bergen–Oslo line. The
venture was destined to fall short of the brilliantly contrived raid on
Entebbe Airport. In fact it never arrived!

HMS *Truant* was one of the new T-class submarines (a sister ship of
the unlucky *Thetis* lost with all her crew on trials in Liverpool Bay). On
Truant's previous patrol Lieutenant-Commander Hutcheson had penetrated the River Ems estuary defences and sunk the German light
cruiser *Karlsruhe*, a daring act of seamanship that won an immediate
award of the DSO. *Truant* had remained submerged for thirty-six hours,
continuously depth charged before finally escaping back to open sea.

With the Scots Guards party *Truant* was not so lucky. Casting off
at dusk from Rosyth, she struck a magnetic mine six hours out from
the Firth of Forth, but managed to limp home with cracked batteries
(she had electric motors), a flooded torpedo compartment forward and
various rivets started amidships.

The officers taking part in operation 'Jack Knife', as it was called,
returned to Keir, while new plots were hatched to try again in a 'River'
class submarine. At this stage I caught up with the conspirators.

Bryan Mayfield took me as an extra hand. The attractions of the
more active rôle were explained to Carton de Wiart. I hope he did
not think me ungrateful. But I won his approval and the short-lived
sniping instructor went back to Scotland, very anxious to 'get lost' until
Bryan had arranged the posting order. It is difficult to describe my
elation. White's is a charitable spot, and if I had stood on my head
nobody would have been in the least surprised.

For a week the raiding party remained in a state of readiness; but
we were overtaken by events. The decision was already taken to evacu-

ate Norway, where the floundering BEF had been forced to inevitable withdrawal.

The proposed submarine trip now ceased to have a purpose. But although the mission never took place, those involved contributed something to a new technique that would become commonplace later in the war, when the navy became partners in a 'new deal'.

Good lessons had been learnt, and we had the makings of a team to put them into practice. But first we required approval, in principle, to find a suitable training area and a staff to run a school – a task made difficult by what was happening in France.

The date 10 May 1940 marked a drastic change in soldiers' fortunes. The winter of discontent was forgotten. The cold war was over; and Europe was ablaze. The shadow of the German army – field-grey uniforms and coal-scuttle helmets, the grey steel of self-propelled guns and grinding tank tracks – half-suspected behind the Maginot Line, erupted across frontiers into blitzkrieg on a fifty-mile front. The concentration of massed troops and armour, with Stuka dive bombers in close support, was preceded by a devastating sweep attack on airfields, marshalling yards and road and rail centres. The grey horde that had threatened Europe was to over-run France in a matter of days. On 19 May Churchill broadcast to the nation for the first time as Prime Minister. By the end of the same week Hitler had virtually conquered Western Europe.

We had a job in hand. Once more I quote Peter Kemp, who supplies the declaration of intent:

The invasion of Norway showed clearly the possibilities of partisan warfare. Paramilitary formations known as 'Independent Companies' – the forerunners of commandos – had been employed in the last stages of the Norwegian fiasco. In spite of angry controversy they were considered to have proved their usefulness. There was, however, no organized instruction in this kind of warfare, no school or centre where troops could be trained in its principles. There was a certain amount of theory. Colin Gubbins, who had escaped from Warsaw with Carton de Wiart's military mission and fought briefly under him in Norway, had written a pamphlet urging the adoption of a guerilla system.

He argued that there must be an untapped reservoir of officers and men with the necessary qualifications and experience to act as instructors, Indian Army officers who had served on the North-West Frontier, 'bush-whackers' from the West African Frontier Force or the King's African Rifles, Polar explorers, ghillies from the Lovat Scouts and the like. There should be no difficulty in finding a suitable location for the project: the Scottish Highlands abounded in what Boswell has described as 'fine and noble prospects', ideal country for training in irregular or amphibious operations: and also half-derelict houses which, in size and architecture, might have been designed as barracks.

That passage, I believe, really marks the beginning of the commando story. Peter goes on:

It was Bill Stirling's idea that six of us, reinforced by a few selected officers and NCOs, should form the nucleus of the new training school: we would begin with cadre courses for junior officers from different units of the army. Mayfield was to be Commandant, Stirling Chief Instructor.

There was no fear of our having to remain stranded for the rest of the war: after the first few cadres were trained there would be a plentiful supply of officers to fill our posts as and when we wished to be released for operational duties: the war was spreading and we should be in a strong position to choose our own operational theatres when the time came.

As the Battle of France opened, provisional agreement was obtained from the War Office for our establishment, the selection of training areas and setting requisitioning machinery in motion. The organization was left to the party who had served in the 5th Scots Guards and thereafter in the depth-charged submarine. Lovat, an officer who owned property in the West Highlands, was sent ahead to requisition all available premises astride the Fort William–Mallaig road and railway line which, in fact, meant six deer forests and their lodges covering a land mass for training purposes of not less than 200,000 acres of wild country. The first cadres would report for training at the end of the month.

We had to hurry.

I must interrupt Peter's narrative to mention a personal friend who won the Sword of Honour at Sandhurst. John Royle, a former 'partner in crime' from the old Cairo days when his regiment (the Highland Light Infantry) garrisoned the Citadel, became the Regimental Sergeant-Major. John had lost his commission for wild behaviour in Egypt's staid society. Reduced to the ranks, he joined the 5th Scots Guards as a private, rising to RSM without batting an eyelid or smiling at his friends. I admired him greatly, John went on to become a major in the Glider Pilot Regiment, with a proud record of service, and was killed leading a fighting patrol at Arnhem.

> It's by Shiel water the track is to the west
> By Ailort and by Morar to the sea.

Peter describes the setting:

Some twenty-five miles west of Fort William, as the crow flies – rather more as the road and railway run – stands Inverailort Castle, a large square building of plain grey stone. It is situated at the head of Lochailort, on the south shore, where the gloom of its natural surroundings matches the chill austerity of its design. The front of the house faces north to the sea loch, whose sombre waters, alternatively wrapped in mist and whipped by rainstorms, blend with the leaden tones of walls and roof. The back is overshadowed by a grim black cliff, surmounted by a thick, forbidding growth of trees which rises well above the height of the roof, blotting out the light of the sun at all times except high noon in summer. The rainfall of Glenfinnan, ten miles to the east, is the highest in the British Isles. But on the few fine days at this time of the year the shores of Lochailort and the Sound of Arisaig reveal a wild, bleak beauty of scoured grey rock, and cold blue water, of light green bracken and shadowed pine, that is strangely moving in its stark simplicity and grandeur.

In May and June of 1940 the same clear weather that favoured the German offensive in France brought warmth and sunshine to Inverailort, where, in the last week of May, we arrived to prepare for the first training course. With the help of the War Office, Stirling had been able to recruit some outstanding officers and NCOs to bring our staff of instructors to full strength. Three stalkers – great rifle shots and expert telescope men – were provided from the Lovat Estate wearing civilian plus fours.

Training in amphibious operations formed an important part of each course and was supervised by a Naval commander* with a motley of seamen to provide boating facilities based at Dorlin and Mallaig. The senior instructor in field-craft was Stirling's cousin Lord Lovat, to whom I was appointed assistant. In view of his epic career – as well as being a superb fighting soldier, he taught me in the time I was with him all I know about night fighting, and movement across country and the principles of natural camouflage. Gavin, now a major, was in charge of demolitions: his assistant was another regular sapper, Mike Calvert, who had the satisfaction of seeing himself reported in the Official War Office casualty list as killed in Norway. Three years later 'Mad Mike', at the age of twenty-five, was one of Wingate's brigadiers. Major Munn, a gunner officer who had served on the North-West Frontier, instructed in map reading and allied subjects. He brought with him two lean, bronzed hard-bitten officers of the Guides. They were supposed to instruct in the practices of irregular warfare, but it turned out that their experience had all been the other way – in suppressing guerillas: they exasperated their pupils by making them climb the hills and march along the crests, where they were visible for miles against the skyline. They left us after a few weeks.

Peter was refused permission to wear his Spanish decorations. At the time this struck me as unkind, for among the permanent staff we had two Arctic explorers who sported polar medals. They drew down the wrath of a cynical sergeant. 'For Christ's sake! Since when do soldiers get a ruddy medal for playing polo?'

* Commander Sir Geoffrey Congreve, DSO, RN.

The midges were bad, there were insufficient tents, and we were short of cooks, rations and transport; but every instructor was determined to succeed in this new venture. Our first students arrived at the beginning of June: twenty-five puzzled subalterns, some of them volunteers, others arbitrarily despatched by their commanding officers. They were supplemented by an equal number of NCOs. Among the new boys was Second Lieutenant David Stirling. Hard work was the order of the day – and night. Each course was to last a fortnight, with a few days' break before the arrival of the next intake. From this small beginning Special Training Schools developed apace, which were later established throughout Great Britain, in the Middle and Far East, Australia and on the North American continent. The work was interesting and I think our efforts were appreciated. We settled down to a strenuous and critical summer.

With the first intake came a tall Cameron Highlander – slightly older than the rest, in civvies without a uniform; he had come through the back door of the Foreign Office by way of Moscow. Fitzroy Maclean wished to join the army, but had been refused permission by his diplomatic chiefs. He defeated the authorities, however, by standing for Parliament and winning a Scottish by-election.* Later he helped Marshal Tito's partisans to drive the Germans out of Yugoslavia, having previously destroyed aircraft in the Western Desert with the new SAS. His first book, *Eastern Approaches*, brilliantly written, is the best account of irregular warfare conducted in a lesser theatre. After the war Fitzroy married my sister Veronica, whose husband, Alan Phipps, RN, had been killed defending Leros against German parachutists.

* On the occasion when General Smuts addressed both Houses in Westminster Hall, Winston Churchill introduced Maclean to South Africa's leader as follows: 'This is perhaps the only man who has ever made a public convenience out of the Mother of Parliaments.'

CHAPTER ELEVEN

TIME TO MOVE ON

O what can ail thee, knight-at-arms,
Alone and palely loitering?
The sedge is withered by the lake
And no birds sing.

Keats was a temperamental fellow who had his black moments. I understood the mood in which he penned that verse: there had been suffering at home during the summer caused by a family bereavement. In August my youngest sister Rose died suddenly at the age of fourteen. A tragic blow to my mother, who was not well herself, it plunged a united family into deep sorrow, for everything that little Rose had time to do during her too short day was gentle and flower-like in quality. She had been well named. Hilaire Belloc, her godfather, wrote this small poem to her memory:

Rose, little Rose, the youngest of the Roses,
My little Rose whom I may never see,
When you shall come to where the heart reposes,
Cut me a Rose and send it down to me.

When you shall come into the High Rose Gardens,
Where Roses bend upon Our Lady's tree,
The place of Plenitudes, the place of Pardons,
Cut me a Rose and send it down to me.

In war it can be hell to hang around.

The season changed again and the autumn weather affected all wild creatures. As rowan berries turned red in Lochaber the migratory birds began to drift south. The smaller passerines – chiff-chaffs, warblers, and a family of redstarts – conspicuous flashes of colour in the birches below Beesdale, vanished overnight. Fieldfares flew high in chattering flocks; then smaller scattered bands of mistle thrushes, ring ouzels and heather linnets (known locally as moss cheepers) left the hill. The black-throated divers on Loch Eilt sought the open sea; when skeins of grey geese passed clamouring to winter quarters on the Solway, the equinox finally broke.

We worked hard to keep warm. The fieldcraft wing became a test of real stamina, with a thirty per cent failure of students to last the course. Gales of wind and rain swept the glens all through October and November. The elements combined to provide new and dangerous conditions for day and night training exercises on land and water. The blue skies and sunshine of that long, momentous summer were over. Another year had peeled off the calendar.

Now a strange restlessness stirred the instructors at Inverailort. The war was real. But had we done enough for King and Country? The old hands were becoming stale and disillusioned. And the west coast weather was depressing. Were we getting bogged down in a rut or turning into trench-dodgers? Here a pause is necessary to review events in distant Europe, learned by hearsay from London.

Daily bulletins from the Ministry of Information supplied guarded platitudes without concealing the gravity of the hour. There was no conviction in either press or radio. Lord Haw-Haw kept the 'Germany Calling' wave-length going, with considerable effect on the uneducated mind. There were other rumours: people spoke of compromise in high places – of peace by negotiation. American citizens had received advice notices in red print to take the first available ship 'out of the beleaguered Island'. Others, not so American, sought refuge in the New World: contemptible creatures, who even lacked the excuse of having children. We heard new, unfamiliar terms – subversive activity, quislings, collaborators, partisans, maquis, and the Fifth Column. The most sanguine optimist began to feel inadequate. It was hard to escape the conclusion that the Highlands were living in Cloud Cuckoo Land.

For those who had played no active part it was difficult to believe, in the silence of the hills, that German armour, with close air support, had first over-run neutral Belgium, then France and Holland in less than thirty days; that the vaunted Maginot Line ('Here the gun will stop the Hun') had proved completely useless and could be driven round at either end; that the day Rommel's and Guderian's panzers broke out of the Ardennes Forest (quite impassable for tanks, so the pundits declared) Allied airfields (with aeroplanes parked, in many cases, in rows upon the runways) had been destroyed by the Luftwaffe, and that many of those which remained were shot down in a belated attack on the pontoon bridges von Rundstedt had thrown across the Meuse.

There was no way of containing the speed of the enemy advance.

The so-called Battle of the Bulge was a newspaper tiger. Weygand replaced Gamelin but it came too late. Guderian drove a wedge through the defences, then turned west to race for the Channel Ports. He reached Dunkirk on 24 May, two days before the BEF, but Hitler, who was expecting an unconditional surrender, ordered him to pull back. Rommel, starting fifty miles farther south crossed the river near Namur and breaking into open country, drove his tank column straight across France heading for Rouen and the Seine. As France fell there came the unkindest cut of all – Mussolini turned to stab his neighbour in the back. Now the Mediterranean became a new theatre of war. After a brief lull in Western Europe – a respite tensed by invasion scares – the war entered a new phase.

The Battle of Britain began with an attack on airfields. Then massed German bomber formations flew wing to wing in daylight up the Thames Estuary to flatten London. This spectacular performance reached a climax on 15 September when 163 enemy planes were destroyed. Our losses were described as negligible, but it is now generally admitted that Spitfire and Hurricane Fighter Squadrons had been reduced to 600 serviceable planes and frontline pilots. If the Germans had continued to strafe airfields in the Home Counties (losses on the ground were never revealed), we would have lost the war. Speaking on the subject Churchill later told Stalin: 'Truth is so valuable it must be protected with a bodyguard of lies.' The air battle had proved a costly business for both sides and ended in stalemate: Hitler turned on Russia – a mistake that drained away the resources of the Reich and saved our bacon – but enemy bombers still retaliated by night. The Blitz had begun in earnest over English cities.

All this was hearsay – we felt cut off from reality. No one liked the sound of the news, and the best men strained to get away. It was as simple as that.

Contrast the excitement in Inverness-shire! The appearance of a solitary Dornier, flying on extra fuel tanks to Fort William, which dropped a stick of bombs on the aluminium factory at the foot of Ben Nevis. In a low-level attack that hit the target fair and square, the pilot, a brave man with a good navigator, escaped unscathed. The nincompoop in charge of the anti-aircraft battery protecting the works had gone off to lunch in town. Unbelievably, in his absence the NCO left in charge – by order of Scottish Command – had no authority to fire the guns. The bombs made holes in the roofs of various buildings, but they failed to explode; all missed the vital pipeline which supplied water to the works. I think Invergordon suffered minor damage to oil storage tanks, and Edinburgh lost part of a distillery, but the Germans never hit the Forth Bridge in two attempts, and the attack was not pressed home. By contrast, a prowling Heinkel dropped a stick across Aberdeen, aiming at no particular target: an outrage that incensed the hard-headed citizens of the granite city. The bomber was shot down. Though the Clyde experienced serious night attacks with loss of life, fires and much property destroyed, Scotland as a whole had little cause to grumble. In 1940 the war in the air had revolved around London and over the Channel and the Home Counties.

That summer, on a weekend pass to Edinburgh, I ran into Colin Pinkney – Philip's brother in the Air Force, another Cambridge wild fowler – as he walked down Princes Street. We had time for a drink and a nostalgic talk about grouse shooting, for the glorious Twelfth was just around the corner. Soon afterwards, No. 603 City of Edinburgh Squadron flew south to Hornchurch outside London. Twenty-four planes left Turnhouse on 10 August. When the Battle of Britain ended, eight fighter pilots came home to tell the tale. The Edinburgh Squadron returned to Scotland before Christmas having destroyed 107 enemy aircraft officially confirmed. Colin was among the walking wounded; he had been shot down, baled out and burnt. That time the luck had held, but he was later killed by the Japs flying over jungle in Malaya.

I shall not touch upon the war at sea: losses among merchantmen showed a steady rise in spite of the convoy system; the Blue Star Line in which I had sailed to South America lost almost their entire fleet; U-boats, armed raiders and pocket battleships provide action stories which are all their own and cannot be touched on here. Civilians tightened their belts as the meat ration dwindled to 4 oz each week per person.

Commando raids got off to a bad start. I have to confess only two attempts were mounted across the Channel to assault the enemy during that first year. Without the benefit of a Combined Planning Authority to provide maps, intelligence, transportation, air cover, signals and specialist weapons, the outcome of each sortie became a comedy of errors.

The first excursion was led by a staff officer, Major Dudley Clarke: a gunner to whose brain the credit of forming a Special Force of army volunteers must be freely given. Within three weeks of Dunkirk, those early optimists, with blackened faces and armed with a new Tommy gun, sailed for France in a variety of craft, hastily assembled at Dover, Folkestone and Newhaven, 'to have a bash at the Boche'. At the 'off' the three raiding parties were apparently pointing in the same direction. In theory they started on time, but the result was a shambles. In mid-Channel, the Folkestone group was mistaken by the RAF for German E-boats, and got machine-gunned. This baptism of fire held up the flotilla's rendezvous at a critical stage of the crossing. In the interval of waiting, part of the rum ration was consumed to cheer everybody up, and the libation must have been overdone. One-third of the force ran aground on sandbanks four miles south of Le Touquet.

The Dover party went ashore nearer Boulogne, where they engaged a German cyclist patrol – an easy target which should have been ambushed as it passed by. (At night it is a simple matter to kill a man walking, with the moon behind him.) In the event, an officer mishandled his submachine-gun at the critical moment: as he belatedly cocked the unfamiliar weapon, the magazine dropped off before he pressed the trigger. Clarke, however, bled for his country, with a bullet (only a glancing blow) along his neck and earlobe which must have stung a bit. So they returned to England without a prisoner – failing to achieve the object of the exercise. The third party, after stalking a seaplane which took off before they could open fire, claimed a dead German in a similar skirmish, but the body was not picked up. Such claims, if the night is dark, must be substantiated. Tendencies to exaggerate go with inexperience, nerves and a lively imagination. The claim, on this occasion, was 'blown' by the assertion that there had been no room in the crash boat and the corpse had been towed astern. The rope conveniently parted before reaching England! Clarke had his severed ear sewn on again on reaching Dover, but the rum drinkers, returning to Folkestone, were accorded a very different reception.

The harbour authorities had not been alerted of the prodigals' return. In fact, they had no inkling of the foray itself. (One can only suppose the departure to France the night before had passed unnoticed.) The unfortunate raiding party was refused permission to enter harbour and was made to lie off the boom, covered by coastal defence guns until their identity was established. This took time; the men were wet and tired. To ward off further chills, they broached more jars of rum – but their troubles were not over. When the troops finally came ashore, a number of soldiers showed signs of 'having drink taken' away above the official ration. The Military Police, mistaking the dishevelled and unsteady warriors for deserters, placed the lot under close arrest and marched them off to the local clink.

The story of the second sortie – to the Channel Islands in mid-July – makes no better reading. This time Guernsey was the target, and the identification and capture of prisoners was again the primary objective.

Because of distance, the local boats lacked capacity (without refuelling) to make the round trip. Transferred from destroyers into what extra craft were available (more than one failed to make the passage), the raiding force stole into Telegraph Bay.

The official report of the operation was said to have infuriated the Prime Minister:

The crews of the speedboats were RNVR volunteers of a most cheerful disposition. The first thing we found as we left the destroyer was that they had forgotten to adjust their compasses which were many degrees wrong: we took a chance and guessed at our landing place. Touch down was approximately on time. The patrol was uneventful. The barracks visited proved to be unoccupied so we were unable to take prisoners. We spoke, however, to residents who were too frightened to talk properly, not being able to believe we were British soldiers and thinking it was some trick of the Gestapo. The trouble started when we tried to re-embark. There had been a miscalculation as to the state of the tide. A heavy swell had got up and the boats would not come nearer in the darkness than 50 yards. After the one small dinghy capsized, and got smashed on the rocks, we all had to swim for it. Three men who had stated in interviews that they could swim proved helpless in the rough water and were left behind.

The Senior Officer, it transpired, was mildly concussed in a fall down stone steps leading to the water's edge, but he still managed to struggle through the surf and was hauled over the stern. The report continued:

One of the speedboats then broke down. We finally got underway making about two knots with one boat towing the other. By this time we were an hour late for our rendezvous with the destroyer. As we emerged from Telegraph Bay we could see her disappearing at speed. By flashing torches we attracted attention and the Captain, very gallantly, returned to pick us up, leaving himself open to air attack, on the way back to England. As we got on board, with Dartmouth next stop, the dawn was already breaking.

Faulty navigation on the raid caused one boat to miss Guernsey altogether and land on a shore with high cliffs and no beaches – which might have been the island of Sark.

A third fighting patrol had trouble with their engine and pressed the destroyer *Saladin*'s whaler into service. It proved so full of leaks that they never reached the island.

The Prime Minister was not amused by this tomfoolery and laid it on the line in no uncertain terms. I must quickly add that no participants in either operation had passed through the Training Centre at Inverailort. The sailors got a reprimand and Churchill ordered an immediate reorganization. There would be no more slackly planned, unco-ordinated efforts mounted by a collection of amateurs – naval or military – against targets of insignificant importance.

With a hostile War Office, limited resources, our poor track record, and the disapproval of every Army Command, it required courage to reinforce the concept of a corps élite with ships, aeroplanes and planners, working independently under separate command. But, to quote Churchill:

> I feel that the Germans have been right in both wars in what use they have made of storm troops: the defeat of France was accomplished by a small number of highly equipped and brilliantly led spearheads. There will be many opportunities for surprise landings by nimble forces accustomed to work like packs of hounds instead of being moved about in the ponderous manner which is appropriate for regular formations ... for every reason therefore we must develop the storm troop or commando idea. I have asked for five thousand parachutists and we must also have at least ten thousand of these 'bands of brothers' capable of lightning action.

Churchill was in no mood to be trifled with, and the Chiefs of Staff gave ground. Admiral of the Fleet Sir Roger Keyes, GCB, KCVO, CMG, DSO, afterwards Baron Keyes of Zeebrugge and Dover, was appointed the new Director of Combined Operations.

Smart the word and prompt the action. By the late autumn of 1940 Keyes was ready to do business, with a competent three-dimensional staff co-ordinating army, navy and air force personnel, set up in Richmond Terrace, Whitehall. This organization became known in service parlance as COHQ (Combined Operations Head Quarters) and, by the more profane, as HMS *Wimbledon* – all rackets and balls.

Aeroplanes and parachutists were not available, and airborne troops branched off on a line of their own; but a brigadier, Charles Haydon (Irish Guards), took command of Keyes' private army and moved his headquarters to Scotland with all the 'brotherly bands' of fighting men who had been on the Channel coast.

The two thousand volunteers from Independent Companies (survivors of the Norwegian *débâcle*), who still squatted dejectedly about the glens in muddy tents, were packed off with a new name – Commandos – to different concentration areas along the Firth of Clyde. Achancarry later became a training and 'trial by ordeal' centre for sorting out new recruits.

Five converted Dutch cross-channel steamers, that had escaped from Holland, were made available for long-distance raiding projects. Much-needed training in assault landing-craft for junior naval officers, coxswains and boat crews was provided at Warsash on the Hamble River. Full-scale practice landings were stepped up in the Firth of

Clyde, with more boats (trawlers) pressed into service for training in the West Highlands at Dorlin and Mallaig. Inverailort had served a useful purpose: the significance of its toughening-up processes, virtually under active service conditions, had not escaped the notice of the War Office. The *coup de main*: ambush and sabotage, forced marches, opposed landings and long swims, unarmed combat and night attacks were ruthlessly exploited. The hills echoed with the detonation of high explosives: bursts of tracer fire flattened the careless patrol – face down in the heather – should they show up against a skyline. In six months several hundred junior leaders had survived the gruelling fortnight course, emerging fitter, more determined to succeed, and with the self-confidence to do so.

The British Army slowly got back on its feet, and West Highland training can take some of the credit. Here I must praise some colourful Allied soldiers, who needed little guidance: a group of Poles joined us that summer after walking across Europe by different escape routes, taking many months to rejoin the Allied cause. Some sprouted wings and learned to fly Spitfires. One hundred and fifty new Polish pilots shot down 193 enemy planes in the Battle of Britain, many of them bombers.

Irregular warfare schools were started overseas: in Egypt, Singapore, Canada and Australia. Various instructors were promoted to command them. Some of the best who came from the fieldcraft wing deserve special mention: Peter Kemp, DSO, Martin Lindsay, DSO, Freddy Chapman, DSO, David Stirling, DSO, to name four of the best. Jim Gavin, DSO, along with Mike Calvert, DSO, both sappers, left the demolition branch. These were outstanding officers who went their separate ways, destined to make a considerable impact in different theatres of war: we did not meet again. David Stirling, who was to destroy 143 enemy aircraft behind the lines in the Western Desert, left No. 8 Commando to form his own SAS regiment.

Auberon Herbert and his tempestuous Poles departed cheering, to die with conspicuous gallantry in the liberation of Europe. They looked upon the lovable, flat-footed and wholly unmilitary Auberon as their father. His own parent, Aubrey Herbert, a Turkish scholar and wayward genius who wrote *From Anzak to Kut*, had passed on some of the same qualities of head and heart.* Aubrey was once offered a Balkan throne.

> Land of Albania: let me bend mine eyes
> On Thee, thou rugged nurse of savage men.

* In *Mr Standfast* and *Green Mantle* John Buchan portrayed his Oxford friend and contemporary, Aubrey Herbert, as the elusive Sandy Arbuthnot.

The predominant Scots Guards régime came to an end. In November, Inverailort was thrown open to the Regular Army; a new Commandant in the person of Colonel Hugh Stockwell took over. We had pushed back training frontiers, but the pioneering days were over. I suggested to Sir Roger that it was time to move on, and he appointed me his personal representative: as Umpire and Military Adviser on exercises (brigade strength) that were to take place on the Islands of Arran.

At the time hopes ran high: it felt good to be doing something useful in the build-up for some great adventure. The troops did their best, but met with frustration at every turn. There followed a period of trial and error, with landings on various islands. The War Office were in part to blame, but it became painfully clear that the navy had little wish to co-operate. Nor was commando diplomacy up to the mark. Quarrels and bad feeling were in evidence at every level.

Sir Roger Keyes, who spent half his time on the train trying to cajole Chiefs of Staff into mounting an operation, set us an example of single-minded determination, turning out in all weathers for night exercises and foul weather landings. The admiral believed in leading from the front. I have seen him in London clothes, wet to the skin, struggle to the wrong beach in a high running sea, and then call for a repeat performance. Nor did he quit until the last man was back aboard his ship; more than once I helped to change his clothes, for his feeble hands were too cold to undo hooks and buttons on a duffle coat. Time spent in his company made us feel twice our size; and that, of course, is what leadership is all about. He was, like most genuine people, a very uncomplicated man, with straightforward beliefs and simple views on life. Old Sir Roger had none of the devious ways which can pass for cleverness today. To let him down was unthinkable. There are some people devoid of humbug (few in number) whose understanding (perhaps inspiration is a better word) seems more important than a firebrand reputation. These qualities were apparent in four men with whom I came into contact for brief periods during the war: Churchill, Smuts, Roger Keyes and Carton de Wiart. They all believed in the old-fashioned virtues of loyalty and patriotism and found it difficult to understand people who did not. Leaders of genius, like those I have mentioned, impart confidence to subordinates. Daily life seemed enriched, for I believe their encouragement made the blood course faster through our veins. They taught me to make soldiering exciting, even fun, carried out with a certain panache that in no way detracted from efficiency. General Alexander was a borderline case, but his men swore by him; likewise Bill Slim in the Far East whom I never met:

both were very great soldiers, but perhaps the former lacked purposeful ambition. In Montgomery and Mountbatten, the inner qualities were not apparent. De Gaulle seemed unfriendly, Stalin was a disappointment. Physically a smaller figure than expected, he appeared entirely lacking in humanity. Shuffling and swag-bellied, the Russian leader was imperturbable certainly, but in a negative sense, with an almost oriental cast of countenance. I met Smuts on only two occasions. He was a mild old man, but his presence filled the room.

Practice landings continued: it was rotten weather, with gale force winds. My diary read: 'December: damp – prolonged and not successful.'

Collective training in the Firth of Clyde lasted into the New Year and then fell apart. The manoeuvres had indicated full-dress rehearsals for something really big: they became a non-event, to be cancelled at the eleventh hour. British phlegm was sustained with a deep sigh and Keyes, a disappointed man, went down with a chill.

Three fortunate Commandos, Nos. 7, 8 and 11 (the first two were later disbanded after taking a hiding in Crete), sailed in February for the Middle East. Those left behind were indignant: this was favouritism! The rest of a disillusioned brigade went into winter quarters, billeted on the civilian population in towns along the coast.

Would we be disbanded? It all appeared pretty hopeless: good men went back to their regiments and morale at every level slumped accordingly. But a week's leave can work wonders in wartime, and on return to Inverailort I found a priority telegram ordering me to report to No. 4 Commando at Troon without delay.

CHAPTER TWELVE

THE LOFOTEN RAID

A FULL DRESS REHEARSAL

To Noroway, to Noroway,
To Noroway o'er the faem.

The Ballad of Sir Patrick Spens

Spring was in the air. On farms along the Ayrshire coast the land was being worked by matched teams of Clydesdale horses, preparing the early potato ground. Some ancient golfers from the town were heading for the Troon links. The Commando to which I had been posted by urgent telegram had disappeared as silently as a herd of elephants into tall grass. No rear party, no orderly room. The house, said to contain regimental headquarters, was locked and barred. Recruiting posters, some dusty files and pin-up nudes confirmed its scruffy identity. Upstairs an open window invited closer inspection. The driver was given a leg up. There followed an extraordinary conversation in a voice muffled by distance:

'Lots of doors up here, sir! Notices all over them, but no keys.'

'Well, read them out.'

A pause.

'I think this must be the orderly room, sir.'

Pause.

'It says "My work is so secret that I don't know what I'm doing."'

The driver moved downstairs and tried again.

'This door is open, sir.'

Then came a chuckle.

'How's this? "We run a tight ship here: lately however we have been getting tight a little too often."'

The Bore War encouraged this military graffiti. Pulling up a sash, the man let us in, but the cupboard was bare. More boozy humour was slipped into another message: 'It is forbidden to leave your stool while the bar is in motion.' With this suggestion (a wise one in wartime, when a man is always hungry or in need of a drink), Mellis, myself and our driver went to Troon Police Station, to ask the whereabouts of the best hotel.

Security at the Marine Hotel was not as good as the cooking. There seemed to be no doubt that the Commando (several senior officers of which were residents) had left the day before in a hurry – their destination, Inveraray on Loch Fyne. Brigade headquarters, in far-away Castle Douglas, did not answer the telephone, and a call on an open line to Combined Operations in London would be a definite breach of security. Fortified by lunch, we hopefully headed back to Arrochar up Loch Lomond's twisting side, turning where the side road struggles out of the country of the MacFarlanes and Colquhouns to the Atlantic seaboard and the heartland of Clan Campbell.

The petrol just held for the climb over Rest and Be Thankful. In the sea loch below lay shipping of all shapes and sizes. The little town of Inveraray, not for the first time in a turbulent history, had become an armed camp, all bustle and stir. Nissen huts, in various stages of construction, were sprouting on the very green where former chiefs – so ran the insulting legend in Gaeldom – had placed itching posts for the Campbell dhunie wassael to rub the lice off their backs, as they lounged in idleness before setting off to lift their neighbours' cattle. No flag bearing the Galley of Lorne flew from the castle battlements on the day we arrived, for Macailen Mhor, with his Argylls, and the rest of the 51st Highland Division, were prisoners in Germany.

The evening saw us slapping across Loch Fyne in a skimming dish provided by Donald Ross, a friend from the Black Isle, to the infantry assault ship, HMS *Queen Emma*, a fast converted channel steamer. *Emma* was in the process of a face-lift: sailors in slings, with pots of paint, were busy on her side. All her brasses shone as for a Spithead review. As she was closed down for the night, we had to climb up a swinging rope ladder. Loaded with weapons, duffle coats, web equipment, tin hats and camp kit, we struggled on board, Mellis using language that made the sailors blush.

No. 4 Commando were strangers, recently arrived from Weymouth, recruited by Percy Legard, a decathlon athlete in the SKINS.* His smart adjutant, Michael Dunning-White, from the 8th Hussars was a good soldier. Sir Roger called the unit his 'Cavalry Commando'; several more officers had been recruited from Yeomanry regiments: dashing men of the 'screw and bolt' variety, who had jibbed against the conversion from horses to tanks.

* Inniskilling Dragoon Guards.

Cavalry or not, Legard could certainly lay on the 'bull'. During a previous visit, the admiral had been impressed by a bicycle spectacular, when two troops, mounted on folding parachute cycles, had put on a trick ride followed by an aquatic parade. The whole unit, in denims, web equipment and waterproofed rifles, headed by the commanding officer, entered the water for a 'swim past'; the admiral took the salute at the end of Weymouth Pier. Still keeping formation, five hundred officers and men rounded the structure and returned, applauded by civilian spectators: breast stroke out to sea, side stroke back to shore. The fire-arms suffered damage, and recently a new colonel and second-in-command had taken over, to restore more orthodox forms of soldiering.

The only friendly faces in the wardroom belonged to Mark Kerr, from the Rifle Brigade, and Basil Fordham, a charming Grenadier who served in Queen Mary's household in pre-war days. The rest appeared hostile, the cavalry having taken hard to the training methods at Inverailort, where certain instructors had earned the reputation of running subalterns off their legs in the high hills. The ship was over-crowded; this intrusion meant one less cabin and one more staff officer – a bad combination.

In the absence of Colonel Lister, busy with a court martial, I dined with Lieutenant-Commander Kershaw, RN, and got brought up-to-date with the news.

A raid with the code-name 'Claymore' was to take place in the Lofoten Islands which lay 850 miles north of the Orkneys near Narvik, in the Arctic Circle. The object: to destroy oil installations and enemy shipping, capture prisoners and Quislings, and bring back local volunteers for the Norwegian forces. The islands were important to the Germans. Quantities of the country's herring and cod fish extracts were processed locally; the oil thus produced was supplied for various needs: notably the manufacture of nitro-glycerine (used in high explosives); some was bottled into vitamins and pick-me-ups for German troops exposed to winter conditions. The Norwegian fishing fleet, which caught and carried these products under duress, would be burned without compunction if surprised in harbour.

Pending further instruction, there would be more practice landings before we sailed for Scapa Flow for a final exercise; there we would pick up our destroyer escort, take on oil in the Faroes, cross the North Sea and get on with the job. The sooner the better, as far as Kershaw was concerned: the troops (an undisciplined crowd) were lousing up his clean ship. He hoped I would do something about it. Kershaw, clearly a testy fellow, displayed a bull neck with mauve complexion. From the security angle he appeared surprisingly free with information. His First Officer (or Number One) confirmed the misdeeds of the soldiery, who had even smuggled a dog on board and were suspected of drinking in their hammocks. Number One was tall, florid and fair of flesh. In the naval fashion of the day, he did not shave the tufts of hair sprouting from his cheekbones. Pacing the desk, he shone like his ship's brasses, with a telescope tucked under one arm, through which he scanned the Argyllshire coast as though suspecting the presence of enemy agents crouching in the heather.

I never ceased to marvel at the flow of nautical terms, twice repeated, bellowed through the Tannoy system to every corner of the ship. 'Do you hear there, do you hear there? Cooks to the galley, cooks to the galley. Do you hear there? Liberty boats away!' etc.

And so on, *ad infinitum*, the tedium varying on important occasions with the preamble of 'Now hear this! Now hear this!' A kindly man, but not convincing, Number One was to become the butt of the soldiery whenever he appeared off duty. Take this example:

The scene – a Wardroom, where troop leaders in battle dress and gym shoes are drinking a noggin before the evening meal. Enter Number One, immaculately clad, who measures out pink gin. Polite enquiries from commandos why night exercises are not encouraged? Guarded answer from First Officer that the navy are already working round the clock, and helping to make and mend the mess left by the army. End to conversation. Enter harassed adjutant, loaded with Part I orders.

'Where can I pin these up, Sir? Rather important for tomorrow's duties.'

First Officer: 'Not in here you don't, there's a perfectly good notice-board in the waist of the ship.'

Adjutant: 'How do I find that, Sir?'

Resentful troop leader, on second pint: 'Stand by to go about you ignorant lubber. Don't hang in stays.'

Second troop leader: 'Avast there; shake a leg and take the starboard companion. The notice-board's slung abaft the stoke hole for any swab to read.'

Exit first officer. The laughter is one-sided and the adjutant decides to have a drink. End of story.

In this juvenile manner we failed to 'square yards' and widened the gap between two services whose habits and behaviour appeared to be irreconcilable.

It was not a pleasant time. Geoffrey Congreve, my opposite number in *Beatrix*, came over from Gourock and reported a happy ship, commanded by Captain Jo Brunton, RN – a first-rate commander, keen as mustard and highly efficient. They intended to move up to Scapa and finish training in the Orkneys. The next afternoon we heard they were gone. Worse news followed. *Beatrix*'s departure coincided with the signal 'Claymore postponed'. Kershaw continued to paint and polish. The troops were told to remove their boots to save the decks, which meant the men (without gym shoes) had to pad about in stocking soles. Drying facilities for wet clothes were non-existent, and morale started to slide.

In wartime, inactivity can become a national hazard; with several hundred young men, confined to narrow quarters from tea-time to reveille, a hazard can become a state of emergency, or something worse. In such situations mediators become whipping boys. I had no intention of filling either rôle; claiming, after the day's training was over, important duties elsewhere, I would depart to *terra firma*.

Captain Dawson, RN, the harbour master, a much respected figure who had served in the First World War and the senior officer on the base, came to the rescue. In peacetime he wrote for *Horse and Hound* and was a familiar sight at National Hunt meetings. Somehow, from somewhere, camp cooks, rations, and blankets were rustled up. The Nissen huts were hardly ready, but troops came ashore before the weekend and lived in camp until we sailed. Better still, Rosie scrounged petrol coupons to drive to Taynuilt for two nights. Suddenly life was very good: on Sunday a picnic in warm sunshine among the ruins of Kilchurn, the finest of Black Colin's seven castles; in the evening a dinner party with Tommy and May Ainsworth at Ardrishaig, the former having exchanged the Leicestershire top-hat and swallow-tails for a khaki bonnet in the Home Guard. We dined by candlelight off lobsters and champagne.

Back to Inveraray, where, among the snoring occupants of the Nissen hut, a signal from the adjutant lay on my valise: 'Parade 07.00 hours for embarkation and departure.' As I lay stretched out on floorboards, with a generous view of the night sky (for neither door nor windows had been fitted) some appropriate verse came to mind just before I dropped off in a haze of bubbly.

> There's but the twinkling of a star
> Between a man of peace and war.

Who wrote this couplet? It appears in Wavell's *Other Men's Flowers*.

I woke with a hangover, my first in the war. The day began badly. The voice of the muezzin, 'when dawn's left hand is in the sky', appeals to some. The clarion call of a bugle will stir the sluggard's blood. But raucous shouts on that hung-over dawn of 'Wakie, Wakee! Rise and Shine! Hands off cocks, pull on socks!' positively hurt, and I made a mental note to do something about it before reaching Berlin.

At 5.30 Mellis staggered in with a sooty lamp to report no gunfire tea. The water supply had failed. In our absence – Mark and Basil, who shared the hut with me, had been on weekend pass – Scottish Command had excelled themselves. During Friday afternoon a German aeroplane appeared over the Gare Loch, to spot ships and troop concentrations in the area. Alarms and excursions were frequent in the west of Scotland even before Rudolph Hess had tumbled from the clouds,* and the Home Guard (in 'Dad's Army' style) reported parachute landings over Loch Lomond and Loch Fyne as often as wild swans chose to settle at night on that open water. But orders were explicit; the general staff had issued the following: 'Inveraray has no defences. None can be made available. Troops will provide their own protection. The new camp [an infantry brigade was also training there] stands out like a sore thumb: no thought has been given to camouflage or elementary safety precaution. The oversight must be rectified.' The punch line in the message ran: 'Herewith one hundred GS picks, one hundred GS shovels for immediate action: the Camp Commandant will take charge of personnel to dig trenches – forthwith – sited clear of the target area – the Pioneer Corps will follow with sandbags when available,' etc. Map references, indicating suitable earthworks, screened by the Park's ancestral oaks, ended the signal.†

The metal heads of pickaxes, and blades of shovels, duly arrived next day – without handles. In consequence, the first working party that set forth on Sunday morning with improvised tools, soon abandoned the tree roots hideout. Whether by accident or design, digging down to make the first trench, spades struck a solid object, severing in the process a rusty pipe that carried the duke's water supply to the field below.

That put an end to the camp kitchens. A water carrier was on the way; meanwhile the management regretted that no washing facilities would be available. The latrines had no water anyway, but, much worse, there was no morning tea. Now it was time to get up, and raining cats and dogs; and so on with the play – without any washing.

Rosie's car, at the missing door, saved humping kit to the pier a mile away. Mark was taking the parade. I stopped at the Argyll Arms to leave a message for Ardrishaig. Inveraray was asleep behind frowsty

* Hess parachuted into Scotland (from his own aeroplane, which he had piloted himself) on 10 May 1941.

† Over the year, Dunbarton did take a series of pastings, the bombers going for oil storage installations as well as shipping. The weekend of 10 May, when Hess parachuted into Scotland, saw the worst night blitz on the West End of London. Bombs fell on the Horse Guards Parade and Whitehall. Westminster Abbey received a direct hit and damage was done to the House of Commons. Flames burnt their way through the roof of Westminster Hall whose timbered oak beams dated back to the days of William Rufus.

blackout curtains. Captain Dawson opened the door, dressed for an
early walk with his alsatian. He gave me a quizzical glance as we
followed Mellis's vanishing tail-light and the baggage.

'Your CO and adjutant started some time ago trying to get out to
Emma. Optimists at this time of day; but look out for squalls when you
step aboard. Have you seen last night's signals?'

'Well, indirectly, sir. I got back late. I gather we sail at short notice,
but so far no details. Or do you mean the balls-up with Scottish
Command and the water main shambles?'

'You'll hear soon enough, my boy,' said the SNO. 'The telephone's
been going all night: I have no staff here for a round-the-clock watch.
Still less a genius to decipher coded telegrams. But let's talk about
racing. I hope for a leave next month, and if Cheltenham is abandoned
there is always Punchestown and, God willing, no rationing in Dublin:
lots of butter over there. In the meantime I'll see you off; then drive
the car back to the hotel. I'd like to ask the Ainsworths and your Mrs
to lunch when they collect it.'

The captain was a fine old fellow. I was glad darkness hid my flowing
beard. The landing-craft were coming across the loch, audible a mile
away in the silence before dawn.

'I thought those new-fangled tin boxes of yours made a stealthy
approach,' said Captain Dawson. 'Come into the office unless you want
to get wet.'

The pier was lashed with rain. Suddenly the sound of music, and
No. 4 Commando in column of troops, whistling an unprintable version
of Colonel Bogey –

> Where was the engine-driver
> When the boiler burst?

– swung into the High Street, bent double under packs, greatcoats
and rifles. A proper bunch of desperadoes. But the spirit was there all
right: the men were ready for action and they were bang on time.

On *Emma* Number One appeared subdued. The same old cabin
looked plushy by electric light. I was washed, shaved and half-way
through breakfast before we weighed anchor.

The wardroom was empty except for Doc Wood – a tall, sinewy
police surgeon (I think from Hastings), soon to get his own field
ambulance. 'One boil lanced by the light of the moon, a crop of sup-
purating spots and a nasty case of impetigo that may spread,' he
announced, dissecting a kipper. The MO spoke with a dry wit and his
subsequent talk on sea-sickness carried conviction:

'The first day out – slung in a hammock when she starts to roll –

you may think you are going to die; on the second day in that stuffy
hold you will wish you had. Stay on deck as long as possible, but
remember to throw up on the right side for the wind. "G.S. Bags. Vomit
I" will be issued when the sea gets rough.'

A steward took the plates away.

'Where is everybody? Don't officers want breakfast? I gather they've
been on cold spam for three meals.'

The MO glanced at his watch. 'Haven't you heard? The balloon is
going up. Colonel Lister's been on the bridge with Commander
Kershaw since crack of dawn. One or both are to address the Commando at 8 o'clock after men have stowed their gear and drawn
hammocks.'

'That's in ten minutes; there must be a flap on.'

And that is how we learnt an exercise in the Orkneys – no mention
of 'Claymore' – had been cancelled and we were returning to the Clyde.
It was a bitter blow. The colonel, a highly strung man, looked ashen.
Breaking the news had not been easy for him. The men were at concert
pitch. They had waited, 'rarin' to go', for almost a year, without, as
somebody said later, seeing a shot or a shit fired in anger. Encouraged
by honeyed words, they had signed on to kill Germans. 'Hardship shall
be your mistress, and danger your constant companion.' What a load
of bull!

Lister was wrong to turn the setback into a personal defeat: indignation clouded his judgement. As a fly on the wall, I felt he did little to
alleviate the rumblings of disgruntled soldiers. A folksy American
President was to say years later, 'If you can't stand the heat, stay out
of the kitchen.' It is good advice on such occasions.

The security aspect was not referred to. Here was a ship full of angry
commandos returning to Glasgow, where careless talk had already cost
lives – all kitted up with arctic vests, double socks and an arsenal of
explosives, reinforced with a naval demolition party and Keyes'
representative!

It had been no secret that we were sailing for the Orkneys. Every
man on board, including the ship's company, could have spelt out
Norway as the objective. I had a shrewd suspicion that, after a few
beers, the Lofoten Islands would be mentioned. At least six senior army
officers knew our exact destination; it is an old adage that two can keep
a secret, but three cannot.

With these disturbing thoughts I sought the bridge; if Lister had
looked pale to the discerning eye, Kershaw's countenance had turned
from mauve to magenta. His immediate problem (and it seemed
irrelevant) was a safe return to the friendly waters of the Clyde, outside

which enemy U-boats were said to prowl in packs, or to lie in wait for the unwary. Corvettes and sloops should be in attendance on our passage. Kershaw had experienced a troubled night of contradictory signals from Combined Operations Headquarters in London and from the commander-in-chief, Western Approaches, on sailing orders. Like Lister, he cared little about disembarkation or security at the other end. Time enough when we got there. The Commando should have returned by road to Troon instead of by sea; if the army could not cope, that was hardly his fault, was it?

Back in the wardroom, over pink gin, Gordon Webb, a bright subaltern, gave his considered opinion there had been a mistake over moon and tides wherever we had been going. 'You cannot pass through shipping lanes or enter coastal waters in large vessels at full moon.' Brigade was criticized for not keeping both Commandos together, the Admiralty could surely have planned better sailing arrangements. Well done, Gordon! That was something we tried to teach at Inverailort: always work as a team. The date was 13 February. A fortnight later *Beatrix* and *Emma*, without a moon but accompanied by destroyer escort, sailed to Scapa Flow.

During the interval a crestfallen Congreve (he had never sailed after all) told us of the vicissitudes which had dogged our fortunes. These are hearsay opinions and not necessarily correct. The reader will have gathered that our admiral always sought offensive action. Briefly, Keyes had enthused the Prime Minister with thoughts of mounting a more formidable raiding project. During the winter the island of Pantelleria in the Mediterranean had been mentioned. The Canaries could have been a new alternative. Churchill liked the idea. Keyes was advised not to disperse his available forces – there were other commandos on the Ayrshire coast that could be added to the expedition – subject to planning approval.

Keyes, the hero of Zeebrugge, had his critics (gunboat tactics were no longer fashionable), but he was a trier. As a strategist, Churchill, 'the former naval person', remained suspect since Gallipoli, and a further load of criticism was added after the *débâcle* in Norway.

Both men passionately believed in Admiral Fisher's dictum, 'The essence of war is surprise; moderation in war is imbecility.' The general public shared these feelings, just as much as the commandos who sought, in the words of a song, 'To accentuate the positive', but it was not to be. The country's fortunes were at low ebb. The War Office could not afford to take a gamble. Chiefs of Staff met on a Monday morning in unsympathetic mood, and turned the new proposals down as totally impractical.

SCAPA FLOW

Before the war, when I was in the Scots Guards, I had taken some
adventurous leave in the Caribbean on a sponge boat, dipping in the
trade wind through blue waters and flying fish off the Exuma Cays.
Three years had passed without hoisting a sail. The nocturnal expeditions with Geoffrey to Skye, Eigg, and Rhum had all been powered
by engines of a Brixham trawler. A two-and-a-half stripe New
Zealander's suggestion that the stiff breeze sweeping the Orkneys
provided the right opportunity to try the ship's whaler seemed a
splendid idea. With snow in the air and the wind rising, there were no
other volunteers, and we were short of hands.

Getting the whale boat away on the sheltered side presented no
problem. As the sail took hold and we dropped astern, the little craft
sprang alive, fighting to be free. The downwind tack was exhilarating.
Only Lieutenant-Commander George's skill saved us from flying
the submarine boom stretched across the narrows at the end of the
Flow.

As we went about and started to get wet, the wind freshened to half a
gale. The thought – would we manage to return? – crossed my mind,
but George's cheerful countenance showed no outward sign of alarm.
Later in the week he conceded the wind had risen in a manner he had
not expected. He was an excellent seaman, and the neat way he circled
Emma to draw the watch's attention would have impressed the Royal
Yacht Squadron in Auckland, of which he happened to be the Commodore; but now there was no shelter, and no margin for error. I lay
in the stern as ballast; the Commodore did all the sailing. As ropes and
tackle came down from *Emma*, the master mariner made a fatal move.
There had been nothing wrong with the timing of our approach; the
fault lay with the lubberly crew. With less than thirty yards to go,
George, dropping the sheets, bounded like a monkey into the bows to
seize the proffered aid, at the same time shouting to me to trip the mast
and sail. The mast came down all right, but in the struggle the sail
whipped free, and the wind did the rest.

Poor George! He might have been wrapped up in a tent, and to
cheers from spectators on *Emma* – I hoped Kershaw was in his cabin –
we swept past with an inch to spare, without scratching the new coat
of paint! But worse was to follow. Wind and tide twirled us, done up in
a bundle, across the fairway. We still carried on sailing, in swoops and
dashes, through the fleet anchorage, to end up all standing against the
broad side of the newly commissioned battleship *King George V*.

The unceremonious bump that followed (for a collision it could scarce be called) will rank for ever as a nautical version of Bateman's Wellington Barracks cartoon: 'The Guardsman who dropped it'. Lieutenant-Commander George, the pride of Auckland, hid his head in shame and positively grovelled at the dreadful deed. The bridge of *King George V*, seen from our lowly craft, appeared away up in the sky looking not unlike the dome of a rococo cathedral, all studded in scarlet and gold; but here the colour scheme was supplied by angry red faces, and by the gold on the caps of the hierarchy of the Home Fleet. They gazed on our discomfiture in awful silence, as we rigged the mast and then fled away, swooping and skimming, like a swallow that had found new wings, back to the channel steamer. Except for one order, 'Into the bows to get hooked up,' the Commander did not address me again until the raid was over. Wet to the skin, with stiffening limbs, we climbed painfully up the Jacob's ladder, then along a greasy pole, back into *Emma* – no easy task with sea boots in a gale.

That night the wind dropped sufficiently for a motor launch to take a few senior officers on a round of visits, first to Geoffrey on *Beatrix* and the rest of the raiding force: No. 3 Commando and its commander, Lieutenant-Colonel John Durnford-Slater. In an age when panegyric is frowned upon, it is difficult to describe this genial sportsman. No one would have been more indignant if presented as a paragon of virtue, and such he was not, though all his faults were of a lovable kind. No beauty in appearance, going bald, stocky and of medium hight, with a jerky, short-stepping kind of action, John Durnford-Slater spoke in a high voice, but the restless energy and drive were immediately apparent. He knew what he wanted, and damned the hindmost into heaps with a cheerful smile. John got the best out of his men and went down well with the navy. A Devon man, he must have had hollow legs, for he could drink all night in the mess, parade the next day as fresh as a daisy, train for the morning and play a good game of rugger in the afternoon. Sea-sickness was his only weakness.

What constitutes a leader? My dictionary says, 'A man of deserved influence; one who shows official initiative; front horse in tandem; a top shoot on tree or a newspaper article'! There must be different answers for different situations. For a soldier, leadership is that power to inspire confidence in battle, or unforeseen crisis, where others fail: the quality possessed by the individual, of whatever rank, whose courage lifts lesser men with the certainty that all will go well as long as he is there. John was such a person. A regular in the Gunners, he had served in India, where, riding with ten necks to spare, he reached the final heat of the Kadir Cup. Living on his pay, with champagne tastes and a beer income, he somehow managed to ride and train his own steeplechase horse. Flutters on the turf were to prove his eventual downfall.

On 1 March a bearded Lieutenant-Commander with a sad face dumped his kit-bag in the refuge of my cabin. We politely tossed for the bunk, being of equal rank. The Commander drew the floor and then disappeared onto the bridge, for he was the navigator who was to steer our course – a difficult task well performed during four nights in a darkened ship. It was not until we were homeward bound that I heard he was one of the survivors from the *Royal Oak*, torpedoed the previous autumn in the Flow, with the loss of eight hundred lives.

THE CROSSING

The next day our force, supplemented by Norwegian army personnel (combatants and interpreters), sailed north, arriving at Staifjord in the Faroes. Here the destroyers refuelled, but nobody went ashore. 'Operation Claymore' was 'ON', and the darkened ship was officially sealed.

In the background, blurred by rain, rotted a splendid regiment. For two years the Lovat Scouts, men of the long stride and keen eye, kept watch and ward over these storm-stayed rocks; banished because of the indecision of a hesitant colonel when they had been required in a previous emergency. There were no Scout bonnets to be seen. I hoped the Faroese girls – rumoured to be of surpassing beauty – were keeping the good fellows warm in their involuntary exile.

In darkness the force sailed once again. Picking up navigational beacons that marked out the channel passage, we entered the West Fjord on 3 March between night and morning. Except for fulmars and stormy petrels that accompanied *Emma* across the wastes of the North Atlantic, those pin-points of light were the only objects seen for two whole days. HM Submarine *Sunfish*, which had preceded us, was in position to provide navigational aid. The tempo quickened and sea-sick soldiers came to life.

Assault briefing, as depicted on stage or screen, harks back to trench warfare. The scene shows a young hero about to go over the top. The foetid atmosphere is vibrant with suspense. Empty bottles, stuffed with candles, give a flicker of light; others, not so empty, cover the tables where the map should be spread. Guns keep up a rumble like distant thunder, making it hard to hear the orders. If this were not enough, the roof collapses at intervals, showering the leading actor in dust and rubble. In real life, the tension varies. A large-scale raiding force, committed to an operation planned and rehearsed with meticulous detail, helped by aerial photographs, transcripts and intelligence reports, should know the answers. Scaled models; charts; navigational aids; details of the state of the tide, the wind, and the moon: all leave the combined force with little to talk about at the eleventh hour. An 'escape officer' gives a pep talk on what to do if one is left behind. (This time each man was presented with a handkerchief map of Norway and a collar stud concealing a small compass.) It remains prudent to synchronize watches and to check signal systems, calling arrangements and meteorological reports, with a word of encouragement to the ship's officers, and not forgetting the issue of a rum ration. All this is best done in the wardroom after the last meal. If, for security reasons, the targets are kept secret until the ship is sealed, then all ranks should be briefed without delay – when they are wide awake, allowing time (this may be considerable) to discuss details, make notes and study maps. Another half-day is needed to draw ammunition, iron rations, escape apparatus and first-aid kits, to prime grenades or prepare demolition charges. The last task is done on deck and in daylight. Never rush the fuse in a hand grenade! After this the soldier's mind grows dim, and thereafter he should be left to his own thoughts – a soporific, compounding anxiety, sea-sickness and boredom in equal quantities.

This council of perfection applies to seasoned veterans. On *Emma* there was little sleep that night. Colonel Lister (the Buffs) had won a Military Cross in the First World War, once held an army boxing championship, and also represented his country in Madison Square Garden. And no rugby international ever put up better performances for England or the navy than Lieutenant-Commander Kershaw. It would be impossible to doubt the physical courage of either man. (Both are long since dead). But that night they both reached a state of nervous anxiety amounting to the jitters.

Although we had sailed too far north for air attack (Norway being covered with deep snow), the possibility of U-boat or surface craft interference was not ruled out. German patrol boats covered shipping lanes, and the risk increased as a following wind fell away to light airs. As we closed the coast, the sea flattened out.

Kershaw was not in favour of a sleeping ship. Lister thought the men should rest, but agreed that reveille should be put forward to 0400 hours, with gunfire tea at 0500 hours; in spite of the intense cold, the troops would parade at 0600 hours, and wait beside their respective landing-craft. All officers, after roll call, would report to the wardroom for further instructions. Boats would be lowered from the davits at 0700 hours, to make landfall at 0800 hours. These proposals caused protest from officers who had experienced similar campaign conditions much farther south. It was pointed out the run-in was over three miles of sheltered water, it would be dark until 0900 hours, and men on deck would freeze from unnecessary exposure.

Why not wait fully dressed, lying in hammocks – better still, go ashore at dawn, fortified with a hot breakfast?

The indecision lasted until midnight; then came a compromise to cut the deep-freeze treatment by one hour. Breakfast was out; so too the proposed landing at first light. The disagreement did not instil confidence in a command which was now divided. Lister sulked, refused to climb down over reveille or a final briefing, and bored all who cared to listen with a series of contradictory orders. Michael Dunning-White, the 8th Hussar adjutant, prompted corrections from the wings.

Kershaw retired to the bridge. Number One saw to it that we got no sleep by constant 'Do you hear theres' on the loud-hailer, making a hullabaloo that carried half-way to Lapland. Moral: it is better to keep your mouth shut and let people think you're a fool than to open it and remove all possible doubt. I have overstated this tale of woe. But it was a poor performance. The era of more Chiefs than Indians was fortunately coming to an end.

Before the end of hostilities we would discover that for young men war is tolerable – even enjoyable; but introduce an element of age, and it becomes impossible.

THE LANDING

Mark Kerr (Rifle Brigade), with Bill Montgomery (Black Watch), were the first away in two of the six landing-craft, each containing thirty-five men. They had some distance farther to go – objective: Brettesnes. While the main force destroyed factories and oil installations in Svolvaer town, their task was to seize or set fire to shipping anchored in the haven farther up the fjord. Both were the right men for the job. Mark, handy with his fists and a hard-hitting bat in the Eton Twenty Two, was an old friend. We had been on the same course at the Small Arms School at Hythe. The summer of '33 was so hot, that we had resisted all the temptations of the London Season, and never left the south coast. Kent is a charming county and there were many diversions. One free Saturday afternoon at Folkestone Races, we noticed a carp of noble proportions cruising in the lily pond which graced the lawn of the Members' Enclosure. By the way it rose to have a look at a ticket, the fish appeared to be hungry. On Sunday we returned with a hand line, a hook, a lump of bread and an empty sandbag. The Commandant, an irascible martinet, who lived in his own red brick quarters, was away for the week-end. Two storeys and a convenient drainpipe presented no problems. On Monday morning, he found a three-pound carp swimming in his bath. How it got there was never discovered.

Montgomery, in the fighting traditions of the Black Watch, drew cutlasses from *Emma*'s armoury for boarding any German Mercantile Marine found in the harbour. As it turned out, Kerr's and Montgomery's considerable success proved a bloodless affair. Sadly, neither officer was to survive the war.

In Svolvaer (the main objective) docks and berthing facilities were limited. Boats could be expected alongside the quays; according to the state of the tide, a climb was envisaged up wall ladders that rose from the waterline. No steps were available, and the flotilla's movements were restricted to a simple plan.

Because of this narrow front, the assult would land in three waves. Three landing-craft in the first flight, followed in the fourth boat by Colonel Lister and his headquarters: adjutant, signals officer, intelligence officer, MO, runners, sapper section with the high explosive, and Sir Roger's personal 'representative', now feeling a trifle self-conscious. After disembarking the assault party, boats would return with all speed to the parent craft and bring on the rest of the Commando (full strength, five hundred all ranks).

That morning No. 4 Commandos wore steel helmets – useless impedimenta of a forgotten age – for the last time as a raiding force. The more sensible men sported cap comforters under freezing head-shrinkers to protect ears from the cold.

Out in the fjord, the destroyer *Afridi* covered the landing. The final approach, still in darkness, was uneventful. A trawler came out of the fjord as we stopped engines. She had been challenged by *Afridi* and seemed in a hurry. She was hailed off our bridge by a liaison officer Max Harper Gow (who spoke fluent Norwegian), calling 'Aloo, aloo' across the inky water in beguiling tones. The boat vanished with all possible speed. Now a sense of unspoken urgency sent the men clattering into the landing-craft. Max got a rocket from the commanding officer: 'Damn it, I didn't ask you to come all this way just to say "Aloo".'

Mark had been gone half-an-hour, and snowy peaks had begun to show against the pale stars, when, numb with cold, the flotilla moved away from *Emma*'s side. Surprise was complete – not a stir in the town, no lights to be seen. So intense was the weather that the first men to spring aloft, scorning gloves as they clambered up the iron rungs of the ladders, had skin torn off their hands.

It was still dark; the time 0830 hours. Except for shouts from deploying sections, 'Fortress Europa' remained silent as the grave. Judging by the smell, and leftovers that shone in the light of a torch, our primary objective, the fish-processing factory, was not far away! Apart from the indignity of tripping over frozen cods' heads which littered the wharf, we had arrived in good shape.

The police station, post office and town hall were taken over, then closed. After an untidy start, a bridgehead was established that controlled the port area; the demolition squads took over.

Sensibly, No. 3 Commando, farther up the archipelago, waited for first light; they suffered the embarrassment of being hauled protesting from their boats (armed to the teeth and with blackened faces) by enthusiastic citizens awakened by gunfire, following an engagement between the destroyer HMS *Somali* and an armed merchantman. The German was knocked out in an attempted escape to open water, was then captured, and later sailed home with the loss of fourteen Germans (killed or wounded).

In our sector, Colonel Lister established headquarters in the harbour master's office. Scouts and patrols disappeared in the darkness to pinpoint installations marked for demolition. The sappers, carrying assorted charges, wrenched open factory doors and set about their tasks. Within an hour, explosions shook the town. Norwegians emerged to witness the destruction.

Before ten o'clock the sun climbed over the mountain. Tension relaxed and the troops started to thaw out. With daylight came the inevitable anticlimax which erodes authority. The numbing cold had curious side-effects. Discipline, essential in a raiding force (perhaps it was never there in our case), disintegrated in the gala reception offered by rejoicing citizens, who ran up flags and brought out refreshments for the short-term liberation.

The commanding officer retired from control HQ to inspect the cellar of the harbour master, a suspected collaborator. He then proceeded to distribute the contents to crowds on the street outside. Informers had a field day paying off old scores; as the morning wore on, fires appeared among wooden buildings that were in no way connected with military objectives.

Troops wandered from roadblocks and rooftops. The telephone exchange, seized as a primary objective, became a popular centre from which insulting telegrams (in basic English) were despatched to Hitler, Goebbels and Goering. Snipers descended from vantage points; posts were abandoned or left undermanned. The temptations – not great in themselves – were no worse than a 'cuppa' laced with aquavit! But what if a motorized column had descended on the town? I fully believe this ill-assorted contingent of volunteers (first known as independent companies, then special service troops, and finally commandoes) would have given a very poor account of themselves and ended up in a prison camp. The same thought occurred to our Norwegian liaison officer, and to his official interpreter, who now delivered a message. He reported a rumour of a not uncommon kind: a suspected German garrison was located round the next corner. When Michael Dunning-White received this information (a Luftwaffe signal station was in fact in barracks on the other end of the fjord) he smartly rang the sales bell in the fish market where our O group were pumping up a Primus to boil tea.

The call for a strong fighting patrol was met in no time. A party of fifty (all ranks) were quickly assembled, while men who could drive collected a column of vehicles. Then the ramshackle convoy, led by a fish lorry with a Bren gun crew mounted on the cab, Michael Dunning-White, the Norwegian officer with revolver drawn and myself crowded in beside him, chugged along the coast to seek opposition and sight a German uniform.

The early twists and turns in the winding road beside the fjord proved uneventful; we should have been ambushed round every corner. The windscreen of the truck had fugged completely by the time we got some action. I remember calling attention to a flock of snow buntings. (The most charming of all Alpine passerines, the cock sings all night in the nesting season on High Cairngorm, where the June sun barely goes below the horizon.) The little birds were feeding on seed in horse manure along the frozen road. As they rose and flew twittering up the hill, I turned down the window and leaned out sideways to follow them, inadvertently saving my leg. For simultaneously all hell broke loose. Michael, at the wheel, was craning his neck at the buntings as we turned a blind corner and ran slap into a column of Germans. The Norwegian officer pulled the trigger of his revolver in shocked surprise. The heavy bullet smashed the dashboard where my knee-cap had been resting, filling the cabin with cordite fumes. The lorry swerved and hit the parapet of a low bridge with a bump which threw the Bren gunner and his Number Two on to the hard-packed snow. I got out of that truck in what could be described as a flowing movement. The speed of the knockabout turn had its effect on the *Herrenvolk*, who put up their hands or scattered up the hillside.

This head-on encounter was quickly explained. Although the leading foemen still wore caps and reefer jackets of the German Merchant Marine, the marchers were only unarmed sailors from ships' companies

at Brettesnes, who had escaped when Mark Kerr made his assault landing. They had walked to safety in the wrong direction.

After a fright on active service the first reaction is either relief or
indignation. On this occasion the adrenalin fairly boiled! I was about
to murder our unfortunate Norwegian, who sat twitching in the
wrecked truck, when one of the Germans, more courageous than his
fellows, dropped a canvas satchel over the bridge into the creek which
ran below.

At Inverailort, Messrs Fairburn and Sykes (two mild old gentlemen,
one looking very like Father Christmas, and both retired Shanghai
policemen) could truss the toughest officer or NCO by what appeared
no more than the pressure of their fingertips. They taught how a sharp
chop with the heel of the hand, in a downward cutting movement (where
the jugular vein joins the collar), was a lethal blow only to be used in a
dire emergency. Two long strides and whacko! The result was disappointing. The German's scarf or turned-up collar may have saved his
neck, but it hurt my hand like hell. I was angry enough to cross-counter
with a left that knocked him over the parapet, to the amusement of
friend and foe, who now got hopelessly mixed on the road, Germans
advising on how to get ashore; commandos less helpful, 'Go to it, Fritz!
Belt up and retrieve that blanketty-blank packet of papers!' These
floated half-submerged in the icy water.

The Bren gunner, his dentures broken in the crash, was all for taking
a pot at a swimming target and had to be forcibly restrained. A weird-
looking member of the intelligence section, bi-lingual Sergeant Goyne,
who in his spare time wrote detective stories and claimed to be a friend
of Edgar Wallace, now stepped into the act. Goyne (who drove to work
every morning in his own Rolls-Royce, after picking up the Regimental
Sergeant-Major) reported that the sailors wished us to kill the swimmer,
now acting as a retriever. 'He is an ardent Nazi, Sir, and nobody wants
him back. The ship's company would be glad to see him removed.'
A nice turn of phrase!

This brush with the enemy lasted five minutes, but after the skipper
and his log were hauled ashore, a half-section detailed to take charge
of the prisoners, and the patrol regrouped (back in the vehicles), we
were running behind the clock.

The skirmish raised morale: we had gained experience and with it
new assurance. In the next encounter we did well. The story is soon
told.

The signal station, a converted police barracks on the forward slope
of a hill with a good field of fire, was flying a swastika flag. It
looked solid and forbidding. From local urchins, whose advice should

always be treated with respect, we gathered that only Luftwaffe technicians lay within. Led by a subaltern, later to distinguish himself at Dieppe, part of the fighting patrol worked its way to a good firing position in bare rocks, and at a range of three hundred yards opened up with Bren guns firing tracer in short bursts at the door and upper windows of the building. This manoeuvre drove the inmates into the arms of a reception committee from the rest of the patrol who slipped round in a detour to cover the rear exit.

Monty Banks, a Second Lieutenant in the Tank Corps, was turned loose with a crowbar in the room containing wireless installations: a well-equipped set-up of high-frequency apparatus, too heavy to move and clearly of considerable importance to the enemy's network in coastal communications. The task soon had Monty sweating on the job, but it did not please our signallers, men whose souls were in their cylinders. 'Nothing like this at home, sir,' said one. 'A treasure-house of beautiful stuff,' whispered the walkie-talkie operator, who was having trouble in raising Svolvaer only five miles away. 'Please don't smash it, sir, I can't keep my hands off these dials and buttons.'

Before we sailed from the Clyde a pep talk had been given by a seedy individual from the Ministry of Economic Warfare, whose function was to assess the potential and expose any weakness in the German war effort. This appraisal covered weapons and equipment, and also raw materials (supposedly in short supply). The information was fed to the press by a Ministry of Information anxious to show that declining powers, ranging from suspected cancer in Hitler's vocal cords to the dud bombs which failed to explode at Fort William, indicated the parlous state of the Third Reich. His talk had begun with the vintage story of a German court martial, sitting in Colmar, which sentenced a local merchant to a fine for repeating in a public place the jest of ordering a sandwich at a Prussian railway buffet and being served with a meat ration ticket between two bread cards!

Those who cared to listen to such nonsense before the raid had been urged to bring back samples of barbed wire, water bottles, gas masks, ammunition, rations, etc. – even a sandbag was of interest – just in case the fabric showed sub-standard quality. This last request was greeted with derision by those who had seen the Maginot Line turned inside out and armies destroyed, with ruthless efficiency, in a blitzkrieg by incomparable troops armed with superior weapons and machinery. The official was hurt and complained to higher authority; the sceptics, it was hinted, were unpatriotic and completely failed to understand the range and subtleties of the war effort.

The lecturer left a profound impression on his audience, though not in the way intended. For his remarks appealed to the fighting man's sense of humour, the most lovable of all his qualities and one which stays with him to the end, however rough the sledding. Also it suggested possibilities of worthwhile looting in the name of science and research, and the police barracks seemed a good place to make a start. The previous occupants, disarmed and shivering outside in the snow, were allowed to collect kit-bags; then, one at a time, to pack for prison under a rough-and-ready supervision. Judging by the cries of protest, this could not have conformed to the rules of the Geneva Convention. Various detachable parts of the wireless installation, code books, logs and so on were stuffed into duffle bags, along with other articles 'of national importance, friend Fritz, to our glorious fatherland, now beset by Herr Hitler; please understand that no personal sacrifice is too great; help me lieberkind to strike another blow for final victory!'

Wrist-watches, fountain-pens, gumboots, scissors, sun-glasses, louse powder, shaving soap, a piece of orange carpet, the office clock, the swastika flag off the roof, and – perhaps the most remarkable of all – a hair net removed from the flaxen head of a prisoner who got out of bed in a hurry, were loaded on to a lorry.

The spoils from the Second Lieutenant's bunk were of a different category. *Vae victis!* The treasures were divided between officers. Michael swiped a Leica camera, Monty Banks took the Luger Parabellum. 'Fairy' Veasey, who came in off the rocks where weather conditions were reported to be colder than a witch's tit (his black moustache was coated with icicles), settled for a pair of binoculars. I unwisely chose a fleece-lined, rawhide Luftwaffe coat – a perfect fit, but too conspicuous to escape detection. Better still, being the senior member of the party, a pair of woolly flying boots were added to my loot. Thus fortified, and with a case of Schnapps divided between the men for comfort on the North Sea crossing ('Anyway, sir, this stuff's too heady to hand over to a chairborne Civil Servant'), the patrol, plus prisoners, returned to base in high good humour.

THE RETURN

The clock was striking the hour in Svolvaer as we pulled up on the harbour front. Our sortie had hardly been noticed in the confusion of re-embarkation. Landing-craft were loaded with a variety of quislings and prisoners, among whom I recognized the bedraggled sportsman on whose neck I bruised my hand with the karate chop too lethal to practise. The population stood on the quayside; as the captives got under way and stood out to *Emma* (where Number One, faintly heard across the fjord, issued instructions), an angry shout, compounded of hisses, shaken fists and other ruderies, followed the Germans and quislings over the still water. The ugly sound, an unexpected display of feeling from the stolid Norwegians, had a noticeable effect on the troops, who perhaps for the first time saw the realities of war in a more brutal context than the comedy described during the morning.

Next to leave, to a thunder of applause, were 150 patriots: young men and a few girls in ski pants and wind-cheaters. The rucksacks strapped on their backs, containing socks, shirts and sweaters, were all they took to start again in a free world. Finally, slung about with plunder, pride of place going to a barrel of aquavit lowered with block and tackle, came the bridgehead party, with O group and commanding officer. Three red Verey lights rose through the smoke that hung over Svolvaer: the signal that withdrawal was complete.

Out on the fjord, *Afridi* showed signs of restlessness before the last man re-embarked. Now every delay seemed interminable. As the small craft waited to be hoisted on to our Mother Duck, Veasey confided that, despite two vests, two pullovers, flannel shirt, leather jerkin, inflatable Jeeves waistcoat and a mackintosh with a woollen lining, he still felt extremely cold.

My last memory of Lofoten, against the backdrop of snow, smoke and burning buildings, was the stirring sight of *Afridi* – gun crews at action stations, a wave of white water creamed off her bows – heeling over in a tight turn as her commander hailed the bridge to 'get the hell out of there and join *Beatrix* and *Somali* in open water'. It was a worthwhile battle picture, unsurpassed in savage beauty and bright light.

After checking with Charlie Head, the adjutant of No. 3 Commando, a signal was blinked out by Aldis Lamp and we got the score. A war diary should be kept by the intelligence officer or adjutant on active service, and the entries written up as in a game book. The commandos produced this joint 'bag' for the half-day 'shoot':

Norwegian volunteers taken back to England	315
English manager, Allen and Hanbury, rescued	1
German prisoners, including Merchant Navy and Luftwaffe personnel	216
Norwegian quislings	60
Ships sunk: total over 22,000 tons	11
Armed enemy trawler manned and sailed home with loss of German crew (14 killed or wounded)	1
Number of factories destroyed	18

Fish, oil and petrol installations burned to the ground
    or destroyed – contents containing 800,000 gallons   7
Photographs and propaganda needed on the home
    front   100% success story

On the debit side there was one officer casualty in No. 4 Commando:
a loud individual whose style of training was noted with distaste at
Scapa Flow. Slipping on the frozen surface, he accidentally wounded
himself with a .45 Colt carelessly stuffed into a trouser pocket. It had
been a bad day for revolvers.

The return was uneventful. The boats were crowded; prisoners
suffered uncomplainingly in the hold, where conditions must have been
indescribable. If water remained on board, it was not used for luxuries
like washing. Away to the southward a covering force, with the cruiser
ships *Nelson*, *King George V*, *Nigeria* and *Dido*, and a destroyer escort,
looked out for *Scharnhorst*, should that battleship have the temerity to
intercept us on this, the first of two Lofoten raids. The proffered bait
was not accepted; nor were our signals from *Emma* to Admiralty,
because of poor transmission.

> Hung in the void, the crescent moon
> Is all that's seen in this white world.
> On polished steel of sword and shield
> The cold dew hangs in icy drops.
> Long may it be ere we return
> So weep not, women left behind.

BACK TO SCAPA

As the convoy came under the lee of the land, the accompanying
destroyers altered course and turned away, crossing our bows at speed,
back to the open sea – all grace and majesty. No signals were exchanged, but wordless partings are seldom forgotten. They were all
'Tribal class', the best in the business. The adjutant spelled out
the closest through his binoculars – *HMS SIKH* – later to be sunk
in the Mediterranean by dive-bombers off Bardia on a commando
raid.

The submarine boom had been rolled back for the Commander-in-
Chief, Home Fleet, who preceded us up the fairway. Tails were high in
the air: the raid was world news. All morning fresh bulletins kept
adding details to our exploits, the cutlass boarding party being
especially popular. Troops and crew stood to attention as *Beatrix* and

Emma, manned overall, glided silently past the tall battleships, whose crews cheered us to our moorings. Strong men shed tears launching a liner in a shipyard, but this was the real thing. Praise from the Royal Navy was especially welcome.

Tugs came puffing out from Kirkwall. The wounded men were taken ashore; Norwegian soldiers shook hands and said goodbye.

That evening, standing politely in spotless wardrooms, commando officers who had slept for the best part of a week in rumpled battledress, 'smelling faintly of cod and dripping in gin' (so described by a wit in the navy), told of various adventures, already fading into unreality: a dream contrasting strangely with silver, soft lights, wax flowers, and senior officers, immaculately clad with broad-striped sleeves, who plied us with drinks and questions.

It is the unexpected and bizarre incidents that are remembered after a successful operation. This was the time to talk about them.

No. 3 had landed a hundredweight of sweets at Stamsund and handed them out to children. Anthony Kimmins, RNVR, a well-known BBC broadcaster, had stood his watch on the crossing, and gone ashore with the other Commando to put in some ski-ing up above the town. He reported the snow conditions to be excellent.

Bill Boucher-Myers (East Lancashire Regiment), a resolute subaltern in No. 4 who had been allotted the only available Tommy gun, considered himself lucky to be still with us. Summoned at short notice from the post office building, he was put in charge of a naval demolition party: their task, to blow up three merchant ships lying in the roads. The landing-craft had frozen solid. Before they got under way, the boarding party saw that *Bremen*, largest of the trio and a vessel of 9,000 tons, was ablaze from stem to stern. The Germans had set fire to her deck cargo – wooden boxes and barrels for fish oil – and the crew were abandoning ship.

The commando party put to sea. After taking the captain prisoner, Bill ordered ratings aboard to deal with the engine room, and chugged on to sort out the rest. No sign of life in the middle vessel, *Hamburg*. Swinging ropes and grapnels, he got men over the stern. They hauled up the explosives and – timely thought – a Bren gun. It was as well they did so. As Bill reached the bridge, he was fired on by the crew of the third ship, who had courageously attempted to cover the withdrawal of the other sailors, now in lifeboats pulling for the shore. The Bren went into action, silencing the rearguard, then engaged the boats and sank the leading craft. The crews surrendered. Demolition charges were placed in the engine rooms and limpet mines laid along the hulls. It was time to return to *Bremen*. The fire had burnt out; sea cocks were opened and machinery systematically destroyed. But the damned ship would not sink!

'She'll never sail again,' declared the chief petty officer, disturbed by the size of the catch that refused to lie down.

Even as he descended the rope ladder, a series of heavy explosions shook the vessel. The boarding party looked at each other in surprise.

'Well, we still have a limpet left,' said Bill. 'We'll hole her side for good measure.' And so they did, below the water line – and that's a cold job for a bare arm inside the Arctic Circle.

As the charge was laid there came another bang overhead and showers of woodwork splashed into the sea.

'Funny thing,' remarked an able seaman, rubbing a bearded chin on which icicles were sprouting, 'We never put a charge in the cutters.' Another blast, and parts of the wheel-house crashed down on their heads.

Then they got the message. 'My God! *Afridi*'s shelling *Bremen* from the fjord. Crank the diesel, cox'n, and beat it to hell out of here!' But the contrary motor had frozen up again – and they had no oars.

Soon after they chugged away *Bremen* finally turned turtle. Bill got a paltry mention in despatches for this wholesale destruction.

Bill Chitty, a splendid provost sergeant who remained John Durnford-Slater's personal bodyguard until he was captured at Dieppe, had been issued with the other Tommy gun (in No. 3 Commando) before the raid. During the mopping-up period, as John harangued a party of quislings, with impassioned words, to cease further collaboration with the enemy, 'Hot shot' Chitty absent-mindedly pressed the wrong button on the Thompson cradled in his arms. The unfamiliar weapon jumped out of control, releasing a shower of bullets into the floor between the CO's legs. John Durnford-Slater never turned a hair, but his adjutant, Charlie Head, swore the Norwegians believed it was part of an intimidation act. The quislings left the building white and shaken. John said to Chitty, 'You must be more careful with those bloody things, they're dangerous.'

Another character, 'Thirty Kroner Jack' – to be so remembered throughout hostilities – was lucky to find a willing lady in the town, and, for services rendered, parted with thirty of his hundred kroner (issued for escape purposes) before returning to his ship. At Brettesnes, the Norwegian National Anthem was sung as the raiders departed. Was it at Stamsund that Jack Churchill belted on a claymore and piped his troop ashore standing in the bows of his landing-craft?

RETROSPECT

The first Lofoten raid was the swan song of Sir Roger Keyes. He retired to make way for a younger man: Lord Mountbatten. After setbacks, let-downs and petty jealousies the war gods had smiled at last, vouchsafing the old sailor this opportunity to prove how a few men, launched boldly in a surprise attack by land, sea or air, could wreck a coast target and should be exploited to advantage. 'Take it from here' was the signal, until such time as new armies, equipped and better trained, again took to the field. The War Office had reservations. No troops were available; the navy and air force could spare neither ships nor planes. The Middle East had top priority.

This was the background – not a friendly one – in which the Lofoten operation was mounted. All ranks were desperately sorry to see our master go. By sheer determination he had put the show on the road. With little help and many disappointments, he first lit, then stoked the fire to fashion a force without the support of proper staff: just a bunch of volunteers, some of doubtful origin. Against the odds he had forged a weapon which in time would earn no small renown from friend and foe.

What was the significance of Lofoten? From the commando angle, the short-term answer was simple: although there had been no fighting, any action was better than no raid at all. At top level Lofoten was hailed as a considerable success and a necessity for survival. Had it failed, or the raid been cancelled, Chiefs of Staff would inevitably have disbanded Sir Roger Keyes' private army. After he had gone the tempo increased and he, poor man, got little credit. In the long term, the constant threat of surprise pinned German forces along the Atlantic Wall, which now stretched from the North Cape to the Pyrenees. These troops could have been deployed in other theatres.

After our sortie came the plunder of Spitzbergen, a second visit to the Lofoten Islands, and the destruction of a German garrison at Vaagso – all mounted within six months. These assaults forced German High Command, bowing before a Hitler directive, to reinforce the northern coastlines of the much vaunted 'Fortress Europa'. (As it happened, the Allies never returned to Scandinavia.) On D-Day, the Norwegian garrison, sorely needed elsewhere, numbered 300,000 men.

It is fruitless, but fascinating, to consider the difference these divisions could have made in Normandy. Either way, 'Corporal Schickelgruber' made a mess of it. The following communiqué came from the horse's mouth: 'Adolf Hitler – January 1942. Norway is the zone of destiny in this war. I demand unconditional obedience to my orders and directives concerning the defence of this area.' Field-Marshal

List immediately made a tour of inspection as the Führer's personal representative; on his recommendation three more divisional commands were established in Norway. More coastal guns were sent north to cover approaches to fjords and harbours, and all-round defensive positions were set up in the interior.

BACK TO BASE

In the murk of a Clydeside evening the raiding force, suffering from sore heads (for Scapa is a hospitable place when the fleet is in) crept back to dingy Gourock – back to the realities of wartime dockland as a big city put up its blackout curtains for the night. 'We don't want to lose you but we feel you ought to go' a Royal Marine band had played as we left the Orkneys, still in festive mood. It seemed good to be alive, and the men were in great heart.

If Colonel Lister continued to live on his nerves, John Durnford-Slater had discovered a cure for sea-sickness. The recipe: quantities of cheese and pickled onions washed down an hour before sailing with a bottle of beer. Should weather conditions become rough this panacea was to be tried out 'on the home run through the Minches by those with queasy stomachs'.

The mirage soon disappeared. Now the sun had set, and the world changed back to drab reality. No sooner were we tied up alongside, and prisoners brought on deck, than contradictory signals tangled up our disembarkation. We moved to another berth, this time at the Tail of the Bank, where transport would be waiting to carry the two Commandos back to Largs and Troon.

Down river, Captain Charles Vaughan was waiting on the quay. A father figure, this ex-drill sergeant, Coldstream Guards, ex-RSM, the Buffs, and ex-Yeoman of the Guard, currently in No. 4 Commando, was shortly to become second-in-command when Mark Kerr returned to the Rifle Brigade. Vaughan ended the war, after thirty-three years' service, as Commandant of the Commando Depot. At Achnacarry he became affectionately known as 'Rommel of the North'. Donald Gilchrist, now an ornament in the Royal Bank of Scotland, became his adjutant, and has written a good book about this remarkable man, whom I saw for the first time standing four-square in the headlights of his pick-up truck. 'Douse those lights, that dopey man on the dock, don't you know there's a war on!' The figure blacked out, and Charles came on board. Like Marshal Ney's, his military precepts were of the essential kind: 'March fast, shoot straight' – not bad advice at that. Because of his bulk, Charles no longer went on training, but the old soldier could still take a big parade by the scruff of the neck and drill men the way that only the Foot Guards can.

It was the same story on the rifle range, where he provided the magic when a soldier was shooting badly. After careful inspection of the Lee Enfield (no question of zeroing sights or asking questions) he would lower his bulk to the prone, or kneeling, position, check the magazine, clips and cartridges, lay more clips beside him, load ten rounds and start to shoot with the slam-slamming rhythm of a machine-gun. With practice, a trained soldier can get fifteen rounds off with reasonable accuracy at one hundred yards in sixty seconds. Eleven or twelve is above average for the conscripted man under war conditions. Charles in his day could fire twenty-five rounds without taking the rifle from his shoulder or his eye off the target, and every bullet in that rapid fire would cut the bull.

He did not go on operations; he was far too valuable. 'I lost my plates of meat [feet] on the Retreat from Mons or I'd start my soldier-ing all over again and come with you!' or 'How many Krauts this time?' was his God speed and welcome home; and go he would, for Charles had been in that Guards brigade which fought the rearguard action across France in the Great Retreat in 1914. After being sur-rounded, and cut off by motorized Prussians at Landrecies, his battalion had been ordered to fight and die like Coldstreamers. They had sur-vived, extricating themselves from an impossible position by the sustained accuracy of rapid small arms fire.

Charles was later to show me every kindness, for Percy Wyndham, my aunt Diana Westmorland's first husband, had been his platoon commander. When Wyndham fell, Lance-Corporal Vaughan carried him to the dressing-station. Coldstreamers have a habit of doing this sort of thing.

Is there a finer campaign story than the occasion when Rommel's Afrika Korps surrounded and partly over-ran Tobruk, forcing the South African General Klopper to capitulate? The Allied troops were ordered to lay down their arms. Not so Tim Sainthill and all that was left of his badly mauled battalion. His signal to Klopper – 'Surrender, un-known term Coldstream vocabulary. Message ends. Over and out' – was the last communication received by the general officer commanding. As dusk was falling, Major Sainthill and three companies of Guardsmen with some brave Indian troops fought their way from the perimeter into the friendly darkness of the desert, escaping down a wadi and back to their own lines. Tobruk went into the 'bag', and Klopper, with thirty thousand men, marched into captivity for the duration of hostilities.

In Glasgow, disembarkation was in disarray. Scottish Command transport, after waiting at the Tail of the Bank, had belatedly gone off to Gourock. There was no way to unload the stores and ammunition. Dejected prisoners and quislings were marched off *Emma* by the provost marshal and his Red Caps; the rejoicing Norwegian patriots erupted down the gangway, greeted by a reception committee of fellow-countrymen. A Ministry of Information spokesman came on board to take a statement for the press, but troops were refused permission to leave ship. This was hard to bear; we had expected to send messages that we were safe and well. A request for a telephone line was refused; Charles Vaughan was interned for the night on security grounds.

Bad orders tend to be disregarded. After lights-out, in the wee sma' hours, when good citizens are all abed and dockland, dreich and deserted, the barrel of Schnapps was rolled out of a sallyport at wharf level and trundled up a plank into Charles Vaughan's pick-up truck. The vehicle was then pushed into a shed on the waterfront, standing next to an office marked 'HM Customs and Excise'. The driver was ordered to sleep in his cab; a sentry with fixed bayonet guarded against all comers. But these precautions were child's play compared to the native cunning of the Glasgow 'wharfie' or 'dock rat'. Before peep o' day a hole had been bored with bit and brace under the truck, through the floorboards and into the barrel. The rest was easy; with a chain of buckets and silent hands not a drop was wasted; nor a sound heard. The barrel was drained to the dregs. The dispossession of our spoils continued.

The chair-borne wallahs started to take charge. After a kit inspection requested by the Ministry of Economic Warfare, all trophies were confiscated. There was no way out. At the top of the heap lay the Luftwaffe signal officer's coat, considered important evidence of the Reich's tailoring trade. It was handed over; but by standing behind chairs and looking half-witted, I strolled off *Emma* after breakfast wearing the fleece-lined flying boots without detection. They were destined to keep my feet warm in Moscow and Leningrad. Judged by the number of respirators missing after the operation, many bottles of Schnapps came ashore in gas masks.

In Troon the civil authorities gave a party for all ranks and a dance in the town hall, with plenty of drink. Before the evening was out Second Lieutenant Monty Banks told the CO he was no use at his job, and furthermore, in Monty's opinion, was treating his wife badly! Banks was sent back to the tank corps on the next train.

Later, the junior officers adjourned to an hotel. There, helped by a piano and more refreshments, feats of strength and agility were performed while the furniture suffered. The 'muffin man' put an end to beer supplies and wrecked the carpet. Gordon Webb proved it was possible to dive over three sofas without castrating himself. All agreed we were the finest fellows in the world.

It took a swipe from Roger Pettiward in a rough game of 'Are you there Moriarty' to break up the party. For this ploy, tightly furled newspapers are gripped by blindfolded opponents. They can elect to fight in pairs or as a team – lying *ventre à terre* or standing toe to toe with face averted to avoid the blows. The damage is considerable either way, but no worse than in 'Russian roulette'.

Roger had served in the Norwegian campaign with a battalion of the Bedfordshire and Hertfordshire Regiment. Older and wiser than the rest, he had helped Peter Fleming to search the Mato Grosso for Fawcett, the missing explorer described in Peter's book *Brazilian Adventure*. Roger had a delightful sense of humour, illustrated by the drawings he sent to *Punch* under the pseudonym of Paul Crum. Not a tidy soldier in the parade-ground sense, he kept his own council as a leader, which is more important. The men admired him greatly and when the red-haired artist fell later, fighting hand-to-hand in the gun pits at Varengeville, he led one of the best troops in the Commando.

At this point in the jollifications it was easier to believe the sky would fall than that any of us might one day be killed. Officers require good nerves; more important, the ability to act correctly and without hesitation in a crisis.

We were a lucky Commando. As raiding became more dangerous, the confidence of the old hands, who scorned the opposition, was something to be marvelled at. Courage takes many forms. It can wear away rapidly when pinned down by shell fire and machine-guns, or by continuous dive-bombing. I knew some, but very few, who appeared unconscious of danger. The obvious answer: such people were either stupid or lacking in imagination. That may be true; but there are those who can discipline themselves to fearless courage and be intelligent as well.

To sort out intentions and achieve results there are certain formulas which transcend individual merit. Pick leaders; win their confidence; work hard to test initiative and adaptability. What is learnt in training is done instinctively in action – without thinking, down to the last gasp. Courage looks after itself, for habit is ten times nature.

Let me give an example of reflex action. Long after the war, Daryl Zanuck asked me to provide technical advice on minor activities during the Normandy landings in the film *The Longest Day*. On arrival at Caen, Zanuck laid on a banquet preceded by a dishy cocktail party. I misjudged the starting time next morning and overslept. A lunatic taxi driver, who drove like the wind, proved to be the proverbial stranger from Paris: on the outskirts of the rebuilt city we missed a turn to find ourselves on the wrong side of the river. I must have dozed off. There was a scream of brakes and an astonished 'Zut alors' from the cabbie: up the road men in field grey uniforms adjusted coal-scuttle helmets. I was out and diving for the ditch before the chauffeur pulled up. That happened ten years after the war. So much for force of habit. The sequel was more surprising. A smiling 'German' dusted me down: 'Come, come, this will never do! Getting a bit rusty aren't we, sir? Keep practising that shoulder roll or somebody's going to get hurt.' The speaker was Corporal Powell. He had been our PT instructor in No. 4 Commando, now turned stunt man.

Did I think I might be killed? When young and strong the answer was 'no'. One accepted the risks of course, but the odds did not stop the right course of action, any more than riding into the last fence on a tired horse or going for the corner flag. Though death had no attractions, and would distress a happy family, I regarded such risk as the luck of the draw. People who enjoy life are seldom troubled by thoughts of departure; having taken the precaution to squeeze the lemon dry they do not grudge to throw the rind away.

My evening was saddened by John Durnford-Slater, who came over from Largs to say that there was no vacancy for a captain in his commando, which I genuinely wished to join. The party continued all night. Charles Vaughan was found asleep, sitting slumped in a telephone-box in the Marine Hotel lounge. By breakfast I had decided that staff duties were not for me – still less a second dose under new management. I had not forgotten Julian Grenfell's lines:

> Oh Lord who made all things to be
> And madest some things very good:
> Please keep the extra A.D.C.
> From horrid sights and scenes of blood.

It was time to move on again. My luck was in: a forty-eight-hour pass and, rare surprise, sleepers available on the London train, a luxury seldom enjoyed in 1941, except by VIPs. At Combined Operations Headquarters in Richmond Terrace, the admiral, with Geoffrey Congreve, was busy at his desk answering congratulations. The success of the raid had exceeded expectations, and he seemed pleased with our performance. Although his shoulders drooped and he looked spent and old, Sir Roger had proved his point, and the War Cabinet had been kind. He was lunching that day with the Prime Minister, and I quickly said goodbye.

I saw Sir Roger only once more, when, on 27 January 1943, flanked by the redoubtable Ginger Boyle (Admiral of the Fleet Lord Cork and Orrery) and myself in scarlet robes over our uniforms, the new Lord Keyes of Zeebrugge and Dover was introduced to the Lords. The House was crowded and the public galleries were full. The ceremony of sitting down, standing up, taking off hats and bowing from different sides of the chamber, with the final handshake at the Woolsack, is normally a subdued affair. The low murmurs of 'hear, hear' from the assembled peers, after the newly created member has a word with the Lord Chancellor, are a formality. Not so on this occasion. There was a war on; it would take a lot of winning and the future appeared grave. There was not much to cheer about, but the reaction was spontaneous: both admirals had made the name of England great at sea, and Keyes, in tears, was given a standing ovation as he left the chamber.

After Pearl Harbor, Lord Keyes somehow got himself to the Pacific theatre, posted as an observer in the United States Navy. He lived to know his eldest son had won the Victoria Cross in the desert, killed in a raid behind the lines, attempting to bushwhack Rommel's headquarters.

Geoffrey Congreve went back to Beaufort to pick up his poodle, which Rosie had been nursing through a bout of distemper. I never saw Geoffrey again. He was killed in a little raid with Philip Pinkney on the French coast; he was the only casualty. There were two VCs already in his family, and he was determined to get a third. Geoffrey was ordered not to go ashore that night, but a machine-gun got him just the same, waiting off the beach in a landing-craft.

Somewhere in a scrapbook I have a photograph of this gallant sailor riding his horse 'Lazy Boots' in the Grand National. He is jumping the big Chair Fence in front of the stands at Aintree, the only bearded jockey in the field, and it shows up well, for his head is thrown back against the sky.

Geoffrey always held his head up. I loved both him and Sir Roger for the same qualities: guts and integrity. 'Without courage there cannot be truth, and without truth there can be no other virtue.'

CHAPTER THIRTEEN

CHOSEN FOR DIEPPE

Singing is the thing to keep you cheery.

HARRY LAUDER

After Keyes retired and Mountbatten took over we went back to
Inveraray to join a 'Puma Force' of three brigades and began training
seriously as part of an expedition planned to seize the Canary Islands.
'Puma' became 'Pilgrim' – then grew, like Topsy, into 'Leapfrog',
becoming unwieldy in the process. The project terminated with a rehearsal in the Orkneys that went adrift in timing and navigation. The
landing took an age, as did the arrival of follow-up equipment, which
became so muddled in the control ship that the expedition was
abandoned.

There followed a period of uncertainty in which our future hung in
the balance. I have said there were growing pains. These took the form
of bickering behind the scenes while chair-borne warriors in Whitehall
engaged in paper warfare. This waste of talent was forced upon us not
by shortages but by a lack of enthusiasm in high places.

The Chiefs of Staff were not our only foes: the regular army considered Special Service troops were a slur on the efficiency of more
conventional soldiers.

We had two friends at court: Sir Roger and Winston Churchill.
The old admiral – he was sixty-eight at the time – went out in a shower
of sparks with a speech in the House of Commons (he was MP for
Portsmouth). The War Cabinet heard they were in for trouble. As a
precaution Sir Roger was sent a copy of the Official Secrets Act, with a
request to return all confidential papers in his possession. The suggestion was disregarded. Putting on the many-ribboned uniform of an
Admiral of the Fleet, he fired his last broadside. Keyes never minced
his words. It was a blistering attack. For a long time he had been at
variance both with the Cabinet and the War Office. The staff of the
three Services, in his opinion, had not been a help but rather a

hindrance. At all times he had been frustrated by official caution; the planners had risked nothing greater than a series of footling imbecilities. It was time a less duck-hearted policy was evolved to make use of the splendid body of young men who had been under his command.

'After long naval experience,' he concluded, 'which had recently embraced the appointment of Director of Combined Operations, I must fully endorse the Prime Minister's comment on the strength of the negative power which controls the war machine in Whitehall. Inter-services committees and sub-committees have become the dictators of policy instead of the servants of those who bore full responsibility; by concentrating on the difficulties and dangers of every amphibious project,' said the admiral, 'the planners have succeeded in thwarting execution until it is too late. Procrastination, the thief of time, is the key-word in the War Office. We continue to lose one opportunity after another in a lifetime of opportunities.'

Back at Troon, I joined No. 4 Commando as a supernumerary captain attached to B Troop, with John Hunter, a gunner, in command, and two subalterns, Gordon Webb and Peter Wilson (who were of the same persuasion), to assist him. The artilleryman's motto, 'Horror ubique' (or 'spreading terror everywhere') was endorsed by Troop Sergeant-Major Woodcock, a Sherwood Forester who came from Nottingham and played cricket for the Minor Counties. Though refusing to take a commission, 'Timber Tool' was to go on deservedly to higher things.

It was a good troop and all ranks handled themselves well. Peter, who had lost a brother at Dunkirk, was killed on D-Day, but most of the men stayed united for the rest of the war. One cause for concern was the commanding officer. Dudley Lister suffered from wife trouble (the old story), with a girl in every port, and it preyed upon his mind. The violent moods – never predictable – began to make inroads on his health and credibility. We were friends, which made the outbursts more embarrassing when I got promoted; a tactful second-in-command should help to leaven the loaf, but Dudley was an officer who never climbed down or apologized.

Charles Vaughan's posting to the depot hastened these withdrawal symptoms. The dependable old soldier had a soothing influence on a man who admitted he lacked self-confidence. Lister proceeded to quarrel, for no good reason, with Mark Kerr; then, in turn, with three out of the six troop leaders – Emett, Montgomerie and John Hunter – all good men who returned to their regiments. Then came a scene with Mr Pettigrew, the urbane manager of the Marine Hotel, over the disappearance of a teapot in which Lister's batman presented the early morning cuppa. This tiff brought about a curious persecution complex: nobody loved him any more! So he took to the wild wood.

Weather on the Firth of Clyde is cold at the best of times, and more so with winter fast approaching. Dudley retreated to a bunker on the golf course, where he pitched his tent beside the fairway, to live in defiance of convention – cooking on a Primus, sleeping in a flea-bag, and freshening up with a plunge into the sea. The golfers naturally complained and the harassed colonel, moving closer to the shore, dug himself into a sand dune which provided shelter from the winter blast. Here he was joined by his shivering batman. It was all very well for Dudley, with his passion for physical fitness, but poor Smith, a well-educated Yorkshire lad whose youthful appearances belied a serious soldier, had a lot to put up with.

The colonel liked to be called 'with hot, sweet tea': Smith's day began early. His next duty: standing by in gumboots with sponge, towel and a torch in case things got lost as his master snorted naked in the waves. That was not all. When they waded ashore the frying pan came out: Dudley expected a big breakfast.

The CO took proper pride in his appearance: he made an imposing figure on parade – which was why he got chosen in the first place. How Smith turned him out so well – working on boots, buttons and Sam Brown by the light of a hurricane lamp before daybreak – remains a mystery. The boy did his training with the rest, though excused afternoon duties to prepare the mess-tin supper. For this Dudley liked 'little surprises', which took the form of baked beans, field mushrooms – even a Welsh rarebit or sardines on toast – wolfed down as darkness fell. But the two men got on well together, for Dudley Lister, in the right mood, had charm and a sense of humour. After he moved elsewhere Smith elected to stay with No. 4 Commando, taking care of my new second-in-command. Wounded in two campaigns and mentioned in despatches, this Admirable Crichton completed his commando service with credit and will be remembered as a stout man at arms.

So the winter of 41/42 passed by. Old Dudley's behaviour had not escaped the notice of the brigadier. We were given a series of collective training exercises. The troop leaders acquitted themselves reasonably well: HQ were seldom in the picture.

Then luck and promotion came my way, first to second-in-command and after the Boulogne Raid (St George's Day, March 1942) to lieutenant-colonel. Imperceptibly this shake-up produced effects which, in the long term, made No. 4 Commando pre-eminent as a fighting unit. I attribute that improvement to the adoption of three basic principles: 1. I cut the dead wood in officers, NCOs and men; then looked for replacements of proven ability, at the same time accelerating promotion inside the Commando at every level of junior leaders, whose talent remained wasted while hanging about in Troon. 2. Individual troops were told to get lost on commando for not less than a week at a time, after submitting a training programme which outlined whatever form of specialized activity most appealed to their respective skills. 3. Short cadre courses began in weapon training and section leading for NCOs who were not infantrymen; this applied to many in the unit, for we were still the Cavalry Commando. But the best news came from the Depot at Achnacarry: to 'Come and choose the cream of the first intake – fifty policemen recently called up, men of intelligence – six feet tall – all drafted as guardsmen with some experience of spit and polish on the barracks square.' Charles Vaughan had not forgotten his old friends: I took the lot; then went out to find some dashing leaders to command them.

General von Hammerstein Egord has been credited with the observation that officers are divided into four categories. There are those who are brilliant and industrious: such men are suitable for the highest staff appointments. Then those who can be brilliant, but are lazy: these will rise to the highest level of command. Use can be made of officers who are stupid and lazy, but those who are stupid and industrious (and plenty came that way) should be ruthlessly eliminated. My selection was confined to the obvious category and I got some beauties, headed by Derek Mills-Roberts, who became my right-hand man.

Dudley was posted to South Wales. There he found a niche for which he was admirably suited: forming No. 10 Inter-Allied Commando, a league of nations whose troops includes Poles, Dutchmen, Belgians, Norwegians, Free French and some very fine Sudeten Germans who enlisted under assumed names. All were dedicated patriots with a personal hate against a common foe. Lest we forget, someone must surely tell their story to help understand the fight for freedom. Most of those heroes rest in lonely graves. In peacetime this love of country can be forgotten; some would have it deliberately destroyed.

I had gained experience under two masters of conflicting character and learnt something from their whimsies. When Dudley fought for the 'golden gloves' in Madison Square Garden, he went into the ring determined not to box, but to inflict grievous bodily harm on his opponent. 'That's how champs are made, old boy.' Unfortunately he lost the fight to a more scientific opponent. He showed the same blind courage during the ability tests set by brigade headquarters which brought about the inevitable downfall. Here is a physical example.

It was a brute of an exercise: a night scheme and speed march across rough country in the north of England to try the stamina and efficiency of commando O groups (regimental headquarters), which included the colonel, second-in-command, adjutant, signal and intelligence officers, medical officer, wireless operators, runners and protection section, in full battle order over a ten-mile course taken on compass bearings, in hill country.

Brigade, at Castle Douglas, rang me up for advice on the route to be set. I suggested crossing the Dungeon of Buchan in Galloway (chancing the bogs of the Silver Flow near by), or the steep hills on the Border – Oh Me Edge, Dour Hill, Hungry Law and the surrounding crags. Then, as an afterthought, I suggested a tilt at the bad ground in Charles Kingsley's charming story in *The Water Babies*, where the little chimney-sweep fled across Harthover Fell and dropped down over Lowthwaite Crag into Vendale, pursued by the squire on his shooting pony, and his keeper with the blood-hound, which puzzled out the cold scent until they reached the cliff 'where the party halted and the hound with a throat like a church bell pulled up defeated; then lifted his mighty voice and told them all he knew'.

Brigade rang up again to say there was no such place! We left it like that. A week later the competitors met at Alston for the contest. A staff officer had taken three hours to walk the course in daylight. No. 4 Commando covered the distance, helped by a moon, moving at the double on the downhill slopes, in two-thirds of that time. We won by twenty minutes, though brigade signals put up a good show in gym shoes, without equipment. They were disqualified.

Dudley registered a stout performance. We had chosen the best hillmen in the Commando to lead the point element; they kept far ahead to blaze the trail, taking quick bearings not to slow the party down. Two Lovat Scouts – followed by Tony Smith, the IO, and Robert Dawson, an iron man on the mountain – set a cruel pace. Doc Wood was a good walker, but the rest were in bad shape when we finished 'spot on' a deserted cottage in the fells. Dudley's legs gave out a mile from home, but a toggle rope was passed beneath his oxter to haul the exhausted man up inclines and through deep heather; supported by his batman, he reeled in with blood in his boots. He was off duty for some time afterwards.

Troon was a station with every training facility, excellent billets, well-disposed civilians and friends on the Clyde coast. No. 3 Commando were at Largs, No. 12 in Ayr, No. 6 at Saltcoats and No. 1, who were strangers, at Irvine and Kilwinning. The summer evenings were enlivened by salmon poaching with Philip Pinkney and Anthony Mildmay, who lived off the fat of the land, staying with the Craigs, the well-known livestock auctioneers in Ayr. Philip, helped by Cool, an Irish water spaniel who retrieved fish, had a Welsh batman, an artist with a cleek; should that fail, Philip carried a home-made, volatile depth-charge in his pocket, or a .410 thrust in season down the leg of his battle-dress trousers. There were adventures fraught with greater hazards: John Durnford-Slater seriously entertained the idea of blowing up the German Embassy in Dublin. The plot was stopped at the eleventh hour by a furious brigadier.

Philip was accident-prone. In a night attack on a guarded ammunition dump, someone failed to alert the defenders of his Commando's intentions. The alarm was raised; volleys of tracer bullets sprayed the area from behind barbed wire, for the guardroom had turned out smartly. That did not stop Philip, who acted quickly before men got killed. He crashed the wire and overpowered the nearest sentry, shouting, 'You bloody fool, we're playing soldiers!' He got a bayonet thrust through the upper arm as he stood off another assailant; the man hesitated before shooting him; then the battle quietened down. The sentry was afterwards promoted to corporal, and Pinkney – strong as a bull – went into casualty ward to have the puncture repaired.

He was soon in trouble again. No. 12 Commando made a diversionary raid (Operation 'Anklet') up in the Arctic Circle (north Norway), to divert attention from the attack mounted successfully by No. 3 at Vaagso; there they sank sixteen thousand tons of shipping, destroyed factories and killed Germans.

Up the coast, Philip's party stayed ashore and were hospitably received by inhabitants who thought the occupation was over. There was toasting in snowed-up homes and Union Jacks fluttered over little houses. Mountbatten, not averse to publicity, had supplied the raiding force with a camera crew and press reporters, to embellish the goodwill story. Great was the consternation in the fjord among civilians who had committed themselves when the troops re-embarked. Greater still the dismay when Lieutenants Pinkney and J. B. Jeffries seized the movie camera and film evidence recorded in what they considered a shabby, almost treacherous deal and chucked the whole lot overboard.

The officers were put under open arrest when they returned and later presented with a bill for £175. I doubt if this was ever paid in full. The last time I saw Philip – before he was shot in Italy – he ignored claims by a cameraman for the price of a new shirt considered necessary to appear in attendance on Mountbatten.

A good soldier – Montgomery was a notable exception – should avoid publicity and keep his mouth shut. The press did not help the commando image with its presentation of reckless, devil-may-care fellows who acted in a slap-happy and generally irresponsible manner. The public lapped it up. Nothing could have been further from the actual truth. Unorthodox, yes; but there were no short cuts to eventual proficiency. I never took photographers, and only one war correspondent, on any operation. Austin, who was a friend and physically up to the task, was given a rifle at Dieppe: later he was to get himself killed, poor fellow. Those who talked to the press got the sack. One imaginative chap described how he crept ashore in carpet slippers and had, on occasion, saved my life on dark nights (very dark) by swinging a loaded cosh. He was a good officer, but he had to go.

Some people do not learn lessons. Before the Pinkney incident No. 6 Commando, then inexperienced, sailed for Norway on Operation 'Kitbag', with cameras and reporters. They never landed. While hand grenades were being primed on the boat deck an inquisitive news sleuth momentarily distracted a man's attention as the short fuses were screwed home. A box on the table was moved to get a closer picture; a live grenade rolled off without its safety pin. Someone dived to snatch it up and throw it clear but missed the port-hole, or hit the rail, and the bomb bounced back; seven soldiers were killed outright and others wounded. The ill-fated expedition returned to base without making landfall. A case of full steam astern!

High explosives have a fascination for men in uniform stronger than the pleasure children take in fireworks. At Kelburn Castle – Lord Glasgow's stately home near Largs – a stag-headed oak tree showed signs of butt rot and decay. It stood in full view of the drawing-room windows, a bow-shot from the house. Could Durnford-Slater's sappers remove the skeletal branches? Certainly: John was delighted to oblige; this was just the kind of practical training his men required – in which, indeed, they excelled. Paper and pencils came out: calculations varied on the amount of 'gelly' necessary to topple the timber; they erred on the generous side; so deep was their concentration that no one thought of opening windows in the castle, three hundred yards away. The deafening bang was accompanied by a tinkle of broken glass. The oak disintegrated, and with it went the windows on the south elevation. But commando soldiers were resourceful. The next day tomato growers, miles away in the Clyde Valley, noticed the best squares of glass appeared to be missing from their greenhouses. The bush telegraph had gone out, rounding up available tradesmen in uniform, and Kelburn was miraculously restored.

A serious folly occurred at Troon during the hanging-about period in the Lister régime. The sapper section was officered by an individual who shall be nameless: untidy and foul-mouthed, it was difficult to understand why he ever wanted a commission. I had twice sent him off parade, to get shaved and to show a hair-cut to the adjutant. He held his place as one of the three rugger blues playing for the commando; they helped boost our ego by winning away matches. On the third offence it was decided to get rid of him. The man could not have been quite normal. The same afternoon, smarting with resentment, he drew, without authority, all available high explosives from the quartermaster's store, loaded them into a truck and proceeded to blow up the most conspicuous landmark in the neighbourhood: a tall factory chimney, some 100 feet high, in a disused brick kiln behind the town. There was hell to pay: children had hysterics; plaster fell from ceilings; cows aborted; a mare broke her leg. Ayrshire was indignant. A court of enquiry was held pending the inevitable court martial. This the accused avoided by transfer.

The force chosen for the St Nazaire raid, then being planned, was short of sappers, needed to destroy dock installations and the U-boat pens then vital to the enemy. The offending officer was added to the demolition team. We were not sorry to see him go. He shared quarters with Stuart Chant-Sempill, a friend who went on to gain a gallantry award and was wounded and taken prisoner. Our own representative sailed with the strongest premonitions of a violent death; during the long sea passage he described to Stuart the exact circumstances in which he would be killed. The sceptics may scoff, but so it turned out exactly as he foretold. I have known of similar cases.

Otherwise we prospered. The year spent in Ayrshire marked a turning-point: No. 4 Commando emerged from being a rabble in arms to turn slowly into a polished weapon. At Troon every man learnt he had to work to keep his place in the team. If we were not experts in the arts of war, it could be said that No. 4 at last measured up to what Churchill described as 'the canine virtues of vigilance, fidelity, courage and love of the chase'. The Dieppe raid was only just round the corner.

I started to enjoy some aspects of the war. Fiona was born the week of my thirtieth birthday, and that is the age when a field officer should (in a shock troop formation) be at his peak. Rosie came down from Beaufort with the two infants, Nanny, a deer-hound (that supplied the jugged hares) and six laying hens. A weekly hamper of garden produce supplemented the nations. We shared a house with Isabel and Loel Guinness, who commanded a fighter squadron (601) at Prestwick.*

* Fiona's arrival gave me a close call. Rosie telephoned from Eilean Aigas to say the baby had started. It was the weekend. Prestwick to Kinloss would take half an hour. Loel had a spare Beaufighter. 'You can start early, but return before dark. This chap is still learning.' It was

In the west the German war machine, turning aside, now rolled its might against Russia. By 1942 Hitler's vaulting ambition showed signs of over-reach. Full advantage was taken of the respite: on the home front new factories went up; tanks and aeroplanes, guns and ammunition poured off the assembly lines. There seemed a chance of winning after all.

As we ran and leap-frogged, up and early on the beach, the rising sun caught the peaks of Arran in a noose of light; the same brightness glinted on the wings of United States Liberators. The big bombers were greeted with a cheer, as they droned in after flying the Atlantic. The Flying Fortresses came later. The Americans called them 'four-fan destruction ships'; they looked enormous and we were glad to see them.

A good story came back from the French colonial empire by way of Roger Courtney, who ran the Folbot and canoe section; he had spent years in coast ports, where pidgin English was spoken. 'Steam chicken topside drop plenty no good shit' was the West African's verdict after an air raid on Dakar.

Recreational training was an integral part of daily 'strength through joy' activities – especially when in Troon's rest billets.*

> Hill you ho boys
> Let her go boys
> Bring her head round
> Now together
> Hill you ho boys
> Pull for home boys
> Sailing back to Mingulay.

There were heavy cutters and ship's whalers in Troon Harbour: good boats in all weathers and preferable to the folding goatleys in which we paddled stealthily ashore on exercises by night. For sporting purposes they took eight- and ten-men crews, the picked pride of each troop. Seated on double thwarts, the Commando competed over a two-mile course, pulling round Seal Island in front of the Marine Hotel.

---

raining, with low cloud over the Firth of Clyde. This changed to dense haar, rolling in from the North Sea as we shot across country. The pilot, who was having trouble with his compass, could not get above it. Then he panicked. We were far out over water before getting a blink of sun coming from the wrong direction. After rapid calculations, we turned and landed thankfully at Donibristle in Fife. Weather grounded the aeroplane for twenty-four hours. I hitch-hiked back to Troon. It took all day. Fiona did not appear for another week.

* The men found their own accommodation, each paying the landlady a subsistence allowance of 6s. 8d. a day (or 36p in present terms), with two meals thrown in. Officers received 13s. 4d. I think we made good impressions. By the end of the war 168 soldiers in the Commando Brigade – many coming from south of the Border – had married local lassies.

A strongly built Hebridean, Nicholson by name, from among the police intake won the races. With him rowed a man from Eriskay (or was it Barra?) who sang in the church choir at Sunday Mass. A Teuton spectator might have been surprised to see happy soldiers stripped to the waist and straining in contention at the oars. The short chopping stroke of the winners was rowed to a Gaelic chorus, tossed back at the helmsman, who stood, wild with excitement, in the stern, as he called the verses that once drove the Biorlinn of Clan Ranald, foaming through the western seas. I asked Nicholson for a translation after he won a crate of beer. The answer was entirely Celtic: 'The words are very strange. They were handed down; my family have always been sailors or fishermen. The lines – well, I do not know their meaning, but when the boat crew hears them it makes us twice as strong.' So be it. With spirit roused, such men are hard to beat and will be found in the forefront of the battle. In the ills that now beset us, the cheers of the whale-boat crew ring back across sunlit water. Is it determination to succeed that is missing on the shop floors of industry? Where is the helmsman when less work and more pay are now the order of the day?

Someone tactlessly asked the actor, James Robertson Justice, who wore naval uniform (while his ship was being repaired), if he happened to be the harbour master. For a big strong fellow – later cast in tough-guy rôles – our little town seemed a curious backwater. Justice (so called before he added the Robertson), a boastful talker in his cups, gave an unsatisfactory reply and had his trousers scientifically removed by a small officer. Thereafter boats were less available.

I have touched on the improvement after troop leaders got their men out on specialist training. Robert Dawson was a mountaineer as well as a fine officer. Educated in Switzerland, he started rock climbing as a boy. He now took his troop to Bethesda in North Wales with David Style, another star among the subalterns. They put the fear of God into the rank and file – most of whom had never abseiled – swinging down a cliff or clawing with fingertips up a rock-face. Tommy Handley, of ITMA fame, ran his BBC programme from Bangor and sent a bus every week to take men to see the show; this kindness helped to soothe their nerves. Such is the force of example, that all emerged within a fortnight as useful climbers. Dawson's thinking was to get us chosen for the Dieppe assignment.

We owned two Jeeps, three pick-up vehicles, four 15-cwt trucks, and plenty of railway warrants. Men raiding a power station at the Clatteringshaws one day in Galloway could find themselves on the next cutting snow steps in the Cairngorms. Other troops specialized in alternative rôles. There was a wide choice: sailing in the Hebrides; forced marches and night attacks on a range of targets by arrangement with the army, the air force or the Home Guard; sabotage and piracy. As Glasgow got blitzed, so we came to know that mighty city. Street fighting over roof-tops in ruined houses, and mouse-holing with a pickaxe from room to room, and window sniping were added to activities. The Lord Provost, Sir Patrick Dollan, headed a socialist council. Today he would be considered a blue-blooded conservative. Soldiers and sailors were shown every kindness and consideration during his term of office.

War-torn, grimy Glasgow had a generous heart (less easily detected in the remote gentility of Morningside), and commandos showed their appreciation. We did our best to add some colour. The smartest men marched round George Square in fund-raising and appeal parades: the strongest performed feats for charity at sports meetings at Hampden and other famous grounds. Our prowess in athletics (much enjoyed by the general public) was exceeded by commando boxing and milling contests. The last-named was a desperate affray, a test of courage that drew full houses and wild applause. A dozen pugnacious assailants from each Commando, not chosen for corresponding size, weight or skills, entered from opposite ends of the ring to fight toe to toe for sixty seconds in a flurry of blows. Then the whistle blew and a new pair of gladiators sprang into the arena. No shaking hands or any kind of sparring. No pause. No referee or seconds. The vanquished were hauled out by their heels. The escalating scrap went on regardless. The volunteers fought for a team and not as individuals and with no thought of defence. It was remarkable how the smallest men, milling furiously in a hurricane punch-up, would fell a more gentle giant, who seemed to take longer to warm up.

All work and no play makes Jack a dull boy. Keeping soldiers interested by variety is as important as serious training. All good men enjoy a laugh. Here we differed from the enemy. German front line troops commanded great respect; with the right leadership they were superb. But the Teuton private lacked initiative, and we possessed two assets they could not imitate: the unconquerable spirit of the British soldier (who is at his best when really up against it), and that unfailing good humour which transcends every misfortune. If the Wehrmacht – the fine flower of the 'master race' – had witnessed the inter-troop Morris dancing and singing on the march competitions (Gordon Webb was disqualified for concealing a gramophone in his rucksack), or attempts to play water polo in Troon's oil and dead cats harbour, they could well have thought, 'Denn wir fahren gegen Engeland' is a pushover!

The Second World War failed to provide great marching songs among Allied armies or, for that matter, any epic poetry. 'Roll out the

9 and 10 Rosie and
I during the first winter of
the war, Weedon 1939

11 Salute to the Faroe Islands, a Lovat Scout piper in marching order

12 An LCA takes off the wounded during a destructive raid into Norway. The return journey could be cold and painful

13 Keyes and Churchill: 'Give us the tools, we will finish the job'

14 and 15 Rival commanders: LEFT Field-Marshal Gerd von Rundstedt, Commander-in-Chief in the West; RIGHT General Adrian Carton de Wiart, VC, 'a happy warrior'

16 Dieppe, the main landing: a scene of carnage with a Canadian division dead, broken-down tanks and burnt-out landing-craft

17 Return from Dieppe: some black faces and walking wounded on Newhaven Pier – Gordon Webb (in sling), Len Coulson and myself (pointing) with Fairy Veasey in the background

Barrel' and 'Waltzing Matilda', of older vintage, are borderline cases that will be remembered; 'Lili Marlene', made famous by the Afrika Korps, stands above them all. For their songs and their snipers the German army get full marks. In a war of movement, it is a paradox that there was so little marching. Singing picks up tired feet, and soldiers have been known to drift miles in their sleep to the tap of a drum. The poet requires detachment not easily found in a motorized column. The Commando had four trucks to move stores and ammunition for five hundred men who all came under starter's orders, so they had to use their feet. But they could not sing the way I wanted. Len Coulson preached a good lay sermon and, being a Geordie, could bellow 'Blaydon Races' – but only at a party. The French Commando sang marching songs on all occasions, like their fathers before them – 'Madelon', 'Saint Cyr garde à vous' and 'Les Artilleurs de Metz'.

Fairy Veasey and Len Coulson, each a fine subaltern, were men of magnificent physique. Coming out of the Central Station to pick them up for an athletic meeting, there was a hold up. The cortège of a civic dignitary was passing by, the hearse loaded with wreaths; the crowded street stood silent as it passed. Both officers towered above the Glaswegians, right arms rigid in a stiff salute. Joining them through the press, I enquired in a whisper who was dead? Veasey's humour never deserted him. He did not see me coming up behind him. The answer came through the handlebar moustache: 'If you ask me, sir, I think it must be the bloke beneath all those flowers!'

Some of the jokes were not so kind. Evelyn Waugh was cordially disliked by every combatant officer in the brigade. I had known him vaguely at Oxford and, while I admired his literary genius, had marked him down as a greedy little man – a eunuch in appearance – who seemed desperately anxious to 'get in' with the right people; then proceeded to mock the gushing hostesses and bright young things who befriended him. If Waugh bit the hand that fed him, he was also the world's biggest snob. He did not always get away with it; Rosa Lewis could spot a social climber at a distance. She was not amused by descriptions of the Cavendish, or of herself in the character of Lottie Crump, in *Vile Bodies*.* At a Daphne Weymouth party Waugh entered and sat down uninvited in the Eleanor Glyn sitting room, the inner sanctum reserved for special friends. 'Take your arse out of my chair and get your body out of my hotel' were the brisk marching orders for this infringement. Belloc, on making Waugh's acquaintance, is said to have remarked, 'That man is possessed', which may be putting it on

---

* Great exception was taken to Waugh's appraisal of the hotel's celebrated game pie: 'full of beaks and shot and inexplicable vertebrae'.

the strong side. But there was a malignant element – it was worse than spite – that I had not previously encountered in a grown man. Having brought about his dismissal, it seems incumbent to give a military appraisal of this misfit personality, who joined through the back door.

Waugh left the Royal Marines after failing to control a platoon who refused to obey orders; veiled sarcasm – which was of a high order – had infuriated his immediate superiors. Befriended by Bob Laycock, he entered the Household Cavalry; then was passed on to our brigade staff with the rank of captain. This promotion was resented by officers who risked their lives cheerfully and had willingly lost seniority to join the commandos. Waugh made no attempt to train or learn his trade; when he emerged from protective custody it was in 'shit order' – that is to say, without equipment, belt, boots or anklets, for which suede shoes, a cigar or a glass of wine (drunk in a staff car) were substituted.

If asked what made a Commando function, which in the absence of a sergeants' mess (the true power-house of any regiment), I would unhesitatingly reply 'The total commitment of all ranks to a best personal performance – good, bad, but never indifferent.' The officers in particular strove to excel; many succeeded. For most, myself included, the war meant a welcome freedom from class, money or position, in a close fellowship of total involvement. Good subalterns are tolerant, and mine were not well off. The newcomer was judged on his merits. If an officer became unpopular the verdict was usually correct. Waugh played his cards badly; condescension is no substitute for common sense. The man misread the strength of public opinion and in the end his weaknesses found him out.

In diaries edited by Michael Davis, the entry dated 13 November 1940 (it marked his first appearance) made no secret of the part he wished to play in a comfortable berth: 'Took the morning train to Largs ... where Bob Laycock failed to recognize me. There is no place for me in the troop so I was made liaison officer with Harry [Stavordale] – a life of untroubled ease.' On the next pages he goes on to say, 'We are the only Mess in Europe which constantly drinks claret, port and brandy at dinner.' This had no bearing on life in a good Commando. Then he had this side swipe at friends, now dead: notably Phil Dunne, Peter Milton, Randolph Churchill, Dermot Daly, Bones Sudely and Toby Milbanke (there was also a defunct hotelier, Basil Bennett, who attended to the requirements of Waugh's high-protein diet): 'The smart set drink a very great deal, play cards for high figures, dine nightly in Glasgow and telephone to their trainers endlessly.... There is no administration or discipline.' The whole thing sounds like a delightful holiday! A nice, loyal fellow – though hardly one to criticize! But Nemesis was stalking at 'Uncle's' elbow.

In December Laycock's Commando, with two others, moved to Arran to train for Operation 'Workshop' (Pantelleria). They lived partly on the island and later on the troopships *Glen Roy* and *Glen Gyle*. I was serving on Roger Keyes' HQ and was present when security reported there had been careless talk by soldiers who paid visits to Glasgow. No. 8 Commando were considered the worst offenders. Waugh, on his own showing, had nothing to do; he was accordingly told to censor all outgoing mail prior to Layforce's departure. Godfrey Nicholson* (who had an impish sense of humour), helped by David Stirling and myself, decided to pull Waugh's indolent leg.

I do not remember the text of the missive addressed to a side street in Zurich. The communication ran something like this (it might have been more subtle):

Herr Schmidt, c/o Vogel Von Bumelzeug,
6 Barkhausen Strasse, Zurich.

Dear Schmidt,
Time is short, and I will soon be on my way to the destination already referred to. It is some weeks since we met and I hope that, according to plan, you have reached this address in a neutral country. The Dutch route is dangerous and must be discouraged. I have said enough. On this chance I am writing. It is not possible for me to see A, but if all goes well I hope to meet B where I am going. He will keep you informed.
Gustav sends greetings.
[Signed] No. 89706,
            BOMBARDIER HILDEBRAND HARDCASTLE
            IN TRANSIT

Next day Waugh looked green and shaken by suppressed emotion. The *Glen* boats were sealed and Red Caps came on board. A search began for Hildebrand: the suspect was nowhere to be found, nor was his name listed among the passengers. Waugh could not laugh at himself and was not amused when the hue and cry died down and the joke became obvious. He sailed with Layforce to the Middle East: once was enough for all concerned. Two years later Bob Laycock went to Italy, leaving me in charge of the brigade. His side-kick stayed behind, in high dudgeon, with the half-promise he would be sent for later.

Then his father died, and I granted Waugh compassionate leave to settle his parent's affairs, with instructions to report on completion. After two weeks Evelyn had not shown up; he was said to be passing the time at White's. Summoned and asked for an explanation, Laycock's

* Sir Godfrey Nicholson, MP.

staff captain had nothing to say: he was handed a railway warrant
and ordered to the Depot, where I suggested some training might not
come amiss. There was a painful scene. Waugh could not contain himself and stormed unannounced into General Haydon's office. Nobody,
said Waugh, could interfere with the arrangements made for his
departure overseas. It was a rash speech. He was sacked on the spot
for insubordination. An appeal to Mountbatten on the grounds of 'unfair treatment' failed abysmally.

I have dealt with our daily round. Intelligence was not overlooked.
Tony Smith had a smart section led by Sergeant Ken Phillott,
an amusing character who still makes the best after-dinner speeches
at reunions. With Lance-Corporal Lansley he used to bluff his way
into the offices of local regiments and remove documents left lying
about marked top secret. The pair also enjoyed putting on German
uniforms to bluff their way, as escaped prisoners, about the country
without detection. Thus attired, at a dance in Ayr, they were asked
if they happened to be Poles by the Military Police! I have a letter
from Phillott that reads:

My record book may be of interest to indicate the variety of specialist
courses I enjoyed, over and above the normal qualification required for a
trained soldier. These included a driver mechanics course; a snipers course;
signal course; marine navigation course; railway engine-drivers course; and a
mountaineering course. And believe it or not I was even sent on a cooks
course. I was already trained as a machine gunner and dispatch rider. On
occasion I took a turn as orderly room clerk. I can say, with truth, that no
other unit could keep up with the officers and men of No. 4 Commando.

Now the Commando was wound up and ready for action: men, like
good horses, take time to reach concert pitch, but they cannot be kept on
the boil. We had not been in line for Vaagso. I smarted a little not to
be chosen for St Nazaire in the spring; Derek pointed out that if not
killed in that encounter, the men of No. 4 Commando would be behind
bars in Stalag VIIIB, instead of Charles Newman (who had won the
Victoria Cross). But it was time to prove something. We heard vague
rumours of a big Canadian operation called 'Rutter' that had been
cancelled earlier in the month. The port of Dieppe was said to have
been the target. Why had we not been given the assignment?

In most Commandos (except for Nos. 3 and 4) wastage was continuous and replacements increasingly difficult to find. Our brigade
strength was one thing on paper, quite another in practice. Layforce
and two of its Commandos, Nos. 7 and 8, had been disbanded in
the Middle East. No. 11 remained in the Mediterranean. No. 5, hard
hit by fever, was in Madagascar. No. 12 had broken up to form a snow

and mountain warfare school, while other personnel (in fact, most of No. 12's best officers) took to boating in the Channel, then left to join the Special Air Service in North Africa. No. 2 was in the bag. We were growing thin on the ground.

That was the picture in late July 1942 when Bob Laycock, who had replaced Haydon, arrived hot-foot from Ardrossan. There was no warning of the visit. He found me perched in the ruins of Dundonald Castle with a group of selected marksmen, firing tracer (after removing the local livestock) at a converging attack when men showed in the fields below.

Brigadier and ADC got pinned down before being recognized by a zealous Bren gunner guarding approaches from the rear. Furious semaphoring ensued from a ditch before the visitors were given the flag of truce. Michael Dunning-White was at hand and saved the day. He had been a Sea Scout at Harrow and read the message for me, 'Hold your fire: this is Sunray. Where is your commander?' Neither side had flags for this discussion. Michael waved back, 'Sunshine right here, dear,' or words to that effect. The bugler blew cease-fire and Laycock climbed the hill.

A big raid was on. He did not say where. Two Commandos were needed to replace parachute battalions chosen to knock out coastal batteries covering the sea approaches to a certain town. 'Can you climb cliffs?' Robert Dawson was summoned as we walked down to the cars. Yes, he had sixty men who could scale anything with a reasonable surface, provided there was no overhang. The word 'chalk' was not mentioned – the worst stuff to negotiate.

I caught the night train with John Durnford-Slater to start planning in London. The Commando would leave within forty-eight hours, destined for Weymouth. There was no time to spare. The job was on. That meeting at Dundonald will be remembered as a small version of the 'Field of the Cloth of Gold'. The bush telegraph travels fast. I had complained about our inability to sing on the march. As they swung back to Troon, the Commando did better than whistle 'Colonel Bogey'. They burst into an improvised version of a cowboy Western. The first verse was printable:

> It's a penny for coffee and tuppence for bread,
> Just three for a rumpsteak and five for a bed,
> Sea breeze from the pier wafts salt water smells
> To festive commandos in south-coast hotels.

CHAPTER FOURTEEN

PLANNING THE RAID

In all difficult operations a moment arrives when brave decisions have to be made if an enterprise is to be carried through successfully. He most prevails who nobly dares.

SIR ROGER KEYES

I gave John breakfast at the Guards' Club in Brook Street, a meal punctuated with the wail of air-raid sirens. The war had come a stage nearer and we got a reminder from the Military Police to carry gas masks walking across to Whitehall.

Combined Operations Headquarters in Richmond Terrace swarmed with red-tabbed gentlemen. The bee-hive illusion was enhanced by busy passages, honeycombed with rooms filled with every branch of the Services, including the powder puff variety, who looked elegant in silk stockings. There was said to be a fair proportion of drones among the inmates. Signing a pass allowed the visitor, having stated his business, to sit around talking to pretty Wrens, or out on the terrace when it wasn't raining. On one occasion I wasted three days in answer to an urgent summons from Wingate, who wanted tests made on the swimming powers of pack mules across the Severn Estuary! As a port of call Combined Ops was not favoured by the serving officer.

This time we pulled a fast one: John was a go-ahead fellow. The course of action had been decided in the train. Bypassing the smaller fry, we made for the room at the top – to find Charles Haydon, now a general. The new Chief of Staff was admired by all. Firm but fair, with a Guardsman's unswerving discipline, he had proved a stern taskmaster in the drive to smarten our brigade in the early days of tribulation.

The general made us welcome, then locked the door, disconnected the telephone and outlined the operation, describing commando objectives but not our destination. He hinted at a bigger picture, but he did not like the overall plan and said so. Allied infantry in considerable strength were involved; there had been interference by rival planners in Home Forces.

He was anxious for commandos to do well. Two batteries had to be destroyed before a main landing could be effected. A lot depended on us: he had staked his reputation that we would succeed.

'Downstairs you will be provided with intelligence and more general information. There are good air photographs, also models of both target areas. For security reasons no maps are, at present, available. They are not necessary in my opinion. You will be given an appreciation of the situation, general intention and method by which it is proposed to carry out the operation. You must conform with overall decisions. As yet, nothing is final. The time factor is against us. I might as well say you will be working with Canadians – good people who are glad to have you and are anxious to cooperate. They know your track record. You will be meeting some of them tomorrow. Now go away and think it out. Working on the flanks of the main target should simplify your problems. At the moment landing-craft are in short supply, which complicates the sea crossing. Just not enough ships to go round. There will be tiresome meetings from which there is no escape. Navy and air force are fully committed and everyone must work together. You stay in London until things get sorted out. This is too big for any slip-ups. Come back if you get bogged down. I want to see your plan in writing as soon as you have made your minds up.'

And so we went downstairs.

I shall move on quickly. Some sparks flew after contradicting a planner (the novelist, Robert Henriques) who tried to do my thinking for me. His suggestions, as it happened, would have done us all in. Henriques had come up in the world. His 'too clever to learn' attitude had not been approved at Inverailort; I had been one of several critics when in charge of the Fieldcraft Wing. He had not gained stature in the interval.

We left Richmond Terrace having won two valuable concessions. First, we would land before daylight; second, each Commando would be independent, fighting its own way: in wireless communication with the force commander, reporting back when the job was done. I had been luckier than John, drawing the good ship *Prinz Albert* to cross the Channel. No. 3 Commando was to be provided with small motor launches, in doubtful condition, called Eurekas. They were destined to be shot up, dispersed or sunk at sea.

Before we left the general he asked for frank opinions. Nothing loath (having gained my cloak of darkness), I came out strongly against the infantry's late start and a daylight attack against, presumably, the port of Dieppe. The general changed colour but stared me out. He was clearly stung. He had a hot temper but, for the moment, it was under control. 'And why the devil do you settle for Dieppe, may I ask?' 'For two reasons, sir; one is hearsay based on supposition: something very similar was cancelled last month. The other is confirmatory. On the back of the picture postcards studied downstairs which show the cliffs I have to climb, there is a give-away line on the reverse side of the pictures, "Les Falaises de Vasterival près de Dieppe". We are landing on a flank and I can understand French, sir.' Haydon reached for the telephone before I finished speaking. 'Thank you, Shimi – I can rely on you both to say nothing to anyone.' We saluted and took our leave. The air turned blue in the room above; John and I went on our different ways, down to the sea and ships.

Later I will give reasons why the raid went wrong. The less said the sooner mended; for posterity it is nevertheless important to stick to facts. I do not wish to join the intoxicated sailor-style of thinking. I refer to the story of a brand snatched from the burning who swore to give up gin; then attended a dock gate mission where he heard for the first time of Christ's crucifixion. Minutes later he met a Jew in the City Road and proceeded to beat him up, shouting 'Y're one of those bastards who murdered our Blessed Lord.' 'But,' cried the Jew, 'that was two thousand years ago.' 'Maybe,' replied the irate seaman, 'but I've only just heard about it.'

Let us move on to Weymouth and rejoin the Commando. Derek Mills-Roberts was a good soldier; he was also full of new ideas. That evening I found him in Portland Harbour, experimenting with his latest invention: 'the unsinkable sea-borne container', designed to specification by Raymond Quilter, a friend in the Grenadiers, whose firm made parachutes for the airborne forces. Four heavyweight commandos in their underpants stood on *Albert*'s rail, gingerly testing the handles on a large cardboard tube of cylindrical design, its noble proportions resembling a watertight umbrella stand with a cork at one end. Derek gave his final instructions in the gravelly voice reserved for special occasions. 'When I say jump you go over the side; swim clear as the container is thrown in. Then try and sink the damn thing. If the lid comes off you dive to the bottom of the harbour and pick 'em all out. I don't want a court of enquiry. Is that clear? Come back up the scrambling net. I shall heave a line to pull in the box of tricks. Any questions? Then stand by. Throw it overboard; now jump!' He turned and saluted. 'I've got four service rifles in there with three hundred rounds of ammunition. That container could be useful in a rough weather landing and save lives if we got sunk off shore.' Even as he spoke the infernal machine dived glistening in a train of air bubbles and disappeared under the hull. 'She's gone to feed the fishes,' quoth an able seaman. But its buoyancy bobbed it up again and all went well.*

Derek had another suggestion. We had lost two good men, drowned off the Arran shore when their Mae Wests had failed in a heavy sea. 'Now the base plates of those three-inch mortars' (I had already telephoned these would be required) 'weigh a ton. If we have a wet landing who is going to swim them in?' I suggested he designed an unsinkable raft. And so we passed a happy day. After supper the adjutant sent for troop leaders. They had a fortnight to learn their parts.

If Waterloo was won on the playing fields of Eton it is truer to say that Operation 'Cauldron' owed its success to Lulworth Cove, once the property of my great-uncle, Herbie Weld Blundell. Here we trained tirelessly in eight rehearsals, working night and day from *Albert*'s landing-craft. Seafarers may be interested in details of the most up-to-date infantry-landing ship.† She carried eight assault landing-craft (LCAs). Her armament consisted of two twelve-pounders (very ancient) and six Oerlikons (anti-aircraft guns). The company mustered thirty-five officers (under the command of Captain Peate, DSC) and 165 ratings, with additional capacity for 350 military personnel (all ranks).

Assault landing-craft were of Thornycroft design: 41 feet long, with a 10-foot beam and a draft of only 2½ feet. Two Ford V8 engines were supposed to give a speed of 10 knots under optimum conditions – though I very much doubt it. Each carried a crew of four (one officer, three ratings), with a maximum load of thirty-five fully armed men. This meant dividing a Commando troop. The craft were cold, uncomfortable and very wet in a running sea, but bullet-proof until the ramps went down.

Captain Peate was a particularly nice sailor. The boat crews were good and trained with a will as No. 4 Commando wound up to concert pitch. Every soldier would meet the events of the day like a trained athlete off his mark to the crack of the starting pistol. We were playing for high stakes. All knew it. But I held the cutting edge.

When intelligence is available, only fools fail to take advantage of such information. Usually a battle is fought at short notice, with little or no plan of action. Here the data had been sorted out and sifted like a jigsaw puzzle. We had an admirable model prepared to scale by RAF Intelligence. We knew the range and the distance to be covered

---

* I believe Derek's invention helped the Polish Resistance in the Pripet Marshes, and later the Americans made use of it in the Philippines.

† HMS *Prinz Albert* (and *Prinz Charles*) were built for the Belgian Government by John Cockerill at Hoboken in 1937. Converted by Harland and Wolff, both ships were commissioned for special service in September 1941. Their 15,000 horse-power Diesel engines gave a maximum speed of 22 knots on a length of 370 feet, with a beam of 46 feet and a draft of 12 feet.

and learnt every fold and feature set out on the ground. The demolition squad could blow gun breeches in their sleep. Wireless communications were tested and counter-tested. Every weapon was fired over measured marks. In camouflaged clothing, snipers zeroed telescopic sights while the Boyes, anti-tank, Bren, Tommy gunners, and riflemen – blazed ammunition on short-range practices. No. 68 grenades were fired on the day with telling effect from cup dischargers on reinforced rifles. The 2-inch-mortar men became so accurate they could drop eighteen out of twenty shots into a 25-foot square at 200 yards. The heavy mortars were fired at extreme range off the beach, controlled by field telephone working 600 yards from the base plate.

We practised crossing wire, throwing rolled rabbit netting over the defensive aprons. We fired Bangalore torpedoes to blast a way through barbed entanglements. We laid smoke to cover withdrawal to the sea. Correct seating in each boat was worked out with meticulous detail. Junior officers changed places with supposedly wounded superiors. Landing-craft were sunk in theory for instant improvisations, carried out with no delay.

On a raid it can be more unpleasant coming back than going in, especially in daylight. Final withdrawal was practised first as a drill, then with stretcher cases, then with fierce opposition, and finally under cover of smoke. All officers and senior NCOs inspected the layout of coastal defence systems. The men were splendid and I was well pleased. If we were put down on time in the right place, eventualities seemed well covered. It was interesting to speculate what might go wrong!

The first hurdle came in answer to the thought. A RNVR lieutenant looked in to say he had been appointed to lead our flotilla on the ten-mile run to a crack in the French cliffs. We knew the officer from Oxford as a cheerful extrovert; now we feared his 'nae bother' attitude. Perhaps I misjudged the optimistic approach, that seemed too casual. Derek did not like it either, and that is an understatement. Any miscalculations – altering course or faulty timing, as the Canadians would find to their cost – spelt lost surprise and sitting ducks over open sights. The Commando touch-down had been put forward to 0450 hours: ten minutes before termination of nautical twilight, or, as the French say, 'entre chien et loup'. Those ten minutes, I believed, were vital. One beach – it happened to be mine – was covered by machine-guns firing on fixed lines, enfilading the wire above the waterline. All defences are jittery and shoot high in the dark. The Commando would take advantage of that fact; but some (including Veasey) had a low cliff to climb, and nobody wanted to be picked off like a fly on a wall. I asked for the best navigator available. Commander Mulleneux, RN, ran us in right on time to help win the battle.

Before departure we made friends with the Canadians. Derek was given a suit of battle-dress much smarter than our own. There were the usual security alarms – not improved by the arrival of six Free Frenchmen sent by Dudley Lister, who swaggered into Weymouth all covered with weapons and insignia. They had uses as interpreters, but first they were told to change into something less conspicuous. We saw a funny film, shown by the security chief, which raised a laugh. The story, predictably, concerned the Second Front. First a clean-cut youth with curly hair, who knew more than was good for him, was seen walking out, 'all innocent-like', with the wrong kind of girl. Despite the Cockney accent and feline appeal, she was an agent – rotten to the core, working for the Führer. The audience whistled and groaned. The film did not explain how she reached London; parachutes for the movie industry were in short supply. But we had a glimpse of the down-at-heel family caught trafficking in Berlin's black market. Unless the daughter's treachery pays dividends all have been earmarked for a Concentration Camp. And a good thing too! There were indiscretions in the bedsitter. Much more was revealed. That was the best part. Next came the careless official from the Air Ministry, who mislaid his brief-case in a restaurant, with dire results. Near the end, the battalion finally got started for the front. The scene changed. It looked a short drive from Aldershot. A hail of lead greeted the troops in the Long Valley. They fell in heaps and never stood a chance. Lastly a brigadier appeared, looking, as Derek said, like some antediluvian relic of the past. He also waddled into the death trap. Tactically, one felt that his demise was well merited.

The public will pay more for laughing than for any other privilege. For Rabelaisian humour, I mean the spontaneous kind, give me the British soldier every time. On the last Saturday at Weymouth B Troop gave a party to which I was invited. Gordon Webb, who had won an MC at Boulogne (and was to add a Bar on this next outing), had booked two rows in a concert hall where an ENSA party was putting on a show for the garrison. The cast were amateurs.* The leading lady was fair, fat and well over forty. She started badly with a guessing game played in single syllables. The first clue was a lily of the valley. The dame, who had an affected manner, started off, 'My first is a li—, my second is a ly—, and my whole is a sweet-smelling flower.' A voice from B Troop: 'And that's another bloody lie.' Worse was to

---

* The organization provided entertainment for troops up and down the country – much appreciated in the dark days of the war.

follow. She reappeared, dressed as the Queen of the Fairies, with all the trimmings. In her hand a tinselled baton. 'And what shall I do with my fairy wand?' she asked the licentious soldiery. That was too much for 'Timber Tool' Woodcock, who bounded to his feet. 'If any man tells the lady what to do with it he goes straight to the guardroom.' This departure was greeted with a roar of laughter. The Fairy Queen blinked and pocketed her secret weapon, nobly taking the jest in good part. The sergeant-major left the hall, slightly bewildered by his sudden acclaim.

It was our last fun party. On Sunday top-secret maps arrived by despatch rider. The messenger bore instructions to ring a London extension by scrambler telephone next day on the stroke of twelve.

Antony Head,* a staff officer (GI) recently posted from a more active rôle, did the talking. The Commando would leave Weymouth in the small hours of Tuesday morning, in a kind of vanishing act. For security reasons, the men (unwarned) would carry on normally, return to their billets after 'closing time', and so to bed. Then stand to before daylight. Transport Command would supply TCVs (subject to weather conditions in the Channel), and the Commando would proceed to Ringwood, there to await orders and stretch legs after a two-hour drive. My call to Combined Operations would be on an open line: I should ask about 'the stores' – a guarded reference to the fighting men. If sea conditions remained favourable I might expect the equivocal answer, 'Well, just as well get them loaded.'

Captain Peate, in *Albert*, would take aboard all heavy weapons, ammunition, Mae Wests, iron rations and explosives. Then, after rigging a dummy funnel (that fell off on passage), he would sail empty to Southampton, where we would rejoin the ship by afternoon. Spotter aircraft paid daily visits to check shipping in the Solent.

It was short notice, and reactions to these precautions were unprintable. Anyone who thought it possible to find and rouse up two hundred men, in widely scattered billets during darkness, ought to have his head examined: apart from waking every citizen or suspect agent in Weymouth, it showed small consideration for personnel unlikely to sleep for the next thirty-six hours. A compromise was agreed. During the afternoon's training programme, troop leaders would announce an early start for a two-day exercise in the opposite direction, the men to parade on the waterfront in marching order, with light packs, at 5 am.

The transparent farce was successfully executed. The quarter-

* Antony, who was a personal friend of Laycock's, became a minister in the Conservative Government during the 1950s.

master's side ran smoothly enough, but the men had to go through a
hard day weapon training in the country. They did not get back before
tea. Activity round *Albert*, including the porterage of the unsinkable
sea-borne container, would not have deceived a child; no doubt there
were outspoken ruderies in the pubs that night. The eleventh-hour
security flap is always counter-productive.

I was in an angry mood. My last words to Head were: 'And what
do you propose in the event of an air raid, with no lines open to
London? Or, if your place gets a direct hit, which is the best thing
that could happen to it? I shall expect a despatch rider at Ringwood
Post Office to double check somebody's damn fool arrangements.'

> O there was horsing, horsing in haste,
> And cracking of whips out over the lee.

Every man paraded the next day on time. One good soldier – who
appeared with a bursting bladder, as the clock struck the hour – relieved
himself (and had his name taken) before the roll call ended. There were
no defaulters. The adjutant saluted, a torch in his left hand. 'All present
and properly shaved, sir.' The transport had arrived, but there was
no sign of the officer in charge. This was intolerable. 'Here he comes,
sir,' chorused the MT drivers, as an individual emerged from the shadows with a greatcoat collar turned up and a Woodbine dangling from
his lip. My nerves were frayed: it was too dark to identify the offender,
but I took the cigarette away with a swipe that put the gentleman on
the floor. 'You step to the rear, my lad, and go into the report for being
late for parade. The senior NCO will ride in the leading vehicle, with
a map reader beside him. Michael! Get the men aboard, then crank
up and let's get started.' And that is how No. 4 Commando set out
to see Dieppe. We noticed empty Canadian transport outside pubs all
the way to the docks. One can only guess what was being said inside.

The thinking of the RAF was impressive. Leigh-Mallory's representative waited in Southampton with six US Rangers. He zeroed watches
as we went on board. Derek had one on each wrist. There was to be
split-second timing in the fighter sorties that strafed the enemy at tree-
top height next day.*

I have given an exhaustive account of the build up before the raid.

---

* Senior officers are seldom killed in battle any more, but flying in the war was a tricky
business which had its ups and downs. Four top names occur to me: Leigh-Mallory and
Admiral Ramsay, who also played an important part, both lost their lives in flying accidents
in 1945. Carton de Wiart came down in the sea on his way to the Far East and was lucky to
survive the experience. Harold Macmillan crashed on land and was badly hurt. None of these
misfortunes was the result of enemy action. Prince George, the Duke of Kent, crashed in Caithness,
piloted by a close friend, Michael Strutt, at the beginning of the war.

It is time someone else took up the story: first Derek Mills-Roberts, then Donald Gilchrist, who landed on separate beaches and each fought a different battle, which started in Tennyson style, 'when the moon was setting, and the dark was over all'.

CHAPTER FIFTEEN

THE DIEPPE RAID:
'FAIR STOOD THE WIND FOR FRANCE'
by Derek Mills-Roberts

At daybreak No. 4 Commando, consisting of 252 all ranks, including Allied personnel, assaulted the six gun battery covering the West approaches to the port of Dieppe.

The position was defended by an approximately equal number of Germans with all the advantages of concrete, wire and land mines, concealed machine gun posts, mortars, dual purpose flak guns mounted in a high tower and full knowledge of the ground. They had had two years to perfect these defences and when the time came they fought with the greatest determination. Yet within a hundred minutes of the landings, the position was overrun, the battery and all its works totally destroyed and at least 150 Germans left dead or wounded in the path of the raiders and the scene of the fighting. Prisoners were also taken. British casualties numbered 45 all ranks, of whom twelve were back on duty within two months.

Operation Cauldron is a classic example of the use of well trained infantry soldiers, bold leadership and the thoroughness of the plan of attack and its swift execution.

OFFICIAL COMMUNICATION

At 1.15 am I went down to the wardroom for breakfast. There is rarely a gladiatorial spirit at the breakfast table. Officers were suffering from that uncertain feeling experienced by everyone coming under starter's orders, or on proceeding to the wicket. The stew needed more salt but no one reached for it and we ate in preoccupied silence. After breakfast I was accosted by the steward with a mess bill for 13s. 4d.! I then got my equipment on and went below to the mess decks for a final check-up.

There was an air of quiet concentration as the men got into their fighting order. As befitted a short battle, it was all weapons and ammunition with no trimmings.

Soon it was time to go up and hear the commanding officer say a few words before the landing. It was crowded in the wardroom. Lovat was brief. He reminded us that this was an operation of prime importance and that, if we failed to destroy our objective, the battery would wreak havoc among the ships of the main convoy. He ended by saying the German soldier was not at his best at night, and that here lay our advantage in the first part of the operation.

When the Commando had dispersed he said, 'D'you think you'll find your crack in the cliffs, Derek?'

'Yes,' I replied, 'no need to worry.' My words did not reflect my real feelings.

The bell rang, the lights went out, and everyone filed slowly to their boat stations. In an unending queue one expects stoppages. Even disciplined paragons of the barrack square are incapable of moving at a controlled speed along constricted corridors, yet those whose mothers doubtless jammed every shopping centre in Liverpool or Glasgow overcame this inherent disability. Slowly and methodically they filed out to their stations and so into the boats.

I was in No. 2 boat on the port side. Like all the others in the flotilla, she carried well over her stipulated weight, mainly in mortar ammunition. The capacity of the davits in a calm sea presented no risk. A low whine came from the electric winches; dropped slowly, with an almost imperceptible movement we found ourselves afloat. The flotilla now moved away from the side of the ship – her dark silhouette looked immense. She had no time to lose getting back across Channel before the dawn broke.

The landing-craft were escorted by an MTB; we had ten miles to go. I settled down and was dozing when distant gunfire roused me. It came away to our left flank, but the noise and the tracer fire soon died down. Minutes later we halted. The MTB had brought us to within three miles of the shore; it was time for the two groups to part. A whispered 'Good luck' from Shimi Lovat, and off we went again.

The lighthouse, our main landmark, was working. Its beams swept the sea and we felt like thieves in an alley when the policeman shines his torch. It is hard to pick out small craft at night; we reassured ourselves with this consoling thought.

We were within a mile, when the lighthouse suddenly doused, and flak lit up the sky. The tracer rose from all angles along high cliffs as a squadron of Brewster Buffaloes roared inshore at low level. Surprise might have been lost and haste was now essential; we moved in at top speed, the plan being to turn left at the cliffs and so seek the gullies which were half a mile to our flank. The overhang seemed to blanket the noise of the flak. It cannot have lasted long, this cruise along the cliff base – but it seemed an age. Robert Dawson, David Style and I looked through our glasses. David Style got it first – yes, there it was, the buttress of the promontory with gullies on either side.

We landed on a narrow beach. It seemed like stealing round to the back door where a noisy party is in progress.

David Style's section moved in fast to reconnoitre the gully on the left. This would be our best bet, as its exit showed no defence works or tracks leading from it on air photographs. The men quickly disembarked and moved up close to follow Style's section when given

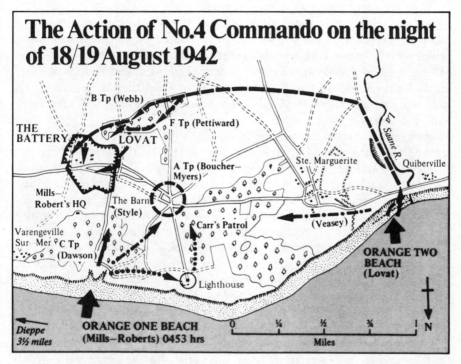

The Action of No.4 Commando on the night of 18/19 August 1942

the word. Waiting was an anticlimax; although Style was a quick worker, the time taken over his reconnaissance of the left gully seemed for ever. The right-hand gully appeared heavily wired up. I hoped the one on the left would prove practicable. The sky was getting lighter. Every minute was precious. David came back. The gully was choked with thick wire and falls of chalk; so it had to be the right-hand crack.

At this moment a patrol passed the message: 'There's someone on top of the cliff.'

The news was worrying: we had to blow a path through the banks of dannert wire which blocked the bottom of the remaining exit, and we wanted no interference.

David placed the Bangalore torpedo in the first bank of wire, lit the fuse and moved out of range. I looked over at Robert Dawson and could see that he too felt the delay interminable. Was the fuse a dud? With a loud report, magnified by the cliffs, it blew. We raced in to spot the damage – a poor gap but we widened it, and placed the second torpedo in the next bank of wire. This one blew a big enough gap to scramble through singly; a small party was told off to widen the breach.

Now we were past the wire and moved up the steep shelving sides of the gully, avoiding the main track like the plague: it was sure to be mined. We reached the top of the promontory and moved in among villas which formed the seaside watering place of Vasterival-sur-Mer. The grass had grown and the gardens looked wild and unkempt. The early light caught shutters which were fast shedding their paint. To our left lay the Hôtel de la Terrace, once, no doubt, an attractive summer hotel.

Robert Dawson's troop, which was to form a defensive perimeter, moved fast into position. Style's section searched the houses between the beach and the battery before moving up. Two leading scouts passed through a hedge and emerged with an old gentleman in a night-shirt which reached down to his bare feet. I felt sorry for the old chap, no doubt a solid citizen of the area; how could he preserve his dignity in such a garment? I explained that we were British soldiers and not Germans – he seemed greatly surprised. Canvas uniforms worn by different armies don't vary much in shape and colour. Our friend, who thought we were a party of local Germans completing a routine patrol, remonstrated over the damage to his hedge.

While this old-timer was being assisted back to his door I noticed a nice-looking girl on the veranda.

'Are you going to shoot Papa?' she enquired philosophically.

I looked at my watch. It was 5.40 am and we were well up to schedule. We had to be in position and ready to engage the battery at 6.15 am; Lovat's assault was timed to start fifteen minutes later. At 6.28 am a squadron of cannon-firing fighter aircraft would rake the target prior to the assault. The timing was precise.

Suddenly the silence was broken by a shattering noise. The battery had opened fire. I got a wireless message from Tony Smith, our intelligence officer, who was on the beach: 'Convoy in sight, apparently within range of enemy guns.'

The battery had to be knocked out by 6.30 am. The convoy appeared to be well ahead of schedule. There was no doubt about it: operation orders were in strict conformity with the main plan. Fifty minutes had been clipped from a very close timetable. It was no good cursing staff work outside our orbit – I had to improvise as fast as possible. No time to search the houses. Now it was imperative to reach the battery at once. It would be fifty minutes before Lovat's party got round by the Saane Valley and be in position for their final assault.

Corporal Smith and I, Ennis the mortar officer and our respective signallers raced through the wood. I had sent a message to David Style – whose section was searching the villas – to join us at once. We heard the battery fire six salvoes in close succession. The noise was deafening. The undergrowth was waist high and heavy going; no one appeared to have entered it for a long time. We heard shooting on our right. Stealth was out and we crashed ahead like a herd of elephants.

We topped a little rise and came face to face with the battery. Ennis and I dropped; so did the others. We worked our way forward to a patch of scrub, some fifty yards in front of the wood and about a hundred yards from the perimeter wire. There was a good view and we heard the words of command distinctly as the battery fired another salvo.

On our right lay a barn-like structure on the edge of the cover: from there we would get a better view. I decided to have a look at it. I crawled back to the wood and ran the ninety yards to the barn. We met David Style and, taking two snipers with us, forced open a door and raced up a flight of wooden stairs to the first floor. From the window we had a magnificent view of the battery as the crews serving the three right-hand guns fired a salvo. One of the snipers got ready and I pointed out his target. He settled himself on a table, taking careful aim. These Bisley chaps are not to be hurried; it reminded me of the awesome time it took General Sir Bindon Blood to fire at Bisley. I remembered seeing him there when I was a schoolboy. At last the rifle cracked; it was a bull's eye and one of the Master Race took a toss into a gun pit. His comrades looked surprised. It was rather like shooting a member of the church congregation from the organ loft. In distance, the barn was 140 yards away from the perimeter wire and about 170 yards from the nearest gun.

David Style's section started to snipe the gun pits with well-controlled bursts. When engaged by a rifle or light automatic, it is not easy to pinpoint the exact spot from where fire is coming. This was the Germans' difficulty; we made it no easier for them. The immediate reaction was lack of movement within the battery itself. The guns had small parapets of sandbags and the crews kept low within them; we could now see no movement between battery buildings. Over on the right the three farthest guns fired again; they had been loaded before the attack: whatever happened they must not be given the chance to load again.

We already knew the locations of seven heavy machine-guns. The Germans had sited these in the double apron fence of barbed wire which lay all round the perimeter. Suddenly a 20 mm gun started firing from a high flak tower on stilts. It possessed an all-round traverse. We could see the streams of phosphorescent shells as they raked the edge of the wood and exploded against the tree-trunks. Fortunately this fire tended to go high.

I went back to Ennis in the scrub; the signal line to his 3-inch mortar was dead and he couldn't get a sound out of it. Suddenly over farm buildings on the left of the battery came the phut, phut, phut of German mortars: soon the area resounded to the crash of shells. The infantry soldier dislikes a mortar intensely, particularly when he is not dug in. With it you can lob explosive bombs into otherwise inaccessible places. A man can use his wits in finding cover from rifle and machine-gun fire, but once a mortar gets his range and locality things become difficult.

Two other detachments were now ready and joined the fray – the anti-tank rifle, a ponderous but powerful weapon with penetrating power, and the handy little 2-inch mortar. The anti-tank rifle was to pierce the steel-plated armour with which we knew the enemy perimeter defences were protected. Gunner McDonough firing this gun, with Private Davis as his No. 2, operated against the flak tower with great effect. It ceased to revolve and gave the appearance of a roundabout checked in full flight. Then he turned his attention to the German machine-guns, ably assisted by the Brens. The aim was deadly and the volume of fire seemed to be checked.

Then our 2-inch mortar detachment started to shoot. Their first round fell midway between the barn and the target; the next was a good one and landed in a stack of cordite, behind No. 1 gun, which ignited with a stupendous crash, followed by shouts of pain. We could see the Germans as they rushed forward with buckets and fire extinguishers; every weapon was directed on this target. The fire grew; meanwhile the big guns remained silent. I sent a report back by Corporal Smith to the signallers.

Messages were relayed to Headquarters Command Ship from our own headquarters; on the beach was a special 'Phantom' link which signalled direct to Admiral Mountbatten, the Chief of Combined Operations, at Uxbridge.

The big guns were silent and any move in the battery was immediately answered by a burst from one or the other fire group. At this stage the German spotting improved, and retaliation swiftly followed. Several men were wounded, and as Private Knowles lay on the track just inside the fringe of the wood, our RAMC Sergeant, Garthwaite, rushed over to dress his wound, and was immediately hit by a mortar bomb. Garthwaite fell mortally wounded. Knowles was brought back safely. One man, Fletcher, had all his equipment and half his clothes blown off by a mortar bomb, while he himself was unhurt.

On the whole we were getting away with it better than expected.

Corporal Smith came back. 'Your message has reached Uxbridge, sir,' he said.

The flak tower opened up again; this time it swept the scrub more accurately. Again the anti-tank rifle returned its fire with effect.

So far so good. But I was desperately anxious to know how Lovat's main assault force were getting on in their wide detour round the flank. We had been unable to get them on the air and did not even know if they had got ashore. Otherwise it would be our task also to carry out the assault on the battery at 6.30 am after the cannon fighters had raked it with their guns.

The German mortaring was becoming heavier, and the forward position more precarious, when I received a signal that we were in touch with Lovat's party. They were actually in their forming-up position behind the battery. There was not long to go, and magazine after magazine was slammed into the Bren guns, with sustained fire by the riflemen. The 3-inch mortar communications had got going and they came into action with a heartening crash.

At 6.25 am we deluged the battery area with smoke and saw the cannon fighters roar in for their two-minute strike at 6.28 am.

Then a Verey-light signal went up for the assault. According to plan, we withdrew to the defensive perimeter round the gully. The Commando was to re-embark from this exit and it had to be held at all costs.

At the Hôtel de la Terrace, Tony Smith, the intelligence officer, with Austin, gave extra assistance to help the more badly wounded back. David Style, who had been hit in the leg, insisted he was fit to carry on and took his section to reconnoitre the left flank. Tony Smith gave a graphic description of the shelling before the guns were silenced: not a ship in the convoy had been hit.

Our new task was to hold the bridgehead while the assault force retired to the beach – Robert Dawson had this well in hand. The withdrawal would be the most likely time for a German counter-attack.

I moved up the road with Corporal Smith – this road lay parallel to our original route through the wood. We pushed on past the barn and on to the battery itself. There was a spasmodic noise of shooting. I saw a couple of German soldiers running in front of a cottage, followed by Pat Porteous and two of his men. Pat shouted that Roger Pettiward had been killed and Lieutenant McDonald badly wounded.

Now German snipers started shooting at the battery from outside the perimeter.

We passed the gun sites, now occupied by the demolition teams; they were blowing them up as fast as they could. Apart from a few dejected prisoners there was no sign of enemy life. It had been a rough party with no quarter asked or given. The many dead – mostly German, but some of our own – lay in and around the gun pits.

Lovat and his adjutant, Michael Dunning-White, were beyond the flak tower. I noticed one very light-haired, blue-eyed German lying staring up at the sky without a sign of a violence upon him. As I turned him over Lovat said, 'Killed by a grenade,' and I saw that a fragment had penetrated his back.

The assault had gone well: it had not been expected from that quarter. The enemy had been looking to their front and our fire group facing the battery. Of the two assault troops, the right-hand one had been lucky, but the left-hand troop had met accurate machine-gun fire, had lost its officers, and been badly knocked about. Captain Porteous took over and led the charge. He and Troop Sergeant-Major Portman had killed one German gun crew and then charged the next gun pit and seized it. Pat Porteous was later awarded a VC for this exploit.

Suddenly a squadron of Messerschmitts swept in low overhead. From the air, troops look alike: instead of taking cover we waved genially to them receiving in return a reassuring wave from the squadron leader: their interference would have been an embarrassment. They were flying very low, but did not realize we were not the rightful occupants.

Time was running out and we had to get back to the beach. There was a company of German infantry in the village of Ste Marguerite about a mile away; should they counter-attack they would use one of three small roads and not waste time going across the fields. Three standing patrols concealed themselves to await developments.

I got word that Bill Boucher-Myers had ambushed and annihilated a reconnaissance patrol moving down from Ste Marguerite. This meant the German commander was still without accurate information. More German infantry were in the village of Quiberville on the other side of the River Saane, about two miles away: no doubt they too would be making preparations to counter-attack. The sooner we got our casualties clear the better.

The stretcher cases were taken to the beach and the walking wounded marched with them. One man, Sergeant Watkins, was lying on a stretcher. He had been shot in the stomach. His face looked grotesquely ill beneath its covering of camouflage paint, but he was quite cheerful. He asked me how the battle had gone and told me that he was all right – the stretcher-bearer silently shook his head.

We made German prisoners take a hand at carrying the stretchers – it is a tiring business.

The bridgehead perimeter extended to the house where the owner in the nightshirt had appeared. There was the old chap in the garden, resplendent in a black coat and striped trousers. He offered me a drink but there was no time – after saying that I was sorry for the damage to his garden, I said goodbye. He and his family appeared impressed by the destruction of the battery, but were obviously nervous, no doubt feeling they were under Nazi observation, which meant trouble later.

Down on the beach the tide had gone out. We expected this; a line of smoke floats was placed on the beach and in the water to cover the withdrawal. The badly wounded were manoeuvred down with difficulty and then ferried to the landing-craft in a canvas collapsible boat.

The beach was now under mortar and rifle fire from the lighthouse area on top of the cliff. This was returned by Bren guns from the landing-craft. When the rest had embarked, our party waded out chest deep and got aboard. We drew out of range and made for the nearest destroyer which lay a couple of miles away. We put our wounded aboard her.

A vast concourse of shipping lay off Dieppe.

We were ordered to return to England. The flotilla made its way through gaps in the minefields marked by small floating platform buoys from which flew diminutive flags.

The main Dieppe approach was shrouded in haze, but we knew from the scanty reports received that the battle was not going well. Speculation was rife, but facts were few.

The night was spent in a transit camp near Newhaven and after dinner we compared notes about the raid. In an operation of any kind you know only what you have seen yourself, and it was interesting to hear the rest of it. Lovat's landing on the Saane Valley beach had drawn intense fire and there were casualties in crossing the barbed-wire apron. Veasey had stormed the cliff pill-box with scaling ladders and silenced the opposition. Once over the wire the assault party had run up the Saane Valley and formed up in the wood as planned. F

Troop had a rough assault, but B had gone in intact and played merry hell in the battery buildings as well as the gunsites. The German commander had been killed after a game of hide-and-seek round the battery office.

Telephone wires between the lighthouse and the battery were cut by Corporal Finney, who climbed up a pylon under fire to do the job. Lance-Corporal Mann, a sniper, dressed in his camouflage suit, had climbed a tree and operated from there with great success.

It will be remembered a German patrol had been successfully ambushed during the withdrawal. It happened as we hoped it would. A peasant woman with a basket of eggs showed her appreciation by giving each man in the Commando patrol a fresh egg. Eggs happened to be scarce in England and these were brought back with care.

The total casualties in the Commando were sixteen killed and forty wounded or missing. We returned to England with the battery office files and prisoners.

Some of the badly wounded men were left to surrender, in the care of medical orderlies: in such cases the movement of a sea passage would have meant inevitable death and hospital was the only alternative. The gallant orderlies knew before the raid that some of their number would have to remain to look after the wounded, and become prisoners of war.

We all felt Shimi Lovat had planned the attack on the battery brilliantly. In the planning stage he had to contend with those who wanted it done differently, but he was strong-minded enough to get his own way. Once the operation had started, he had led and controlled it perfectly.

CHAPTER SIXTEEN

DIEPPE: 'A SUBALTERN'S IMPRESSION OF HIS FIRST OPERATION'
by Donald Gilchrist

And oh! the lad was deathly proud!

BELLOC

I awoke to the steady throbbing of the ship's engines. As I washed and dressed, I told myself I would at least make a well-groomed corpse.

The mess decks gave the impression of a Hallowe'en party, not a major operation. Everywhere men were applying grease-paint to their faces. It varied in colour. Some had achieved the desired nigger minstrel effect. Others resembled Red Indians.

A Sioux brave in khaki approached me. 'You haven't got your face blackened, sir. I'll fix it for you.'

He smeared me with a greasy substance, the smell of which made me sick.

'Where the hell did you get that?' I demanded.

'Right from the cook's pan, sir,' he grinned.

The ship's engines slowed. Every light was dowsed. There came a shrill blast on the bosun's pipe – our signal to embark. It was before dawn.

Next to me was a veteran commando; he wore spectacles with metal rims. A Tommy gun was cradled lovingly in his arms. Across the muzzle was a strip of adhesive tape to prevent water getting in when he splashed ashore. Gunner Hurd, my runner, was on my other side.

The minutes passed in whispers.

'There's the lighthouse.'

The speaker was B Troop's captain, Gordon Webb. The lighthouse was our landmark. Its grey, pencil shape could just be made out in the half-light. There was an excited buzz of conversation.

'Shut up and get down,' Webb rapped out.

The LCA was at half-speed now, silently slipping towards the shore. A grey band of searchlight swept the horizon. Was it going to be a dry landing? Was it going to be unopposed?

It wasn't.

A stream of tracer bullets suddenly leapt out at us, thudding, searing, probing. A mortar thumped as the LCA grated on the shingle. Webb cursed and clutched his shoulder, hit by a mortar fragment. The ramp clattered down and we swept up the beach, Webb still with us. On either side, other LCAs were grounding, other ramps going down, other commandos were racing ashore.

We were in occupied France. This was the Dieppe raid. I was in action for the first time. A keen young junior officer. Proud to be a section leader in No. 4 Commando. Determined to be a credit to my country and an inspiration to my men. And then my trousers started to come down. I still shudder to think of it.

My section was one of the first over that barbed wire, right on the heels of Sergeant Watkins and his reconnaissance party. We were advancing in the face of fairly heavy mortar and machine-gun fire. Under such circumstances the first men over a barbed-wire entanglement are expected to trample on it, wrench at it, and roll about on it to flatten a passage for those following behind. Picked men wore leather jerkins for this purpose and other carried rolls of rabbit netting.

In my enthusiasm I did everything a rogue elephant would have done except trumpet. In protest against these unnatural strains, British buttons sawed themselves free. And down slid my trousers. It was a moment of mental anguish. Clutching my trousers in one hand, Tommy gun in the other, I raced inland. Those buttons had reduced morale to rock bottom. If any of my comrades had noticed my predicament, and had said one wrong word – I'd have shot him dead and then burst into tears.

Happily for me, the hour produced the man – in the shape of a poker-faced commando who voiced the words needed to give back my confidence.

We ran shoulder to shoulder as a solid stream of German machine-gun tracer bullets whizzed past at head height. We ran like half-shut knives, our bodies bent forward, as if we were forcing our way against a strong wind. The private panted, 'Jesus Christ, sir, this is as bad as Achnacarry!' At Achnacarry, the Commando Depot, we'd spent half our time shambling along like Charles Laughton in *The Hunchback of Notre Dame* under a hail of live ammunition. This *was* as bad as Achnacarry. With a wrench and twist, I made my trousers fast around my waist. I was dressed, if not properly dressed. To hell with all trousers – I was in the war again.

A moment later word was passed along to halt, and close up. The tall, debonair figure of No. 4 Commando's CO, Lord Lovat, appeared.

He was informally attired in corduroy slacks. A rifle hung rakishly from the crook of his arm. He looked as if he was out for a pleasant day's shooting on the moors.

Giving us a reproachful glance he demanded, 'Why isn't someone up with the recce group giving encouragement?' At Achnacarry too, we were frequently expected to be in two places at once.

Military experts have described the part played by No. 4 Commando in the Dieppe raid as one of the most successful operations of the war. Since I was there, I'd be the last person to argue the point. But who am I to justify the statement?

That would require a rounded, overall picture. This isn't the way I remember the destruction of the six-gun battery at Varengeville. It just isn't possible for a man taking part in a battle to be everywhere, see everything, and remember it. He only grasps incidents he himself took part in – and not all of those – and it is the trivial events which remain in mind, such as the descent of trousers.

One memory is of the first German I ever saw killed. Boston Havocs were screaming over and around the enemy position, marking it for us. Like an ungainly, grey-uniformed grouse, a German broke cover and ran across a field. He wasn't fast enough. One of the Havocs swooped. A burst of cannon fire lifted him off his feet, hurling him for several yards before he tumbled head over heels and lay still.

Through a thick hedge we could see an anti-aircraft flak tower overlooking the battery position. A few Germans were moving about on top of it.

The troop closed up, and we crawled to the rear of the position. The Germans were unaware they had visitors entering by the back door.

Gordon Webb gave the order to fire. Rifles cracked. We watched, amazed, as a German soldier on top of the tower toppled over the edge and slowly fell to the ground some sixty feet below – like an Indian from a cliff in a Western picture.

The marksman turned to Gordon and said, 'Now do I get back my bloody proficiency pay, sir?'

I recognized the man. A few weeks before he'd had his pay reduced by a sixpence for a bad score on the range.

Webb said, 'I want a few of you to go round and knock out the right-hand gun.' He looked at me. I got up to find Sergeant Watkins and a few others already on their feet: Marshall, Keeley and Hurd.

We cut across a hedge, raced through some trees, and darted between buildings. Before us, not seventy yards away, was the battery position, German heads bobbing up and down. We began to stalk –

we'd learned how – walking upright, stiff-legged, our weapons at the ready. Suddenly, we froze.

A German soldier appeared, carrying a box of grenades. He must have sensed danger at his back. He stopped, turned cautiously, and looked open-mouthed at us, his face distorted. Then, still holding the box in both hands, he started to jump up and down like a baby in a temper, as he shouted dementedly, 'Kommando, Kommando!'

A sarcastic voice said, 'I'll give him flippin' Commando!' A rifle spat, and the German fell over his grenades.

Our section doubled forward. Close to us the hedge rustled. The Bren leapt in Marshall's hands. I glanced sideways in surprise at a grey form settled in a huddled heap. Up went hand grenades to explode in the right-hand gun pit. Groans were coupled with a Cockney voice, 'Every time a coconut!'

We were ready to go in.

With fixed bayonets F Troop attacked, yelling like banshees. In too came B Troop, led by Gordon Webb. His right hand was dangling, useless – but he had a revolver in his left. Razor-sharp, Sheffield steel tore the guts out of the Varengeville battery.

Screams, smoke, the smell of burning cordite. Mad moments soon over.

A rifle shot from the buildings behind the hedge. Lying in a yard was a wounded commando soldier. From the gloom of a barn emerged the German who had cut him down. He jumped up and crashed his boots on the prone face.

Our weapons came up. A corporal raised his hand. We held our fire. The corporal took aim and squeezed the trigger. The German clutched the pit of his stomach as if trying to claw the bullet out. He tried to scream, but couldn't. Four pairs of eyes in faces blackened for action stared at his suffering. They were eyes of stone. No gloating, no pity for an enemy who knew no code and had no compassion.

We doubled across the yard to where the two wounded lay side by side. For our comrade – morphine. For the beast – a bayonet thrust.

Over at the guns, Sergeant Jimmy MacKay was fitting special explosive charges into the breech of each gun. In a satisfied tone he was saying, 'They fit like a glove – just like a glove.'

Lovat's next words brought me to my feet.

'Set them on fire,' he ordered, with a gesture at the surrounding buildings. 'Burn the lot.'

Not the words of a commanding officer in the British army. They were the order of a Highland Chief bent on the total destruction of the enemy.

As we withdrew to the beach, a black pall of smoke covered the corpse of the battery at Varengeville.

On the shore Lovat yelled at the navy to come closer. 'No reason why I should get my feet wet,' he said.

Aboard the landing-craft, heading out from the beach, some of F Troop were eating apples.

Aircraft whined overhead. A Bren gunner had a shot at one marked with a swastika, using his pal as a mounting.

We watched a dog-fight between planes – one German, the other American. Smoke belched from the American plane, and its pilot baled out. He drifted down; our LCA changed course to rescue him.

The Yank pilot hit the sea alongside our boat, and was hauled aboard by his parachute simultaneously. The commandos sat him down. Somebody stuck a tin of self-heating soup into his hand. The American gazed at it incredulously. A voice enquired, 'How's that for service, bud?'

We were allowed to send a telegram home. Faced with this uncensored opportunity, I was at a loss for words.

CHAPTER SEVENTEEN

THE RETURN

Home is the sailor, home from the sea,
And the hunter home from the hill.

ROBERT LOUIS STEVENSON

Enough said. Close-quarter fighting is a messy business, leaving the survivors drained of emotion and mentally exhausted.

After standing down, the battle over, some men shake; others turn to stone. One thought blurs the mind: a vast relief steals over the senses, transcending conscious feeling. 'It's over.' In a detached way I felt neither pleased nor sorry to have destroyed the enemy. The sense of elation, a grasp of victory, hunger or thirst, and physical fatigue – these came later. So we returned to England – each to his thoughts, wrapped in a dream.

The seventy-mile crossing passed without incident. Squadrons of fighter planes passed high in relays to engage the Luftwaffe, their shadows reflected on the burnished water. Two pilots – one from each side – were picked out of the drink. The sea was like a mill-pond, shimmering in a mirage; cliffs along the Sussex coast, bathed in sunshine, looked warm and welcoming. As we approached Newhaven the LCAs closed up and took station in line ahead; sailors put their caps on and recumbent men got shaken from drugged sleep.

'In Sussex by the sea.' How good it looked! Over the red roofs of the town, huddled round the gap in the chalk cliffs, a scatter of fantailed pigeons wheeled and tumbled in a happy display of acrobatics. The quay swarmed with people. It was a sweet moment and the dogs of war receded slowly into the background.

The anxious crowd – civilians, general staff, auxiliary services and ambulance drivers – who lined the harbour wall neither knew our identity nor were aware which side had won; the battle had petered out across the Channel and the assembly gazed down at the troops in silence. Active and passive – a stark contrast in styles. Time stood

still that August evening as boats made fast below and commando
soldiers climbed stiffly up iron ladders. The badly wounded had been
transferred to HMS *Calpe*, under air attack off Dieppe. Her sister ship,
the Hunt Class destroyer HMS *Berkeley*, lay crippled by dive-bombers,
on fire and sinking near by. There had been an ugly pause while we
circled the control ship *Calpe* after I reported 'Mission Complete':
were we going back into Dieppe? General Roberts took his time to
order us home. Sadly, the dead and dying, with some fine medical
orderlies, stayed behind in France. But the sprinkling of bandaged
commandos, patched up on the crossing, climbed nimbly enough on to
the dock: for reasons now forgotten, the prisoners had been blindfolded
on sight of land.

Our appearance could not have been reassuring. All commandos
were soaked to the skin (they had waded up to the armpits to re-
embark), and war-painted faces, streaked with sweat and powder
burns, clashed strangely with starched aprons and pretty VADs. The
green beret was not yet issued: we had raided in stocking caps, few
of which remained in place: I lost mine on the wire, early on. Only
Bill Boucher-Myers, as befitted a former lance-jack in the Coldstream
Guards (we served together at Caterham), presented a smart appearance – right down to his collar and tie.

Identification badges were always removed before a raid, but it didn't
take long for a whisper to turn into a buzz; then a shout went up: 'It's
the commandos, God bless them, and they've knocked out the German
guns!'

Next day the papers carried a photograph across the free world of a
rugged soldier with a rueful smile and a bloody bandage round
his head. The four-word caption read, 'I was at Dieppe.' But I
only remember a fair-headed boy with a white skin who lost his trousers
by a bomb blast yet survived without a scratch. At once a rush of girls
found this hero a blanket; blushing, he wound it round his waist into a
kilt. Tension immediately relaxed: 'Hasn't he got lovely legs!' chanted
the back-slapping crowd.

As the news spread across the harbour, cigarettes and hot tea were
handed out from mobile canteens to tired men, who rested on their arms
while sergeants called the roll and troop leaders reported losses to the
adjutant.

Brigadier Laycock arrived with a grave face. The news from France
was bad. Beside him a civilian with sharp, inquisitive features; it was
Walter Knickerbocker, the American journalist. I introduced them to
Mackinnon, the senior boat officer, who had landed us with distinction
twelve hours before and brought us safely home. 'Go easy with the

press,' said Laycock behind his hand. 'The Canadians have taken losses.' Then he addressed me. 'You're for London and Mountbatten: we start immediately.'

I handed rifle and revolver over to Mellis, with instructions to find the quickest way to Weymouth and somehow send me up a change of clothes for the morning. I asked Michael to ring me from the transit camp, giving the casualty list, and we were off.

The staff car stopped at East Grinstead for more news of the battle. I missed the chance of a bite in a vain effort to ring Scotland. The innards were starting to growl by the time we reached London, around nine o'clock; except for a stale rock cake, that sat uneasily on a pint of beer, shared with Derek the previous afternoon, there had been no time for food. But there was no respite: a meeting held at Richmond Terrace was shunted over to the War Office; with difficulty the night porter was eventually roused to unlock the Guards' Club after I had made my full report.

The operation had been pieced together. It was a grim story: the Canadians had lost the greater part of a division; both sea and air losses had been severe. No. 3 Commando had given a good account of themselves.

John Durnford-Slater and his men suffered a rough passage in the twenty Eurekas, which he mistrusted from the start. These boats had a poor reputation. Before the craft were moved to Newhaven (conspicuously concealed under a canvas awning, which drew reconnaissance visits from the Luftwaffe) the Commando had circled round the Isle of Wight to test old engines. There had been one breakdown. It could have been worse: the sea was smooth but progress slow.

Charlie Head, always an optimist, told me he also felt they were in for trouble: prophetic words. As the Commando waited to embark, queued up behind Canadian infantrymen, he got the first alarm. The civilians in Newhaven lined the streets, sensing the importance of the occasion, and watched the troops file past. As Charlie shuffled by – a lofty figure – a fat lady appraised him in a clear voice: 'There goes a tall one. He's the fella for me.' Charlie was taken aback and claimed he spent the next twenty-four hours on hands and knees!

Their luck gave out on the crossing. By 2 am three launches dropped far astern with engine trouble. Ten miles from the French coast the remaining craft ran into E-boats – the terror of the shipping lanes. Star shells at sea turn night into day; at their slow rate of knots the commandos were sitting targets, as wooden hulls were raked with 40 mm cannon fire. The escorting destroyer had disappeared to deal with trouble elsewhere; the motor launch that led the flotilla had her armament shot away, with bridge destroyed and boilers pierced, bringing her to a standstill. All the officers were killed or wounded.* It was a one-sided contest.

The flotilla scattered to escape: six boats were sunk with severe losses. One troop leader swam the several miles to the French coast and was taken prisoner. Charlie Head went down, but managed to rejoin his commanding officer. John, hit by splinters, took off his boots and prepared for a swim. But Commando tradition of maintaining the impetus of the advance lasted to the end. No turning tail for home.

Four of the landing-craft drove straight ahead in the early stage of the sea battle. It was getting light as they beached in front of the Goebbels battery. Three boats landed sixty men, under Captain Wills, in the east cliff gully. They were met by withering machine-gun and mortar fire, but fought inland, reaching the top before the Luftwaffe took a hand at dive-bombing the exposed party. It was not until ammunition was exhausted, and the last man unable to use his weapon, that the survivors were taken prisoner.

But the best performance of the day was put up by Peter Young, a fighting soldier of intelligence and determination. Peter landed with a boat-load of eighteen men, including signals and headquarters personnel, in the west gully. Working inland without heavy weapons of any kind, they successfully engaged two hundred Germans in the Goebbels battery with small arms fire from the surrounding orchards and fields of standing corn. Though the Germans turned the big guns on Young's group, they could not depress them sufficiently to do any damage; the shells whistled over the heads of the raiders, bursting a mile away. For the next three hours the Commando party continued to harass the battery, sniping accurately, constantly changing position and beating off counter-attacks with their Bren so skilfully that the heavy guns never fired on the anchorage four miles away. Young finally withdrew his section by the way they came, suffering only minor casualties, and was picked up by Lieutenant Buckee, RNVR (his boat officer), who pluckily stood off the beach all day under air attack and fire from the cliffs above. A brave feat of arms, for which both soldier and sailor were to receive immediate awards of the DSO.

And so to bed – if one had been available. But the Club was full. I was filthy, and without a penny; there could be no admission elsewhere. So I went in. There were bathrooms and hot water; the old servant got

---

* In Sergeant Collins's boat the naval crew were also killed and the engine put out of action. The compass on the craft was shot away. But Collins, using his army prismatic compass and a groundsheet as lug sail to supplement the defective motor, fetched up at Newhaven the next day. A considerable feat of navigation.

out a new cake of soap – a wartime luxury – that did not lather, and, shaking his head, sorrowfully removed various garments to clean and dry. It was lucky no lamp-black had to be scrubbed away, for I fell asleep in the bath; then spent the night wrapped in towels in the library – shivering from a recurring nightmare, where tracer bullets probed the darkness, and leaden feet pounded desperately on slopes of slippery shingle that rolled back, like shifting walnut shells. It had been a disturbing day. But was it over? Something troubled and hurt more than chilling dreams. With the false dawn came the realization that I must break the news to the families of the men who had died at Varengeville.

CHAPTER EIGHTEEN

DIEPPE: A POST-MORTEM

Aptitude for manoeuvre is the supreme skill in a general – it is the most useful and rarest of gifts by which genius is estimated.

NAPOLEON

Field-Marshal Montgomery left his caravans to the nation. Many stories have been told about Monty; the one I like best concerns a caravan.

The scene: the Western Desert, soon after his arrival to replace 'Strafer' Gott, who had been killed in an air crash the day he assumed command. The war hung in the balance: after initial success against the Italians, the Eighth Army had suffered more than its fair share of defeats under various commanders – out-manoeuvred and often out-thought by Rommel's Afrika Korps. We had lost the initiative: understandably, confidence was shaken after the Axis powers triumphed in Europe. All the pressures rested on our new commander. The Italians were cock-a-hoop; the war had turned their way and Mussolini's white charger, suitable caparisoned, had already been sent ahead for the conquering hero's entry into Cairo.

To restore morale was essential. The new man had to be known and recognized by troops in the field. To his credit, Montgomery succeeded, but certain gimmicks (multiple cap badges, odd hats, church parades and PT) were not popular with the old hands. Austerity can be overdone; only a damn fool makes himself uncomfortable on a campaign, but the comfortable *modus vivendi* is more suitable for a winning side, which was not the Eighth Army's lot at this stage of the war.

Monty had one caravan. It was said to be sparsely furnished. The new general made it a rule to go to bed early. Up with the lark and in bed with the WRN? Well, not exactly, but Monty retired soon after sundown, when most officers, the day's work over, were starting to brew up cocoa, laced with something stronger, to keep out desert chills.

One day, General Alexander visited Montgomery's quarters. He knocked and entered. Just a camp bed, a trestle table and a hard chair. Maps covered the bare walls, except for a studio photograph prominently displayed, of Erwin Rommel – the man he had to beat.

'I believe in studying my opponent,' was the testy answer to some ragging. 'Quite right, Monty,' said General Alex, 'but it's Erwin's arse – not his face – you should be looking at!'

The future field-marshal will not be remembered for his sense of humour, but that day he made a point. Every commander in the field has a weakness which, if recognized, can be exploited. Did army intelligence study such things sufficiently? Before the Dieppe raid the answer was emphatically 'NO'. Were our own shortcomings apparent to the enemy? The answer is affirmative.

Now I enjoy reading about shadowy opponents whose names – if ever mentioned – remained meaningless and beyond our comprehension (apart from Rommel's) at a time when it really mattered.

Years later (in 1964), opening Simon Fraser University, I dined with the Seaforth Highlanders in Vancouver. Cecil Merritt, who commanded the South Saskatchewans at Dieppe (and was captured, with most of his battalion, at Pourville), was another guest. Cecil had won a VC. There was much to talk about. *In vino veritas:* what I remember sounded very wrong. Merritt confirmed that, prior to departure for the operation, he knew little about the Germans, or how their general was likely to react. In the event Cecil was cut off by a cyclist company held in reserve behind the town. There were screeds of operational orders, but after perfunctory briefings the Canadians simply expected to engage elements of the 571st Infantry Regiment. That was all – and that was meaningless. Did Ham Roberts, their GOC, know what he was up against before starting? I suggest the answer is no.

It is as important to study an opponent in a pitched battle as on a raid. To capture or destroy the other fellow it helps to read his mind. But was this ever attempted?

The planning approach to Dieppe may have been casual. I write, looking back, as a commander who had few problems. My task was fundamental: in and out – smash and grab. Recesses in the War Office no doubt held dossiers giving appraisals of Haase and von Rundstedt; but were they accurate or ever used? I suggest the briefing was slack, and the intelligence system inadequate. This is not surprising. Previous attempts to spring surprise had ended in failure.* Wrong conclusions based on insufficient study, led to slapdash efforts that went astray. The

* An exception was the capture by Paddy Leigh Fermor of General Kreipe, who commanded the German garrison in Crete.

raiding party cannot be blamed on such occasions. To out-think is to out-play, and that applies even to stud poker!

We were not conversant with the set-up on the other side? In 1941–2 fighting patrols crossed the Channel to snatch and identify bedraggled prisoners – never high-ranking – dead or alive. Although exciting, these Commando raids, carried out in conjunction with other branches of the service, proved unproductive and were usually time-wasting. In the spring, after a long wait at Dover, I led a mixed force of Canadians (Fusiliers Mont Royal) and commandos on the Boulogne raid. It was an over-rated affair which achieved limited success, though the Canadians did not think so. (The navy put them down on a sand-bank, where they stuck until the tide turned.) But we got home in one piece, that is to say, with all weapons and wounded, and gained experi-ence – meaning that I stepped over my head into the sea on a cold night and missed an easy chance with a revolver at Germans who blundered into the outpost line. It was pitch dark, with no moon, and the muzzle blast spoiled good shooting. Mellis blazed away, his rifle wobbling beside my ear – no harm done to the enemy, but I was deafened for a week! The patrol might well have surrendered, but after the rude reception they fell flat and crawled hastily away. No. 4 Commando carried out all its objectives; yet I doubt if our masters were much the wiser. The raid at least proved it was possible to achieve reasonable surprise with shock troops in a defended area, and some undeserved awards were handed out.

Boulogne boosted morale but two men could have been more usefully employed in a stealthy reconnaissance of the Dieppe coast in Kayak canoes: indeed, plotting beaches and supplying a mapped description of local defences before stirring up the hornets' nest will always save lives. But there were minefields in the Narrows, and suitable craft were not available to make the Channel crossing. Later in the war the COPP (Combined Operation Piloting Party) carried out important surveys to obviate such mistakes.

Someone got their sums wrong before Dieppe. A post-mortem is unpleasant, and it is easy to be wise after the event, but it would be fair criticism to say we either lacked reliable agents or seriously under-estimated German High Command. The Abwehr, in fact, broke down certain coded signals between London and French Resistance groups. False information was systematically fed to intelligence, and our strate-gists misread the situation.* Planners relied heavily on air photographs to pin-point the massive build-up of town and harbour fortifications.

* It was said a French operator, caught originally through intercepted carrier pigeon messages, was then forced to work for the Germans.

Such pictures failed to show tunnels were being dug into the chalk head-lands covering the sea front and principal landing places; or the anti-tank guns that could rake the town esplanade; all skilfully concealed in daylight.

There can be little doubt the success at Bruneval (just along the coast), and to a lesser extent at Boulogne, also the impact of the Commando assault on St Nazaire, increased enemy strength for our next objective. Field engineers of the Todt organization worked for months thereafter, pouring concrete into cliff excavations, using forced labour and prisoners of war. Reinforcing the precipitous terrain round Dieppe, once considered an unlikely spot for Allied interference, went on un-detected. Strong-points were hollowed out and stored with food and ammunition. Light guns and heavy machine-guns were mounted on rails run from protecting cave mouths – to be fired and then withdrawn to cover. Camouflaged pits were sited at every coign of vantage for snipers to fire smokeless cartridges.

Minor victories may have turned heads in COHQ. The mood of the general public had also changed: the war would soon be three years old; the Canadians had left home to fight a battle; and there was irre-sponsible talk of a Second Front.

But the odds were stacked against the operation. Whenever weather, moon and tide favoured assault across the water, German vigilance increased. What is gained on the swings is lost on the roundabouts. Land-mines and wire were added to the beach defences, and cliff stair-cases were blocked off or destroyed. The German Army Commander issued the following warning:

DIEPPE
GERMAN ORDERS TO XVTH ARMY
ISSUED 10 AUGUST 1942

The information in our hands makes it clear that the Anglo-Americans will be forced, in spite of themselves, by the wretched predicament of the Russians to undertake some operation in the West in the near future. They must do something –
(a) In order to keep their Russian allies fighting.
(b) For home front reasons.
I have repeatedly brought this to the attention of the troops and I ask that my orders on this matter be kept constantly before them so that the idea sinks in thoroughly and they expect henceforward nothing but surprises. The troops must grasp the fact that when it happens it will be a very sticky business.
Bombing and strafing from the air, shelling from the sea, commandos and

assault boats, parachutists and air-landing troops, hostile civilians, sabotage and murder – all of these they will have to face with steady nerves if they are not to go under.

On no account must the troops let themselves get rattled. Fear is not to be thought of.

When the muck begins to fly the troops must wipe their eyes and ears, grip their weapons more firmly and fight as they have never fought before.

THEM OR US

must be the watchword for each man.

The Führer has given the German armed forces tasks of every kind in the past and all have been carried out. The tasks which now confront us will also be carried out. My men will not prove the worst. I have looked into your eyes. I know that we are German men.

YOU WILL GLADLY DO YOUR SIMPLE DUTY TO THE DEATH. DO THIS AND YOU WILL REMAIN VICTORIOUS. LONG LIVE OUR PEOPLES AND FATHERLAND. LONG LIVE THEÜHRER, ADOLF HITLER.

[signed] Your Commander,
COLONEL-GENERAL HAASE

The 'Them or Us' order, issued on 10 August, was found on the notice-board of the Hess Battery Office at Varengeville-sur-Mer and brought back to England by No. 4 Commando after they over-ran the enemy position and destroyed the guns.

That is the background to the story. Where did the raid go wrong?

For thirty years the truth has been wrapped in pink ribbons, with saccharin added to sweeten the taste. Justifications (by those responsible) were freely publicized after the operation. Time rolls on, but the buck-passing is periodically renewed with explanations that draw still farther from the facts. Of all the wasted words, few come from those who took an active part. In the recent best-seller, *A Man Called Intrepid*, five inaccurate statements dealing with No. 4 Commando's rôle appear on as many pages. The author claims Dieppe was an outstanding success. He is wrong. At best the raid was an exceedingly bitter experience, learnt the hard way.

There were many reasons why 'Jubilee' miscarried. It is sufficient to provide six, with some added asides. But, for a start, I suggest our Allied and political background – the plight of Russia and the clamour for a Second Front – be carefully considered.

Sir John Colville, Churchill's private secretary, in his book *Footsteps in Time*, makes it clear that if Roosevelt and Marshall demanded a Second Front (the shrill voices of such irresponsible figures as Professor Laski and Aneurin Bevan had chorused British approval), both

Churchill and Alan Brooke opposed it entirely. Both asked the USA for increased technical assistance, especially more tanks for North Africa. American support for the more dangerous suggestion arose from chivalry towards a hard-pressed ally ·and not from ideological or military considerations.* A. P. Herbert answered the graffiti, 'A Second Front Now', daubed on London's walls, with some weak verse:

> Let's have no nonsense from the friends of Joe:
> We laud – we love him, but this nonsense: no.
> In 1940 when we bore the brunt
> We could have done, boys, with a Second Front.

Churchill recalls in Volume IV of *The Second World War* how Stalin, at the end of his tether and facing annihilation on all fronts, sent Molotov, his Foreign Secretary, to Washington with the thinly veiled warning to President Roosevelt that, if Hitler threw in more troops against the Red Army, the contest would soon be over.

The threat was clear. If Russian armies could not hold out, events might force Stalin to seek a separate peace with Germany, as Lenin had done in 1917.

'On the other hand,' added Molotov, 'if Great Britain and the United States were to create a new front to draw off 40 German divisions, the results could be very different.'

What Molotov said to President Roosevelt is above my head. The antennae of a serving officer are not attuned to a high enough frequency to receive messages from such exalted wavelengths. This misfortune does not blind men on the job, who fire the gun; they perceive the more obvious mistakes committed at ground level.

In my opinion the raid ran into trouble from the outset for the following reasons, some have been mentioned earlier in the chapter:

1 Though Canadian forces demanded action (and would not be denied), they were short of combat experience. Because of the severe fighting expected in the encounter phase of the battle, the assault should have been assigned to seasoned troops accustomed to seaborne operations. The navy also had weaknesses. Landing-craft arrived thirty minutes late at Puys and Pourville, thus losing the benefit and surprise

---

* John Colville gives an appraisal by Harry Hopkins (Roosevelt's adviser in the White House), when he visited London, of current American opinion of the war in Europe. There were four groups, each with equally strong views: 1. A small party of Nazis and Communists, sheltering behind Lindbergh, who declared for a negotiated peace and wanted a German victory. 2. A group, represented by Joe Kennedy, which said, 'Help Britain but make damn sure you don't get involved in any danger of war.' 3. A majority group which supported the President's determination to send the maximum assistance at whatever risk. 4. Almost fifteen per cent of the country's thinking people (including the senior ministers Knox, Stimson and most of the armed forces) who were in favour of immediate war.

of darkness. Punctuality cannot be expected from numbers of small vessels, dispersed along the south coast, supposedly to arrive together at the other end. Nor were the boat crews all up to standard.

2 The choice of force commanders, both naval and military, was open to criticism. Neither had directed – still less controlled – a confrontation of such magnitude. I say 'controlled' advisedly, because wireless communication between units and from ship to shore broke down within the first half-hour. Captain Hughes Hallett, RN (then on Combined Ops staff), the naval brain who planned the approach to the French coast, was highly thought of. No taxi driver can be criticized for getting his fares to their destination without serious mishap. After touch-down it was up to the soldiers to carry on. But punctuality was a prerequisite for arrival and, thereafter, to ensure a safe return. Neither came off, and the sailor who later put much thought into the making of Mulberry Harbour was let down by tardy craft on the sea crossing. Ham Roberts, the Canadian GOC, who told his troops, 'Your general will be the first through the minefield,' was a stout fellow, afterwards relegated to obscurity in a 'dug-out' post. But he made a convenient scapegoat to carry the can. Roberts behaved like a gentleman throughout. Hughes Hallett, it so happened, was given the unenviable task of replacing Admiral Baillie-Grohman, posted elsewhere at very short notice. As the coast was shrouded in thick smoke all morning, short of landing, cutlass in hand, and rolling up the enemy flank, Lord Nelson himself would have been stymied. The plan was rigid and, when it came to the attack on the town, suicidal.

But it would be idle to pretend that our Allied Forces were in the same league as Field-Marshal von Rundstedt, Commander-in-Chief in the West, or the opposing army commander, Colonel-General Curt Haase, or of the troops under his orders. From whence came the unbounded optimism?

The considerable success of the raid on St Nazaire (carried out at night and without air support) probably had its effect on the Chief of Combined Operations. Mountbatten was told to put on a show. His orders came from 'on high', but I consider he was over-bold and prepared to take unjustified risks in carrying them out. In fairness, the choice of targets was limited to a short distance and within range of RAF fighter cover.

In battle, the right things often happen for the wrong reasons – but never when a plan is basically unsound. Was Dieppe a worthwhile effort? It is difficult to think so. There were no military objectives to justify landing a division in France, and why use tanks and why attack in daylight? Heavy bombers could have destroyed the batteries, harbour facilities and most of the shipping, and hit the radar station and air-fields behind the town. These were the only targets of any significance, nor would an alerted civil population have been at risk. Our losses in the air, on land and at sea were in inverse ratio to the enemy's, and that is an understatement. These sorry figures are not the stuff of which victories are made. Of 4,963 Canadians who sailed from Southampton, only 2,210 returned next day. A total of 466 naval and commando personnel were killed, captured or reported missing. One destroyer and 33 landing-craft were lost at sea. The RAF had 105 fighter planes shot down, something they could ill afford owing to severe shortage. The German casualties were 600 men all told, which included the pilots of 91 planes the Allies claimed to have shot down. (The enemy admitted to only 48.)

3 Mountbatten's judgement was not improved by sketchy intelligence that failed to provide information on the elaborate dispositions of the enemy. The planners misjudged both the sagacity and precautions taken by German High Command. Who assessed the capability of the rival commander? The 2nd Canadian Division were unlucky to place themselves at the receiving end of a mind-reading exercise. But Mountbatten's suggestion of a flank attack was a good deal sounder than the plan put forward by General Paget and Home Forces, who took final responsibility. In his memoirs Montgomery says the frontal assault was a thoroughly bad idea; it is the best-kept secret since the birth of Moses that Monty presided over one of the most important meetings before the operation. He had no criticism to offer on that occasion.

4 The Canadian forces chosen for 'Jubilee' were inexperienced beginners; all innocent of heart and unaware of withering fire power and the ferocity of war. In contrast, von Rundstedt had worked out weather conditions so well that our arrival was ticked off to the day.* The Luftwaffe were less accurate: their meteorological reports discounted the possibility of a dawn landing or airborne descent because of early mist in the third week of August. This low cloud base before dawn is of frequent occurrence in the Channel, and gave commandos the opportunity to carry the batteries, a task originally allotted to paratroops.

5 Postponement of the raid from July to August seemed unwise, presenting security risks which, in my opinion, underline the incautious approach and general air of optimism. I quote James Leasor, who wrote in *Green Beach*: 'With only a few brief periods remaining to favour an amphibious landing no other operation of equal size could be

* *Special order issued by Field-Marshal von Rundstedt:* 'The night of 18/19 August can be regarded as suitable for enemy raiding operations. Commanders of coastal defences to maintain troops at "the threatened danger alert".'

mounted on a new target. Mountbatten then made a new proposal which he felt was so daring that he decided it should not be committed to paper. The Chiefs of Staff argued against this suggestion. . . Not even the First Lord of the Admiralty was made privy to the plot.' The Dieppe raid was on again.

6 Alterations to the Combined Operations plan (a sensible pincer movement to pinch out the town – later changed, at General Paget's insistence, to a Home Forces responsibility) were to prove disastrous. I refer to a frontal attack on Dieppe town and harbour. The misuse of tanks (landed central to the main defences, on the wrong side of the sea wall, on loose shingle on which their tracks could not gain a purchase) was an act of folly.*

The original intention – to flatten cliff and waterfront strong-points by a preliminary bombardment from sea and air – was vetoed by Roberts, backed by Crerar, his superior general (high-ranking but no infantry soldier), and possibly Paget, not from compassion for French civilians, but because the resultant debris might impede tanks in built-up areas. I read elsewhere that the air commander, Leigh-Mallory, while not approving of the frontal assault (which he rightly held would pin men down), nevertheless suggested better use be made of bombers in diversionary attacks elsewhere. The observation credited to Mallory, that preliminary bombing would alert the enemy, seems hardly credible.

Were these considerations overlooked or wrongly assessed? There are indications that both errors were committed. Generals and planners, like other people in authority, can be mediocre like anybody else.

This *coup de main* called for the best at every level timed to a split-second standard of excellence. No amateur side can take on pros unless miracles are expected; only a foolhardy commander launches a frontal attack with untried troops, unsupported, in daylight against veterans (who had never known defeat), dug in and prepared behind concrete, wires and mined approaches – an enemy with every psychological advantage, who had recently been told by Hitler that troops who held the Channel ports controlled the rest of Europe. It was a bad plan, and it had no chance of success.

Bernard Fergusson, in *The Watery Maze* (a full account of Combined Operations), has added some balm of Gilead: 'Dieppe posed the problems for all to see: it lifted discussions about invasion out of the academic into the practical sphere.' We learnt a lesson which Hughes

---

* Of the twenty-seven Churchill tanks put ashore only eleven got up on to the esplanade, where they were knocked out in turn by concealed guns that covered exits leading to the town.

Hallett summed up rather neatly: 'The next time we bring our harbour with us.'

What of the opposition? They deserve a mention. I read, as it were for the first time, that Field-Marshal Gerd von Rundstedt, a soldier of thirty-five years' experience, noted for his grasp of detail, had been appointed Commander-in-Chief in the West and made responsible for coastal defences; that in campaigns he had led an army group to destroy Poland and performed brilliantly in France, launching the surprise attack that crossed the Meuse after emerging in strength from the supposedly impassable forests of the Ardennes.

After the Allied collapse at Dunkirk, von Rundstedt had fought with distinction on the Russian front. He was a meticulous disciplinarian, respected by his staff, although he drove them hard. I know the type. There are three kinds of general: active, passive and bloody-minded. The field-marshal, who came into the last category, was efficient. He spoke good English, learnt from a former London nanny; his family held a long tradition of public service. Like most professional soldiers in the Reich, he disliked Hitler and mistrusted politicians. In the early days of Germany's revival he approved the liquidation of Ernst Röhm, the homosexual leader of the Brownshirts and principal rival in Hitler's rise to power. A photograph shows a junker of the old school, stern and correct, his face marked with duelling scars, a tightly buttoned figure. He was sixty-six and probably wore stays. He is shown inspecting the carnage on the Dieppe beach with an air of cold detachment.

Von Rundstedt ranks as a first-class soldier, and Colonel-General Curt Haase, who commanded the army group in Normandy, was no fool. His Order of the Day, 'Them or Us', issued a week before the raid, shows how well he read his opposite number.

Haase's XVth Army consisted of seventeen divisions divided into three corps; of these the 81st held the area between Caen and Dieppe, with its headquarters in Rouen. In and around Dieppe itself were soldiers of the 302nd Infantry Division, with their headquarters at the crossroads on the Route Nationale – half-an-hour's drive behind the town. These troops were rated as only good-to-moderate. The garrison itself was composed of the 571st Infantry Regiment, of which two battalions, numbering 1,500 men, were stationed in the town.

Further battalions of the same regiment, and two heavy six-gun batteries, were in positions covering either side of the port. This last fact we knew about. The guns were primary commando objectives: they commanded all sea approaches. (Not suspected was the presence of cyclist reserves, held in a state of readiness behind the town.)

Inland at Amiens lay the 10th Panzer Division – a unit with a formidable reputation. After over-running Poland, 10th Panzers led the thrust across France that finally captured Calais. They then turned on Russia and, after a year's fighting in the advance on that broad front, were finally withdrawn with all their vehicles worn out. After the battle for Smolensk (April 1942) the surviving tank crews limped back to France to rest and re-equip. Their morale stood high.

There were no flies on the SS Adolf Hitler Brigade, under its rough-tough Nazi commander, Sepp Dietrich – a favourite of the Führer's who enjoyed political patronage as an original party member. But Dietrich was a good soldier and his brigade second to none. Having fought through Poland and France, they had recently returned from heavy slogging against the Russians in the Donets Basin. Resting at Vernon, they were available to reach the Dieppe battle by late afternoon.

That, briefly, was what we were up against on 19 August 1942. Who chose Dieppe in the first place? The protagonists will argue that, however unfortunate the results, the raid conclusively proved it was impossible to seize or make use of harbour facilities on D-Day, and a Second Front massacre was thereby averted. This is true. The Normandy landings and the Mulberry Harbour caissons later justified COSSAC planning decisions to cross the widest part of the Channel and land with a ready-made port on an unlikely stretch of exposed coast.

Planners for this sad Canadian venture had insisted on continuous fighter cover wherever the target, but the tactics employed remain a first-class blunder.

CONCLUSION

I must take up a position in the final analysis. It is easy to oversimplify the case for or against the Dieppe raid; I will sum up objectively, holding to the facts.

The critic, be he soldier or civilian, can pick obvious holes both in the planning and execution of 'Jubilee'. The background complexities that surrounded the operation are less apparent. The raid had to go ahead for higher political considerations, and I have little doubt that the Prime Minister's order to proceed was mandatory, and the reasons compelling.*

* I link other events to this chain of mishaps by suggesting impatience on the part of Combined Operations. Naval susceptibility had been outraged by the loss of the battleships *Prince of Wales* and *Repulse* (10 December 1941), sunk by the Japanese in the defence of Singapore. More blows were to follow. On 11–12 February 1942, in foggy weather, *Scharnhorst*, *Gneisenau* and *Prinz Eugen*, the pride of the German fleet, steamed almost unpunished the length of the English Channel

I hold the raid itself was a disaster, and the changed plan nothing short of suicidal. Combined Operations had no say in the choice of troops or military commander. Home Forces, with no previous experience of this type of forced landing from the sea, changed the plan to one of frontal assault and wrecked the operation.

I have been asked about the security aspects of the raid, with particular reference to postponement from 5 July ('Rutter') to remounting the same operation ('Jubilee') on 19 August. Surely the concentration of troops and shipping opposite the target area helped to give the show away? My answer will come as a surprise, but need not cause chagrin to those who defend the safety precautions.

Many thousands of servicemen knew the 'Rutter' target. Ham Roberts announced on the loud-hailer where he was going in July. James Hill, a first-class soldier who commanded the 1st Parachute Battalion (originally detailed to destroy the coastal batteries), was astonished on arrival in the Isle of Wight to hear the Canadian rank and file talk openly about what they were doing. We got much the same latrine rumour in Scotland after 'Rutter' cancellation. There could have been agents on the south coast; I do not believe loose talk made the slightest difference. We were expected to arrive at a certain date – and so we did. The raid was signposted all the way.

When the Luftwaffe dropped bombs on 'Rutter' ships concentrated off the Isle of Wight, the Germans got the message, and vigilance continued. Postponement gave the enemy time to strengthen their positions. The assessment of Colonel-General Haase (responsible for coastal defence from Ostend to Normandy) was remarkable. Troops under command were given 'Special Alert' danger periods whenever conditions appeared suitable for Allied landings. The dates were 27 July to 3 August, and 10–19 August. The Germans are an intelligent and methodical people. They were fully aware of sea conditions in the Channel. Von Rundstedt's Special Order could not have been more accurate.

It is true the Luftwaffe (because of weather reports) were taken by surprise. Not all posts were fully manned in coastal defence. Shipping proceeded normally up the Channel. But where it mattered, on the beaches, the enemy were waiting and there were no mistakes.

'Operation "Jubilee" was mounted the night of 18 August, when a

and escaped back to the safety of their home waters. There had been warnings of radar jamming along the South Coast, but nobody in the Home Forces had taken sufficient notice. This bold action was a severe jolt to naval credibility. Every fighting sailor shared the view that whatever the Germans achieved, the Royal Navy could go one better. The enemy lay seventy miles across the water – it was time to take them down a peg.

force of six thousand Canadians and eight hundred commandos sailed from different harbours on the south coast to converge at first light in an assault on Dieppe and its immediate surroundings.' (Official Communiqué.) The force included the following:

Canadian units

South Saskatchewan Regiment
Queen's Own Cameron Highlanders of Canada
Royal Hamilton Light Infantry
Essex Scottish
Fusiliers Mont Royal
14th Calgary Tank Regiment
Royal Regiment of Canada
Black Watch of Canada
Calgary Highlanders
Toronto Scottish Regiment
Detachments of Signals, Engineers, Artillery, Medical, Ordnance, Provost and Intelligence, with a representative party of US Rangers and Free French

British units

No. 3 and No. 4 Army Commandos
No. 40 Royal Marine Commando

Perhaps 3,500 sailors and airmen also took part in the raid.

CHAPTER NINETEEN

FINAL PREPARATIONS FOR D-DAY: SOME CHANGES IN COMMAND

For though, with men of high degree,
The proudest of the proud was he,
Yet, train'd in camps, he knew the art
To win the soldier's hardy heart.
They love a captain to obey
Boisterous as March, yet fresh as May;
With open hand, and brow as free,
Lover of wine and minstrelsy;
Ever the first to scale a tower,
As venturous in a lady's bower: –
Such buxom chief shall lead his host
From India's fires to Zembla's frost.

SIR WALTER SCOTT, 'Marmion'

History tells us that 1942, in the Churchillian phrase, marked the 'end of the beginning'. After the Japanese attack on Pearl Harbor, the United States, caught unawares, rejected isolation and entered the war with a will. While the conflagration spread alarmingly across southeast Asia and down into the Pacific, the noose in Europe slowly tightened on the Third Reich, now fighting unwisely on two fronts. We owed that growing ascendancy to errors which cost Hitler all he had won and in the end destroyed him. The Russian obsession which drained armed might and manpower became counter-productive after the first winter snows.

In the long term the war looked brighter. There was positive proof in the Mediterranean, where the weaker partner cracked first. Mussolini had promised eight million bayonets; but was Italy ever a serious opponent? An essentially happy people, they had no heart for the war. The decisive Battle of El Alamein began in late October.* It marked the start-line for the relentless advance of the Eighth Army across the

---

* Desert warfare has an exhilarating quality that comes from freedom in the limitless glare of sun and sand. My cousin Andrew Maxwell wrote this account of the Scots Guards in action in the Western Desert (see Appendix V, page 382).

Libyan desert to Tunis. In November the First Army, with superbly equipped US forces, landed in North Africa to push along the coast and cut the enemy retreat. No help reached Rommel. His communications were severed by land and sea. By the following spring the Axis had suffered overwhelming defeat; in high summer the invasion of Sicily began. Landings in Italy soon followed.

Commandos, after previous failure in the Mediterranean, were in demand and more appreciated.* The new reputation proved a mixed blessing: the army did not hesitate to make full use of expendable troops.

After Dieppe, raiding across the Channel ceased. There was one last fright: a nasty project on the Cherbourg Peninsula, where Green Berets were to be cast ashore as decoy ducks to bring out the Luftwaffe, who needed correction and were said to be lying doggo. It was another joint effort with No. 3 Commando. John Durnford-Slater, whose nerves were better than mine, did not like the plan at all. No. 3 had lost a third of their strength during the sea battle in the Dieppe crossing, and men had to be replaced. The Commando were not fully operational. When we returned from London, John greeted his adjutant with the comment, 'By the way, Charlie, have you made your will? If not, you had better see a solicitor in the morning.'

The plan fell through. Not long afterwards, John, again at full strength, sailed for Gibraltar. Nos. 1 and 6 Commandos had already left to reinforce General Anderson in Algeria. Raiding had become campaigning. Like Uriah the Hittite, that ancient shock-trooper, the brave are to be found in the forefront of the battle; casualties escalated alarmingly. Few replacements were available, for reasons best known to Whitehall. There were further changes when the War Office cracked down on recruiting on a volunteer basis. We were lucky in the 'Cavalry Commando' and got the last horse, so to speak, before the stable door was shut tight.

The losses sustained at Dieppe were made good by another draft of police – first-class material, with discipline and a good education behind them. All measured up to the original proclamation that called for 'bold men with fine physique – of the hunter class – able to swim and navigate boats: wanted immediately for hazardous operations'. These stout bobbies were to exert considerable influence as non-commissioned officers. Some rose swiftly from the ranks.

After Dieppe we moved reluctantly from Troon to Shaftesbury, to spend the winter in drab surroundings and unwelcoming billets. Then we went to Winchester: a pleasant station, with friendly people, the

---

* Layforce were never given a fair deal in the Middle East; Crete had proved a *débâcle* for which they were not to blame.

right size for quarters, in reach of London, set in a happy valley with suitable training areas. At Winchester No. 4 Commando reached its peak in morale, efficiency and numbers. We were over strength (after those wounded at Dieppe returned to duty). Nobody asked questions, for a good unit is a law unto itself. It was also time for promotions, and with such a strong hand I had no discards: the best were chosen unflinchingly as the first to go. Derek Mills-Roberts left at short notice to take over No. 6, who had lost a moderate colonel and suffered losses in severe fighting over Tunisia's Goubellat Plain. Bill Boucher-Myers, a regular soldier, was reluctantly sent to the Staff College. Doc Wood, promoted to colonel, got his field ambulance. Michael Dunning-White, the adjutant, became brigade major. Max Harper Gow stepped up to DA and QMG. Robert Dawson took over No. 4. Fairy Veasey left to join No. 3 Commando in Sicily. Sergeant Woodcock, promoted to regimental sergeant-major, was sent to join Derek. In the fullness of time I was given command of the brigade. Tony Smith, the intelligence officer, came with me, promoted to captain, while Joe Lawrence, also moving up, took charge of brigade transport. Bob Laycock succeeded Mountbatten, who left for Ceylon in South-East Asia Command.

There came an ugly moment when, at Orde Wingate's suggestion, our special talents appeared to be urgently required for the Far East. This I resisted with the strong argument that if a Second Front was imminent, our services should have local priority. There were new men to be trained who lacked combat experience for D-Day.

It was a wrench to take leave of old friends in Winchester. They were comrades rather than good soldiers. Every man and officer had given of his best, and I knew them all by name. There were no weak links – crime sheets and desertion had been eliminated. Unorthodox, certainly, but loyal to the core. All ranks lived hard and played hard and enjoyed life to the full. The Commando believed in itself and team spirit is a priceless asset. At games or in action, on the football field or in an opposed landing, in athletics or the boxing ring – the men were seldom, if ever, defeated. For No. 4 Commando was imbued with a fierce resolve to be the best. 'Who dares wins' was to be their slogan to the end.

I should mention every individual man by name, but I can do no better than provide the regimental roll call giving the sources from whence they (and the other soldiers of the 1st Commando Brigade) came (see Appendix IV). In many cases the titles no longer appear in the Army List. No. 4 Commando distinguished themselves at Lofoten, Boulogne, Dieppe, on D-Day, and later in Germany and the Low Countries. The names have a splendid ring. Every soldier – worthy of his salt –

remains imbued with the traditions of his regiment and the example of his predecessors, but No. 4 Commando meant something even more important.

Sailors talk about happy ships as well as lucky ones. The Commando was blessed with a fair share of both advantages.

Those were the great days – and I had been living on the peaks. Now it was over and close contact had gone for ever. It meant starting again from the beginning. There was another reason which made it hard to sever the strong ties. With the police intake, more officer replacements were required, and, despite difficulties, I was lucky to find a batch of good ones – the last of the summer wine.

Welding new leaders into the old organization had brought its own reward in Winchester. Alastair Thorburn, Murdoch McDougall and Hutchie Burt were stalwart Scots – the last an old soldier who had risen from the ranks to captain in the Norfolk Regiment. He reverted gladly to second lieutenant when discovered at a battle school. Hutchie was to prove as dashing an officer as he was once a sergeant in left-flank company of the Scots Guards, in which we had served together. This son of Glasgow, with red hair and fierce moustache, came from the old school – not above taking a troublesome man behind the wall and hammering sense into him in a bare knuckle exchange of views.

McDougall wore the kilt like Donald Gilchrist when they went training in the wilds. A gentle giant with feet so big that the army had to provide a specially made size in boots, he was a German scholar, and his later services proved invaluable. There were other promising juniors. Bob Jennery, a cavalryman like Joe Lawrence, later became brigade signal officer. Sandy Carlos Clarke, a close friend (with whose sister, Diana Walwyn, I shared a horse in training at Lambourn), was shot up in France and again in the Low Countries. Wellesley Colley, killed on D-Day, and Jimmy James also joined at this time. Peter Beckett, Denny Rewcastle and Ken Wright were further good subalterns. Ken, the intelligence officer, was a capable fellow of dispassionate judgement and a clear thinker; both he and Murdoch were destined to make names for themselves in Germany sorting out the fate of 250,000 displaced persons in the Ruhr. Patterson (a Scot practising in Northern Ireland) proved a plucky new MO.

The man I remember best and not strictly a soldier, was David Haig Thomas.

A rowing Blue (three years in the Cambridge boat, like his father before him), Olympic oar, naturalist and explorer, a more attractive and original volunteer never attempted to put equipment on; and David's was usually upside down. He had the supreme gift of making soldiering fun. His enthusiasm was of an infectious kind that turned aside the rockets freely given for various misdemeanours.

Ancient Winchester is built round a cathedral whose chaste carvings of sculptured stone are much to be admired. I personally prefer the plainer spire of Salisbury, that rises to a pinnacle above the plain – so high that it is known to draw peregrines in passage. The difficulties of making the ascent of our tower, said to lack handholds, must have been discussed by unit experts. All agreed a certain pitch rendered this impossible. Haig Thomas listened in respectful silence. He had recently joined the Commando. Nobody knew the boy could climb.

When Winchester woke one morning, a green beret, tightly stretched over a china utensil, crowned the dizzy height – a speck in the blue sky and a desperate climb in the dark. No steeplejacks were available and David, the smiling culprit, was told to go up again, take it down, then have tea with the Bishop, Dean and Chapter to explain the escapade, which he did with no difficulty, disarming the clergy with inimitable charm.

His versatility was remarkable and for that reason I steered him clear of the more blunting kind of military routine. But whether he was rolling a kayak canoe on a Scottish loch, demonstrating the construction of an igloo on snow-fields in the Cairngorms, treating his section to samples of edible seaweed and raw shellfish off the Cornish coast, or simply giving a lecture on bee-keeping or butterflies to his men, David was in a class by himself. He was a most popular officer – never more so than on the day his section, returning from Deeside, caught the train at Aberdeen, each warrior with a salmon wrapped in regulation groundsheet. David had supplied the lot.

He was killed on D-Day. When the time came to open a Second Front, the overall plan required the Commando Brigade to join with an Airborne Division dropped overnight in battle. A man of proven ability was required to liaise in what might become a mix-up between red and green berets. (I opposed the idea, for, though there could be inter-parachute confusion in the darkness before our landing, during the encounter phase of the contest fighting would be too bitter for liaison opportunities; but I was over-ruled.)

David was the first man to volunteer. He asked for one favour: to take Ryder, his servant, with him. Coming to say goodbye before jumping practice, he told me he had disliked three years wasted in the army: but the time spent in No. 4 Commando had been the happiest of his life.

David dropped with James Hill's 3rd Parachute Brigade. There was cloud over the moon; the sticks were blown off course by wind, and badly scattered. Many fell on the wrong side of the flooded valley of the River Dives; some were lost, killed or captured, others drowned in their harness; a very nasty start. Collecting stragglers, the two Commandos floundered out of the marsh back to high ground and the correct Dropping Zone that commands the valley of the River Orne.

As dawn was breaking, David's party mistakenly took a road where they were ambushed by a standing patrol, and David, in the lead, was killed instantly. His servant and the paras then lay up in no man's land. Two nights later, the outliers got back (through opposition) into Le Plein, the tactical features on which my brigade had dug in. Ryder brought with him the bone cosh made from a walrus flipper that my friend was carrying. The lucky talisman, a gift of Baffinland Eskimos from Cambridge Exploration days, had been splintered with Schmeisser bullets. In due course it was returned to his family.

Along with Geoffrey Congreve, Roger Pettiward and Philip Pinkney, David's death, though I did not know him as well as the others, was the saddest personal loss of the war. All four were splendid Englishmen.

David will be remembered by his friends for setting an example to inspire lesser men – the enthusiast, *sans peur et sans reproche*. In a life so spent there can be no room for tears.

> Nor needs he any hearse to bear him hence
> Who goes to join the men of Agincourt.

I left No. 4 in Robert Dawson's capable hands. The move to Cowdray Park, where I set up the 1st Army Commando Brigade Headquarters, was performed with a heavy heart. We were amateurs and no doubt made mistakes which came from lack of experience; but keenness in the end makes up for deficiencies. The work of junior staff officers often goes unsung. Here I must pay tribute to Max Harper Gow's administrative skills. Michael Dunning-White was not far behind him. They won the respect of those they served by practice and precedent. Duds were commonplace at higher levels: regulars, in spite of staff training, could make a hash of things. Max and Michael took the work off my shoulders, tackling incomprehensible administrative and documentary detail. Tony Smith became a third member of the triumvirate who bore me up.

Bob Laycock offered congratulations on my promotion. 'Is it necessary,' I asked, 'to wear a brass hat or put red tabs on my battle-dress?' He replied, 'Yes, you must wear the red flannel of inefficiency against which you have railed repeatedly and denounced so often.'* Bob gave a hint of what the future held.

---

* There had been trouble after Dieppe for striking the transport officer, who reported late with the convoy for Southampton, and, on return from the raid, when the Commando had waited two nights in the clothes they stood up in for a train to get them back to Troon. Michael rang

Fighting had been severe in the Mediterranean. Derek reappeared from North Africa and John got back at the end of October. They had done extremely well, but both formations were below half-strength and John was in need of a rest. This fine unit had put up their best perform-ance at Termoli, seizing the Italian port after defeating the Ger-man parachute brigade that held it, and standing off a Panzer division until belated help arrived. John was a personal friend. He went down extremely well with higher authority, being more tolerant than myself. He was now detached for special duties. Peter Young, of the Bedfordshire and Herfordshire Regiment, took over No. 3 Commando.

Suddenly our services were in demand from South-East Asia to Italy and the Adriatic, where a new campaign had started in the Balkans. Planning for the Second Front had begun. Now there were not enough shock troops to go round. The recent casualties forced a drastic decision. If special troops were required, they would have to be found from other sources. Late in 1943, a Royal Marine division was brought in to fill the gap. The new arrivals were commanded by Major-General Robert Sturges, a good man, ready to admit the obvious limitations of an untried formation of five thousand officers and men.

But first, at Mountbatten's request, I attended a conference, with Derek and Robert Dawson, to meet the marines and give our views on the amalgamation. It was not a tactful discussion. I liked Sturges immediately and he later proved a loyal supporter who did not take sides and showed no favours. 'Jumbo' Leicester and David Fellowes, the highly-tried G1, became good friends. Their best marine was Campbell Hardy, who went to Burma. The Royal Marine colonels, collectively, were not so amiable. Their outlook was, understandably, stiffer and more hide-bound than our own. Great traditions lay behind them. We had none. They could drill and counter-march better than most of the regular army, with bands in white solar topees; while our brigade boasted one solitary piper. But few of these marines had manned gun turrets, or seen action, in the days of capital ships.* None had Commando battle experience or knew what close fighting was about. I personally found the patronizing, 'old pro' attitude to 'upstarts' very hard to swallow. There were more pompous characters on the staff at Group Headquarters. But I will let John Durnford-Slater describe the set-up:

---

up to tell me of their plight. Gross incompetence! All ranks received railway warrants and a three-day pass to return independently to Scotland, with a holiday thrown in. Laycock had stood by me in the row that followed.

    * Major Ian De'Ath, of No. 45 Commando, was a notable exception, having fought in the secondary armament of HMS *Ajax* in the Battle of the River Plate.

When I reached Britain, there were changes afoot in the organization. Until then there had only been one Royal Marine Commando, No. 40, good people who suffered at Dieppe and were now fighting in Italy. They also were volunteers. Now a conscript and regular division was to take over what remained of the army commandos. I was dead against the idea. Perhaps I was prejudiced, for I was proud of what we stood for. In spite of Mountbatten, whose idea this was, I felt convinced that units of conscripted marines could not to be expected to maintain the high standard of shock troops.

Later I went to Group Headquarters at Pinner.

'Your job will be to look after the interests of the army commandos,' Bob Sturges told me. 'Also I want you to help run the planning for D-Day. Two brigades will be involved.'

I decided to visit them at once.

I found the first brigade, commanded by Lord Lovat, was in a highly efficient state. It included No. 4 and No. 6 Army Commandos. No. 3 was on its way back from Italy to join him. The fourth unit was No. 45 Royal Marine Commando. These men came from a recently formed battalion and had a long way to go to catch up with the others, but I thought that under Lord Lovat and in such good company they would be able to make the grade.

At Cowdray, in the rarefied atmosphere of typewriters and telephonists, I asked staff officers to spare me unnecessary paper in out-trays and in-baskets. Then I got to work on collective training for the Second Front.

John introduced me to General Morgan, the chief military planner at COSSAC (Chief of Staff Supreme Allied Command). Although I was barking up the wrong tree, he seemed impressed by rough weather landings on exposed coasts – experiments backed up by hair-raising demonstrations. In one such display a bunch of trusties, dropped by parachute on the playing fields of Roedean School (near Brighton), lowered ropes over the chalk cliff to No. 4 Commando, landing on the beach 160 feet below, and hauled them up in less than thirty minutes. Fortified, after supping on roast goose, the general, complaining of vertigo, crawled to the brink for closer inspection, his heels firmly gripped by retainers – a happy memory of the war. There was another display off the Cornish coast in a big Atlantic swell: to scale the tall sea cliffs known as the Brandies at St Ives. It proved too dangerous to be practical.

In November 1943 I was let into the D-Day picture. The essential needs to establish a Second Front were basic: (1) a port capable of handling large ships should be included in the area, (2) the sea voyage must be as short as possible, and (3) fighter aircraft should be available at all times to cover operations from bases on the South Coast. It was agreed that Normandy offered the best choice. The Caen area was finally selected for the bridgehead. Though the enemy lay in strength on the front of the British landing, the bulk of his mobile reserves were said to be located north of the Seine.* The beach did not appear heavily defended while the coastal plain could be rapidly transformed into runways. On one flank, narrow lanes, river systems and poor lateral communications appeared unsuitable for the rapid deployment of massed Panzer divisions known to be concentrated in the Pas de Calais.

In January 1944 Generals Montgomery and Dempsey returned from Italy to take over Twenty-first Army Group and Second Army respectively. I never met Eisenhower, but we were soon given the once-over by Monty. After an inspection on the cricket ground at Hove, I received the following message:

2 March 1944
To: Nos 3, 4, 6 and 45 (Royal Marine) Commandos
From: I.SS Brigade   881/13/053
                                                              231050
ALL COMMANDOS TO READ FOLLOWING SPECIAL ORDER
TO: Brigadier the Lord Lovat, DSO, MC
Ref: The Commander in Chief's Inspection
Monday 21 Feb. '44
    I wish to congratulate all ranks of I.SS Brigade parading on Monday for the C. in C. on the smartness of their turn out and soldierly bearing. General Montgomery considered this Brigade the best he had seen since returning to England. I am confident that this good impression will benefit us later and more than justify the high level of turn out, keenness and efficiency.
    All informed.

                          F. DE GUINGAND
                    [signed] Chief of Staff.

General Morgan continued to take an interest in our activities, welcoming an excuse for a break from the planner's room. But he was not liked by Montgomery, who, it has been said, was an over-cautious strategist, jealous of his new authority. In the months ahead there were to be some awkward moments at top planning level: sufficient here to say that opinions were at variance. Unfortunately, the C-in-C left no one in doubt about it when his memoirs were published: 'Morgan considered himself a god: since I had discarded many of his plans – he had put me at the other end of the celestial ladder.' An unworthy comment. Monty was to say unkind things about

* At a conservative guess, six Panzer divisions, in support of forty infantry divisions, were believed to lie opposite the front chosen for the British force.

Auchinleck that were even more unfortunate. It was General Morgan who taught me what I needed to know about the brigade's rôle and enemy dispositions. With a kind, friendly manner, he did not believe in barked-out directives.

It is fair to say the London planners lacked battle experience, and their conception of re-entry into Europe tended to be theoretical rather than practical. Montgomery (whose headquarters was in St Paul's School) welcomed positive thinking among commanders at every level in the field. He gave me a fair hearing on two occasions: with the help no doubt of Freddy de Guingand I got what I asked for in the initial stages of the Normandy landing.

The C-in-C has been criticized for the snail's pace adopted in certain operational movements, but he must be given full credit for overall strategy on D-Day and for the extension of the lodgement area of two armies (commanded by Generals Dempsey and Omar Bradley) which, in 'Overlord', landed simultaneously on widely separated fronts.

It is said when Winston asked Monty, on appointment, how he proposed to set about the invasion, he replied curtly that he would sleep on the matter. Next morning the method and principles to be employed were set out in sub-headings on a single sheet of foolscap paper. Montgomery had his faults, but he translated the build-up for a battle into black and white, then reduced it to simplicity. The precepts he laid down for re-entry into Europe are worth remembering. The appraisal of 'Overlord' sent to the Prime Minister read as follows:

A The initial landing must be made on the widest possible front.
B Army corps must be able to develop their operations through their own beaches and other corps must *not* land through them to cause resulting confusion.
C British and American landings must be kept separate.
D After the initial landings the operation must be conducted in such a way that a good port is secured for both the English and the Americans.

The type of plan required should therefore be on the following lines:
(i) The British Expeditionary Force to land on a front of two and possibly three corps. One American army likewise.
(ii) Follow up divisions to come in (land) behind the corps already ashore.
(iii) All available assault craft (steel-plated) to be used for landing the first waves of the attack. Successive flights to be then poured in rapidly in any type of craft available.
(iv) The air battle must be won before the operation is launched.
(v) We must aim at success in the land battle by the speed and violence of the initial opening assault.

(vi) If the flanks are secured you will be well placed in the battle proper.

To sum up – it is essential to relate strategy to what is tactically possible – with the forces at our disposal. To this end it is necessary to decide the development of operations before the final blow is delivered. There must be a direct relationship between the two. Simplicity is vital to the planning of operations. Once complications are allowed to creep in – the outcome is in danger.

The planned invasion of France consisted of a number of Allied landings stretching from the Cherbourg Peninsula to the Caen Canal, preceded by a massive sea and air bombardment.

My brigade came under command of General Gale's 6th Airborne Division. The corps commander, General Sir John Crocker, afterwards wrote, 'No troops were ever given a more momentous task than the men who landed by sea and air in the early hours of D-Day to seize vital ground on the eastern flank of the Allied invasion.'

I learnt all about the enemy. The 711th Infantry Division (strength fifteen thousand) held a short sector of coast between Le Havre and the River Orne, protected frontally by the Atlantic Wall and heavy coastal batteries. The 716th Infantry Division, of similar size, lay west of the Orne with headquarters in Caen. Two of its eight regiments were said to include Russian soldiers: the 352nd Infantry Division, farther inland, was available for counter-attack, while two armoured divisions – the 12th SS Panzer (Hitler Jugend) and the 21st Panzer, each twenty thousand strong and extremely mobile – were available to reach the battle area within twelve to twenty-four hours.

General Dempsey, who commanded Second Army, was well disposed: he had charming manners and was respected by No. 3 Commando, who had fought under him in Italy. He must have been tolerant as well: in a final Cowdray Park inspection he was bucked off a hard-pulling chaser (which I used as an umpire's hack) when the brigade presented arms, drawn up in hollow square. The crash of a general salute performed by well-drilled troops is always impressive, but a jack-knife dive by the army commander was even more magnificent. To his credit, Dempsey saw the long day through with a painful shoulder and no complaint.

During the last preparations, commandos stretched from Worthing to Eastbourne along the Sussex coast, training to concert pitch. Our services were often in demand. Derek, who could best be relied on for public performances, gave a demonstration of fighting patrols showing German methods of infiltration.* It was enacted on the Downs near

* At this time, although enemy invasion was no longer expected, a hostile landing was always a possibility.

Brighton. No. 6 Commando, who brought back captured weapons from Tunisia, had also kept German uniforms. When fully disguised with belts of ammunition slung round their necks, Derek's men looked the part. They were still sunburnt from the African campaign, and this added to the impression they gave of being enemy troops. It was an impressive display, right down to the donkeys impounded from nearby stables as baggage bearers to carry heavy mortars and an anti-tank gun. During rehearsals denim suits had been worn; on the day itself everyone taking part paraded in German uniforms.

After the actors had been dismissed they returned, as was customary, to various lodgings in the town. One man was surprised to see his landlady rush in from the garden and bolt the door against him. Two others entered a pub and ordered a pint of beer; the taproom emptied rapidly, except for a determined-looking character behind the bar, obviously resolved to sell his life, and the local, dearly to the enemy. Other soldiers waited at the bus stop, when, to their dismay, the double-decker put on speed and went past, with anxious faces peering from the windows. After several incidents of this kind the patrolmen reached their digs; then brigade started to receive complaints. It was our fault – the uniforms had been overlooked. Understandably, the shock had caused alarm and despondency.

Life in Sussex was never dull. The arrangement by which commandos lived in billets worked very well. The men became members of the families they were boarding with. There was little trouble, but there were occasional incidents. Derek described another laugh:

One morning I wanted to see the adjutant, David Powell, about something, and, noticing his door was ajar, I looked in. David had been badly shot up in North Africa. His eyesight was not one hundred per cent, and he concealed the fact, as far as possible, by wearing a monocle screwed tightly into his eye. Under magnification, he was now examining with care an enormous expanse of female bosom exposed for his inspection. Not wishing to intrude on this homely scene, I decided to use the telephone. He came at once and explained the position.

One of our storemen had apparently had an unfortunate love affair. The girl of his choice, a relative of his landlady, after reciprocating his advances for some time, suddenly transferred her affections from the dour and silent Scot to a lighthearted Canadian soldier. At first he brooded in silence, but later confided in friends, who took him out and bought refreshment to ease his troubled mind. Towards the end of a convivial evening, they decided the storeman should see the lady and settle things once and for all. The rejected suitor pushed open the gate of the sedate suburban dwelling and entered the front door. He was immediately felled by a blow from a frying-pan.

His friends outside heard the commotion; one of them rushed into the house. The lights were out. There was a blind struggle in which he got hold of the frying-pan. 'Come on, you bloody swine,' he shouted, 'where are you?'

'I'm here,' said a gruff voice in the dark, and the storeman's friend lunged heavily with the improvised weapon.

Unfortunately his wily opponent dodged quickly to one side, and the frying-pan landed with some force on the ample bosom of the former sweetheart.

The matter was eventually settled. The wielder of the pan lost two stripes and all concerned performed a number of arduous fatigues – which was our equivalent of CB. Disciplinary lapses were rare and normally dealt with by returning unsatisfactory men to their regiments. However, these commandos were excellent chaps and I could ill afford to lose them.

Before Normandy, I gave the corporal his stripes back. I remember saying as we embarked, 'I'm glad to see you properly dressed again, corporal.'

'Yes, sir,' was the slightly subdued answer.

'I hope you've got all your weapons,' I continued, 'not forgetting the frying-pan? That should win the day.'

'Yes indeed, sir,' he cheerfully replied.

During the war various foreign royalties had found temporary homes in the UK. The active members were attached to Allied staffs. Prince Felix of Luxembourg, with his charming family of small children and the Grand Duchess, rented Eskadale (a house on the estate) for the duration. I thought he would get on well with Derek, who has the following reminiscence:

One morning I heard that we were to be visited by HRH the Grand Duke of Luxembourg. He was to spend a day with us as he wanted to see the commando training. When he arrived and duly saw the troops at work, I noticed in the entourage, a smart commando soldier, wearing a bugle lanyard in the Luxembourg colours. His Royal Highness was particularly pleased.

I whispered to David Powell, 'Where on earth did that come from?'

'Sergeant-Major McLeod's dressing-gown cord, sir,' he answered softly.

After the morning's training, I laid on a good lunch in my digs which HRH thoroughly enjoyed.

At Cowdray, Jim Mellis, who had served me faithfully in peace and war since we first joined the Scots Guards in 1932, unexpectedly went down with a duodenal ulcer. He was invalided out of the army, staying on as 'the Laird's man' for ten more years in civvy street. He now heads the Post Office in Beauly. Mellis had been a tower of strength and utterly dependable. He was sorely missed. I then drew a dud replacement who could not go the pace and was returned to duty. Private Salsbury, discovered before Normandy, was an immediate success. This smart, apple-cheeked young soldier, who came from Bedford, was a clever mimic and much loved by my children for his cheerfulness and winning ways. His family owned a tobacconist's business. I have a photograph taken on his wedding day, resplendent in tail-coat and top-hat.

THE BRIGADE PLAN FOR D-DAY

The intention

The 6th Airborne Division and 1st Commando Brigade would be responsible for holding the left flank of the Allied bridgehead.

The method

The brigade, consisting of four Commandos, to land on the extreme left of BEF on Queen Beach (Sword) and cut inland to join forces with two brigades dropped inland overnight by glider and parachute. No. 4 Commando to destroy a battery and garrison in Ouistreham then later rejoin. The rest of the brigade, landing thiry minutes after No. 4, to fight through enemy defences to reach and reinforce brigades of 6th Airborne Division, meeting astride bridges spanning the River Orne and Caen Canal at Bénouville. Glider regiments of the Air Landing Brigade would arrive late the same evening descending in country cleared of the enemy.

We had been assigned a formidable task.

I told commanding officers to emphasize our considerable psychological advantage. The Germans were concentrated behind the 'Wall', in strongly defended localities dotted over the area. Concrete emplacements are difficult to destroy; but they impose a defensively motivated outlook on the occupants: the feeling 'It's bullet-proof indoors and here we'll stay and not sally forth.' The view inside a pill-box remains restricted while those without can see more clearly. I intended to take full advantage of this fact. After dealing with immediate defences we faced a running fight across country. Nothing was left to chance. The assault was worked out in great detail. Chiefs of Staff considered the plan over-bold. Three hours to get through to the bridges appeared on the short side.* I stood on my word – having said it was possible – but I had to debunk the pessimists.

To prove the point I suggested a dress rehearsal with identical landing-craft and crews carrying the same personnel they would put ashore on D-Day. The selected beach (cleared of mines, but with the wire left standing and backed by similar terrain) lay between Angmering and Littlehampton, near Arundel on the Sussex coast.

---

* The distances across country were: the beach to the River Orne, 4½ miles; from the Orne Bridge to the village of Le Plein, a further 1½ miles on the route followed by the brigade on D-Day.

I chose Derek to break through and spearhead the advance. Peter Young, who had done brilliantly in Italy, was asked to follow him. To ring the bell on these occasions, one has to cheat a little. The men put ashore were picked veterans. Lieutenant-Commander Rupert Curtis, RNVR, and his flotilla LCI(s),* proved equal to the hour. It was not our first practice landing. We had seen cumbersome performances before, both real and simulated; they invariably went wrong. This was to be the best ever – the all-out attack executed at full throttle. NCOs were warned to run the course in light raiding order. I asked for a blinder with no holds barred, and signalled we must show Home Forces what Commando speed was all about. My word, how they responded to the spirit of that five-mile dash!

All the top brass came down from London. I provided the running commentary on a loud-hailer, first advising Generals Morgan, Bob Laycock and Freddy de Guingand, along with Dempsey's Chief of Staff, to stand clear of the beach and not to part with the keys of their cars. Sergeant Rice, my driver (now a rich man in Canada), and two stick orderlies, stood guard over the Command vehicles.

The Channel was calm that morning. The craft carrying No. 6 Commando ran in, well controlled, in line abreast, hitting the sandy beach in a dry landing. Derek had borrowed two-inch mortars to thicken up his smoke; men mounted in the bows opened up still well out to sea. The little bombs rained down just wide of the spectators, obliterating the landing. There were lucky escapes, but not one got hurt. Some of the higher staff, fearful of mishap, shouted for a cease-fire, but I continued the commentary as No. 6 swept inland through the crowd. 'Rude but high-ranking' was Bill Coade's comment as he cannoned into a furious G1 from Southern Command, coughing in the fog, who was trying to find me. 'Lovat, call off your bloodhounds – somebody is going to get killed.' But the pace was too good to enquire.

Peter Young came in on the second wave and the treatment was repeated. No. 3 Commando went one better. 'Finding was keeping' on this occasion. Section leaders became pirates as they crossed the sea wall and doubled inland. Their CO had asked if he could make use of car park facilities. This was a demonstration; to end arguments, permission was granted. Staff cars, Jeeps, motorcycles – everything in sight was commandeered; protesting drivers were tipped out of their seats or relieved of keys. Fully thirty vehicles (one belonging to a high-up in the Military Police, handling traffic control) roared out of the car park and headed across country for Arundel.

There were big drains to cross in the rubber dinghies; these had me

* LCI(s): Landing-Craft: Infantry (small). Each craft carried eighty-five men, plus crew.

worried, but without heavy packs the task was simple. No umpires caught up to interfere, and by the time the leading troop crossed the River Arun upstream of the castle, Derek had won the game. The whole Commando had reached the Arundel gap in the Downs beyond the river when I called a halt. Generals Morgan, Bob Laycock, Freddy, and John Durnford-Slater, were delighted. The planners were impressed. Everybody else was extremely angry: Southern Command asked for my head and put in a claim for damaged vehicles. No further doubts were expressed over brigade's speed off the mark: tactically, the exercise had been completely successful.

I saw a brave deed performed that day. A rubber dinghy overturned in the Arun where it runs fast and deep. A man got into difficulties, and David Powell, in full equipment (handicapped by his eye blinded in North Africa, along with other wounds), dived in and saved the soldier.

No. 4 Commando stepped up their training in street fighting. Then they went north on an exercise in the Moray Firth with the 3rd Division, whose 8th Brigade were to land before Robert Dawson on D-Day. I sent Derek as an observer. They both gave gloomy accounts of hesitant and badly led troops.

On the South Downs and close country behind them the brigade sweated away, burning off the fat. Fifty volunteers from each Commando were sent on a parachute course – just in case of emergency. Every man became fully conversant with all types of enemy weapon; No. 6 Commando had built a range of their own for this purpose, in a valley where shooting was possible at moving targets. Pill-boxes and other defences were set up, resembling country over which the battle would be fought. I built a de luxe model for the assault course at Cowdray, provided with a bullet-proof glass panel to experience sensations of being shot at or smoked out. It proved a success. Even the seasoned troops back from Italy and North Africa found they were out of practice and had to work as hard as newcomers; every commando soldier struggled to keep his place in the team. They knew sound training would save lives. In war, I kept repeating, the worst casualties are caused through incompetence, and not by the enemy. We were going to keep the losses down. For the more senior officers I had a sterner message: 'A good Commando may be destroyed but it cannot be defeated.'

Finally the day came the free world had been waiting for. Morale in No. 1 brigade stood high. On 6 June, commando soldiers went into battle, elated to the extent that the hazards ahead appeared of little consequence, and fear was a forgotten emotion.

CHAPTER TWENTY

OPERATION OVERLORD

Linked in the serried phalanx tight
Groom fought like noble: squire like knight
As fearlessly and well.

'Marmion'

By the morning of D-Day, and indeed long before, all four Commandos had reached the highest pitch of efficiency. Battle experience combined with intensive training had produced a military machine as perfect as any which history can show. From the individual commando to the tall, immaculate brigadier, every soldier landed that day washed, shaved, and with clean boots. An atmosphere of absolute efficiency prevailed.

To this a sense of exhilaration produced by the solemnity of the hour and the words of their commander were added. 'It was truly inspiring,' says one of them, an officer of Marines, who listened that day to Lovat as he addressed his men before embarkation. There was no nonsense, no cheap appeal to patriotism. He spoke simply, but imbued each man with the spirit of this great task.

During the afternoon of the 5th June, the commandos embarked at Southampton 'in a grotesque gala atmosphere', records Captain A. D. C. Smith, 'more like a regatta than a page of history, with gay music from the ships' loud hailers and more than the usual quota of jocular farewells bandied between friends....' It was a perfect summer's evening, the Isle of Wight lay green and friendly and tantalizingly peaceful behind the tapestry of warships. At 1800 hours the commandos sailed down the Solent before 'setting out to war'. In the first of twenty-two craft in line ahead, Lord Lovat's piper was playing in the bows. ... It was exhilarating, glorious and heartbreaking when the crews and troops on ships began to cheer ... the cheers came faintly across the water, gradually taken up by ship after ship.

I am told the Admiral threw his hat in the air. ... I never loved England so truly as at that moment.

From *The Green Beret*

Certainly a great occasion, but above all a moment of release. The invasion build-up had grown interminable: a mounting tension, to some unbearable, that weighed like a millstone until we stepped at last into the boats. Tony Smith's description of the final departure is accurate enough; in retrospect that euphoria must be qualified. It had not been beer and skittles all the way. The last week, as we waited for the starter's flag, stands etched in memory.

I had my own doubts about the Second Front. Despite superior technical resources in weapons and equipment, and increasing air strength, the general standard of efficiency – both British and American – left much to be desired. In the spring of 1944 few regiments in Home Forces had faced a German tank, nor, in cold hindsight, had General Paget been the best of trainers. Some of the best soldiers in our experienced army were still fighting in Italy, while the navy waited impatiently for the return of salted landing-craft from the Mediterranean. There were top-level differences of opinion: Montgomery was said to be at variance with the planners, and maybe with Eisenhower.

Against this background came a disastrous invasion rehearsal in the Moray Firth. To avoid mistakes I envisaged a shorter sea crossing – given favourable conditions – in late May, bound for the Pas de Calais; so did the enemy, who had massed their Panzer divisions north of the Seine! As a precaution, an O group of key personnel from brigade moved down to Seaford; it looked a good central spring-board (with the necessary accommodation not hard to find, for Sussex landladies were invariably kind). My destination theory proved entirely wrong: rear details stayed where they were, enjoying the comforts of John Cowdray's stately home. I was visiting them overnight when the embarkation signal flashed over England: 'Top Priority. From Movement Control, Southern Command.' The message was short and to the point: 'Commandos will be collected and moved under separate arrangements at xxx hours as detailed: proceeding to C18 Transit Camp, Southampton. Acknowledge and alert all units concerned.'

Brigade headquarters was waiting packed and ready. Despatch riders roared out of town with sealed orders for commanding officers. The wheels of invasion began to turn. Our waterproofed Jeeps had already gone ahead, loaded into ships for France. Midhurst remained outwardly calm. The rear party took over files and switchboard; John Cowdray (who had lost a limb at Dunkirk) gave me breakfast before departure. The fighting men had far to travel, and there was time to say goodbye to Rosie and the children. Then the convoy rolled.

It was a morning of sunlight and high wind and the sea must have been rough in the Channel. Whistling 'Lilli Burlero', formations of the

1st Commando Brigade debussed in the suburbs of Southampton. They did not arrive together but a distinctive appearance was common to them all. There was, perhaps, a hint of arrogance in that first impression, concealing the qualities of individual leaders – each a law unto himself. They looked good and they damn well knew it. The assembly area, scene of some confusion, already swarmed with bewildered infantry, sitting round piled arms with that far-from-home and nowhere-to-go cast of countenance that characterizes troops in transit.

When dog meets dog in uniform a spirit of rivalry forms a convenient lamp-post.

There were shouts of 'Cheer up lads! Get off your knees and keep those peckers up!' as lithe, laughing men sprang down from troop-carrying vehicles; the steel helmets, garnished with camouflaged trimmings (as worn by Home Forces) were always greeted with derision. But Max was ahead with guides; indulging in no further ruderies, commandos fell in and, directed by Red Caps stationed along the route, made quick time into camp.

The men marched at ease in column of troops and the earlier stiffness was soon gone. Some had arrived by train, others came hard-arsed from billets on the Sussex coast – packed in open trucks, with Bren guns mounted on each vehicle against fighter attack; but no Germans appeared that day to strafe sitting targets on roads that were jammed for miles as armies crawled across England, nose to tail, to their ports of embarkation.

The commandos were in fine fettle – and small wonder. Hell's bells, we had joined up, gone places and grown tall together (for almost four years, to be exact). And this was a special, almost festive, occasion. A battle honour to be shared between old friends. It had meant a long haul for most of that great brigade, chosen to spearhead the BEF's re-entry into Western Europe. If the men had seen much fighting in different theatres they surely knew that now they faced the clincher. The challenge was cheerfully accepted. To weld the spearhead into an integrated force, No. 6 Commando had been recalled – battle-scarred – from First Army in North Africa. No. 3, with faces still browned by Italian skies and much raiding in the Mediterranean, were also in time for D-Day. Now as veterans – proud, deep-chested and superbly fit – they stepped out jauntily under Bergen rucksacks. Some wheeled bicycles and all were hung about with weapons; the big fellows carried portable flame-throwers, rubber dinghies, scaling ladders and anti-tank missiles. Though they marched at ease, the men bore the look of eagles; sunshine caught the flash of regimental badges on green berets and toe-caps of polished boots as, ignoring a fussing provost marshal, they swung through guarded wire gates and across a sandy common to lines of tents dispersed among the trees.

No. 45 Royal Marine Commando (a new formation commanded by Lieutenant-Colonel Ries) were already in camp. They had not yet seen enemy action and they lacked the indefinable quality of the old guard that I have just described – the spirit of the bold lads who joined up for the hell of it, because they liked fighting and the roving life. But those young marines carried the traditions of a famous corps: if they lacked experience and cunning, they had done well enough in training and there was a fine discipline at every level. They appeared eminently suited for boating across rivers in the target areas. They were later to prove their ability in aquatic operations, helping the advance over water obstacles through Germany (the Maas, Rhine, Weser, Aller and Elbe).

Today there was no need for inspections or salutes. Brigade signals, in shirt-sleeves, were checking wireless sets when Robert Dawson arrived from Bexhill with No. 4 – the 'Cavalry', now better known as the VC Commando.* With him came the irrepressible Free French under Philippe Kieffer: two hundred Fusiliers Marins, some the heroes of Bir Hakeim. The Gallic panache added colour to the scene. The sailor types were curiously tattooed. One brown-eyed fellow – I suspect he had served a prison sentence – bore the legend 'Pas de Chance' enscrolled upon his forehead, and that was the name to which he answered on parade. I knew some by sight, others by reputation. Maurice Chauvet had sailed in a rowing boat from Morocco to seek freedom. He was picked up by a Spanish vessel, taken to Algeciras and incarcerated in the infamous prison of Miranda de Ebro, suffering there eighteen months of deprivation with other Frenchmen. More had deserted from battleships after the fleet capitulated to Vichy.† Achnacarry is a far cry from Oran, but the escapists got there to offer themselves for special training. The Frenchmen stayed the course and won their green berets – none more deservedly than the weakened Chauvet, who completed the ten-mile march from the station with a carbuncle on his heel. When they drew off his boot the grey army sock was a welter of blood. In attendance on this gallant company marched their doctor, a patriot who was to be killed tending the wounded on D-Day. They sang 'Sambre et Meuse' and eyes were bright, for they were going home. Each wore the Cross of Lorraine with his shoulder titles. 'L'audace et

* Victoria Crosses were won by Captains Porteous and Gardiner (by the latter after rejoining his old unit, the Royal Tank Corps, in the Western Desert).

† In France no man was considered a soldier until he had fought in North Africa or Indo-China. See the Appendix for a further account of these French troops by Robert Dawson.

toujours de l'audace' summed them up. The *élan* with which the returning exiles stormed the beaches brought heavy casualties later in the
week.

I have already sung the praises of No. 4 Commando. A formidable
unit, specializing in night attacks on enemy installations, they had
distinguished themselves in forays from the North Cape to the Pyrenees,
fighting as individuals or in small parties and, when required, at full
strength from larger vessels. Lofoten, Boulogne, St Nazaire, Dieppe,
the Channel Islands and the Low Countries had smoked behind them.
For me they were tops, but I knew them best of all. The size of a
Commando at full strength (some five hundred officers and men, all
fighting effectives) is smaller than that of an infantry battalion. From
small beginnings, No. 4 was to end the war with two VCs, five DSOs
and nine MCs. Among the officers, two became brigadiers, and three
'boy colonels' in their middle twenties emerged to take over other commands. Except for Tony Lewis (Dorset Regiment), none were regular
soldiers. At least one senior officer was wanted, dead or alive, with a
price tag on his head. At this point in time they had lost few men.
Speed and daring made them razor-sharp, but they had not fought in a
campaign.

The rank and file were as good as the officers. Some carried
recoil-less K-guns (strictly RAF property) and the new PIAT* anti-
tank weapon, to increase fire power in assault. On that day Craig and
Donkin pulled a trek cart drawing the heavy mortars. Both were miners,
two great-hearted soldiers, small men who could hold their place on a
forced march in any company. Craig, a wee castle of a man from
Ayrshire, was to fall on D-Day and old Donkin (nearly forty, with
many children) died at Flushing in one of the last bloody raids of the
war.

So much for the troops. Some naval signals, with a bearded FOB
officer, asked leave to come aboard. A new padre made an unexpected
appearance and Bob Laycock sent down Charlie Birkin, an observer
to act in liaison with COHQ. Cameramen and the press were barred.
'Overlord' was set supposedly for 5 June. That night the camp was
officially sealed: it was the end of May.

In Southampton the holiday spirit prevailed. There was a NAAFI,
run by American GIs, selling confectionery and soft drinks (candy's
dandy but liquor's quicker); there was also a cinema tent and enough
space for football played in gym-shoes – Hyde Park style – and stump
cricket. Ropes were obtained to provide a Tarzan course among the tree-
tops.

* PIAT Projector Infantry Anti Tank – also useful in house-to-house fighting.

Officers busied themselves with final arrangements and the check-ups that save lives; dinghies were inflated, bayonets sharpened, automatic springs tested, magazines oiled, waterproof wrapping wound round all weapons, escape maps were issued with ammunition, rations and first-aid kits. Hand grenades were not primed until the day of departure. The intelligence section made a sand table to supplement air photographs showing half-hidden details of enemy fortifications, wire and suspected minefields. Place-names and destination were withheld until embarkation, but Frenchmen identified the destination at a glance. Sappers and signals attended to the mysteries of their trade. The briefings were a formality: everybody knew his job.

The brigade major was ordered to take a rest. He had worked round the clock for ten days without a break. Michael Dunning-White accordingly retired to his flea-bag and silk sheets. Dreamless shut-eye is the soldier's friend, and the officer in question needed it. With two wireless operators and Johnstone, his trusty batman (a furniture remover in peacetime), Michael had to face some unpleasant chores after touch-down on the beach.

The sun shone; there were no parades. After breakfast roll call, some PT, a fox-hole 'dig or die' competition, a rifle inspection before lunch, and then the day was clear. The US caterers did not approve of the commissariat ('If it's cold, it's soup. If it's hot, it's beer'), but the British enjoyed this get together of old friends. It was their last relaxation before weeks in the front line.

FINAL ORDERS

From Tactical Headquarters, Southwick House, near Portsmouth, General Montgomery called his captains to a last conference: an unpleasant drive against the traffic, and for some a waste of time, being a repeat of the Brighton visit. I had a great respect for the future field-marshal. When he got to know us better, commandos were to win his particular regard; but on this occasion he missed an opportunity of remaining modest and to the point, and his lengthy address largely concerned himself:

> 'He either fears his fate too much,
> Or his deserts are small,
> That puts it not unto the touch,
> To win or lose it all.'

It sounded inappropriate, with Germany reeling from defeat in North Africa, Mussolini overthrown and the Wehrmacht committed on three fronts (Russia, Italy and the Balkans), with another coming up round

the corner. Our armies threatened the gates of Rome. We digested
some impressive facts: 'The whole of the Allied air power is available
to see us on shore. Its strength is terrific: 4,500 fighters and fighter
bombers; 6,000 bombers of all types.' Montgomery did not mention
the juggernaut weight of 21st Army Group* or our naval strength at
sea. I seem to remember it mustered 5,000 craft of varying sizes. It
looked as if the invasion was getting off to a good start!

The more impressionable members of the audience tiptoed from the
room. I left feeling that I had been listening to a headmaster taking a
backward class: concise in his approach, but egotistical to a fault and,
on this occasion, something of a *poseur* into the bargain. There was a
piece of good news. De Guingand, the power behind the throne (Chief
of Staff and a former schoolfellow), drew me aside. 'If they can land
and clear Sword Beach you will have swimming tanks in the assault
after all. But intelligence says Rommel has thickened up the underwater obstacles. Frogmen go in early, but no chance to clean them
up in time. The Ouistreham end of Sword looks a hot potato; the
town is strongly garrisoned, but please knock out the battery in double
quick time.' I disliked the sound of those underwater obstacles. The
ones appearing on the surface looked bad enough.

In camp there were last-minute flaps. The axis of advance – a line
of trees chosen to provide useful cover – had been cut down, according
to new air photographs. Then Curtis, the flotilla leader taking us across,
had his navigator posted elsewhere without warning. Speculation on
who had given this lunatic order was pointless: something had to be
done quickly. Admiral Ramsay, the Naval Commander-in-Chief, was
not available, but Derek, as second-in-command, got cracking. First
putting on his blue Irish Guards forage cap and best uniform (collected by special messenger), swifter than Malise with the Cross of Fire
he sought out Naval Operations Headquarters. In conversation Derek
could hit like a heavyweight. His case was convincing. He talked
round the sailors, and the navigator was returned. As Lady Hamilton
once murmured from the depths of her hammock, 'Thank God for the
Nelson touch.' I personally felt like Napoleon after Marengo, extricated
from near disaster by Desaix! The Americans were not so lucky with
their navigators, who put troops ashore in the wrong places.†

On Saturday night Southampton had an air raid. The Ack-Ack

---

* This consisted of the First US Army under General Bradley and the BEF under General
Dempsey. Two more armies would land later: the Canadians under General Crerar and the
Third American Army under General Patton.

† It is not always realized that the lodgement area in Normandy covered different landings
on five beaches along a coast stretching fifty miles: Utah, Omaha, Gold, Juno and Sword. Juno
was on the Canadians' frontage; Sword on the BEF's.

gunners pumped up everything they had and the defences loosed off a smoke screen, a ground-level device which smothered the camp. Next morning Salsbury, dumping a canvas bucket of shaving water, reported damage from falling shrapnel in the marquees.

I agreed with two willow wrens, nesting in a grass tussock by the tent flap, that we had suffered a disturbed night. The small, shy birds, now going about their appointed tasks in spite of clumping feet, helped to instil renewed confidence. I had felt the same way on discovering ladybirds up a sleeve before raiding Boulogne. Time was running out before touch-down. Then perceptions are keenest and the most uni-maginative people seek for signs and portents.

On Sunday René de Naurois, his decorations a splash of colour on a white surplice, said Mass for three hundred men kneeling on the grass. At the Interdenominational Church Parade a favourite hymn, that has since become our own, was sung with feeling:

> Eternal Father, strong to save ...
> O hear us when we cry to thee
> For those in peril on the sea.

It goes well with male voices, but the new padre preached a rotten sermon about death and destruction which caused surprise. There are few atheists to be found before a battle, or later in shell-holes. Tension was building up, and charity perhaps a trifle thin on the ground. There were a number of complaints; the cleric was suspended and told to return from whence he came. Poor fellow! A spark can cause the prairie fire! It was mistaken zeal from a man, lacking combat experience, who did not know his congregation, and doubly unfortunate in that it con-flicted with my own 'God speed' before departure. The incident was forgotten but the dismissal was taken badly. On the last day in camp the unfortunate man took his own life. A sad business, with barely time for regrets, for troops were belting up amid the dust and shouting as embarkation transport came grinding in to Southampton to take us away. Max, a most humane officer and the soundest of administrators, cleared up the pitiful remains. The padre was put down as a 'battle casualty'.

In the right mood soldiers appreciate a word of encouragement, but not a lecture as seconds get out of the ring. The commandos were no exception: as professionals they knew the score and with that know-ledge did not accept half-truths or doubtful leadership. It was mutual trust that raised men above themselves in the encounter; now on the start line something was expected and I faced a critical audience. What I said took two minutes. It was simple enough, the message plain. I weighed each word, then drove it home, concentrating on the task ahead and simple facts – how to pace the battle, which in the event came true. First I spoke in English, then in colloquial French, which was greeted with a yell: 'Vous allez rentrer chez vous. Vous serez les premiers militaires français en uniforme à casser la gueule des salauds en France même. A chacun son Boche. Vous allez nous montrer ce que vous savez faire. Demain matin on les aura.'

The English version was different. I touched on past achievements, then congratulated the commandos for being chosen to fight together for the first time as a Brigade – a proper striking force, the fine cutting edge of the BEF. They knew their ability from past experience; they could expect a physical encounter in which they had no equal; the greater the opportunity, the greater our chances of success. They knew their job and I knew they would not fail. Put it this way: 'The bigger the challenge, the better we play.' History might say that there were giants in those days: we were going to prove it tomorrow. Our task appeared a tough assignment, but we held the advantage both in initiative and fire support. It was better to attack than defend. The enemy coastline would be flattened and defences pulverized before we arrived. Three 'maids of all work' infantry battalions, landing at zero hour, would clear the beach, then find us exits through the minefields above the tide-mark – as easy as kiss your hand. It was all laid on a plate.

The brigade major would be ahead, waiting to guide the way. Spearheaded by No. 6 Commando, the brigade would breach the Atlantic Wall and then press through demoralized enemy positions. The momentum of our advance inland must be kept going at the double to reach the airborne division. Dropped overnight, the paras would be holding high ground over the canal and river bridges north-east of Caen: we were not going to let good men down.

The break-out meant fierce fighting by direct assault on a narrow front, without getting involved on exposed flanks or stopped by any holdups. Each commando would leap-frog through the one in trouble, and there would be no pause to slow the speed of our thrusting attack.

I allowed three hours to reach the Orne bridges and all afternoon to make good the high ground beyond. We would be dug in before dark to meet inevitable counter-attacks. No. 4 Commando and the French (their task to destroy the enemy guns and garrison in Ouistreham at first light) would rejoin the brigade to firmly anchor the left flank of the landing. It was going to be a long day, but given a reasonable crossing in fast launches, we should be fresh for the encounter phase of the battle. After that, stamina would decide. Michael's signallers would be in touch with airborne forces. If the bridges had been seized intact, the rubber dinghies – which weighed men down – could be cast aside. The brigade was going to make history and I had complete confidence in every man taking part! I ended with the suggestion, 'If you wish to live to a ripe old age – keep moving tomorrow.' And so we stood across the sea to France.

THE PASSAGE

There was a knifing wind in the Channel. Rupert Curtis described the sea as 'lumpy' when I joined him on the bridge.

The weather had changed since we left the shelter of the Isle of Wight; so flat was it in Stokes Bay that a New Zealand boat officer's* gramophone, playing 'Hearts of Oak' carried over and beyond our ships as the brigade supped off biscuits, American plum duff and self-heating soup.

Waiting for the darkness, Derek clambered over from his motor launch to mine. Immediate worries were over, with time to unwind before touch-down. Now the parcel was in the post and out of my hands. The navy found a half-bottle of gin. Later, amid unseemly merriment, Peter Young, another bachelor, was summoned to attend the party – not, as expected, for a fond farewell, but to study some remarkable information in a paperback discovered below decks. The guidance in Dr Marie Stopes's Marital Advice Bureau for Young Couples setting out on a honeymoon was full of surprises. Soft beds and hard battles had something in common after all! The officers returned to their commands, shaken by the revelations – hopping across lengthening shadows on gently heaving decks. An amplified band played 'Life on the Ocean Wave'; I took my boots off for a last kip in Rupert's bunk and slept well. I can snore through any form of disturbance, provided I go to bed with a quiet mind.

That happened yesterday. Now we were in mid-Channel. 'C'est le jour – le jour de la Liberation.' The paling stars spelt out 'Invasion'. It was blowing half a gale and getting light enough to see Curtis, now with his steel helmet on. He had reddish hair and a serious face. A quiet-spoken, dependable man, keenly aware of the importance of the occasion, Rupert was to be awarded the DSC for the work he did that day. I imagine he felt lonely on the crossing: twenty-two boats pitching in line ahead; seven hours of eye-strain darkness, keeping station in rough weather up the swept passage through the minefields. 'Twenty

* Lieutenant D. G. M. Glover, DSC, RNZN.

miles from the coast and twelve to lowering point,' he shouted against the wind. I nodded respectfully, trying a shivering smile with eyes on the duffle coat. The navigator had done his job well – on course and ahead of the clock. Nautical twilight was past and the sea changing colour to oystershell in the grey dawn when an Aldis lamp blinked on our port bow. 'Good morning, commandos, and the best of British luck.' Curtis and his yeoman spelt out the signal. We made a suitable reply: 'Thanks; think we are going to bloody well need it.' Rupert ran up his battle ensign. War was becoming personal again.

At 0530 hours *Warspite* and *Ramillies* opened fire. The men came up from below to stretch legs, and sea-sick soldiers gulped in fresh air. The cruiser *Frobisher* joined the battleships on our port quarter. Muzzle blast from the turrets of the ironclads lit the dawn with a yellow glare as fifteen-inch guns hurled their one-ton shells into the batteries round Le Havre. The fearsome salvoes screamed over like trains coming out of a tunnel.

Tony Smith – veteran of the Norway, Dieppe and Boulogne sorties, a philosopher with a dry sense of humour – thought it was rather noisy and overdone – with the big guns pointing in the wrong direction. Broad daylight now; Tony, like myself, preferred night raiding to this kind of public exposure. 'Fear nowt, Tony!' I said. 'That is just an overture. Listen to the racket starting on the starboard side; old *Ajax** with destroyers are way on ahead, shooting up defences. *Quincy* and *Nevada*, with French and Canadians to match, *Georges Leygues*, *Qu'Appelle*, *Restigouch* and *Saskatchewan*. HMS *Enterprise* and *Black Prince* are there as well. At 6 o'clock some real noise will startle the *Herrenvolk*.'

In France, belatedly, the BBC warned citizens *en claire*:

This is London calling. I bring you urgent instructions from the Supreme Commander. The lives of many of you depend on the speed with which you obey. It is addressed to all who live within thirty-five kilometres of the coast. Leave your towns at once – stay off the roads – go on foot, and taking nothing with you that is difficult to carry. Do not gather into groups which may be mistaken for enemy troops.

The Resistance had more timely notice. The first alert among daily bulletins came on 1 June: just one line of Verlaine's *Chanson d'automne*. 'Listen carefully,' said the announcer in London to Frenchmen crouching over radios. 'Here are some personal items of information.' They

---

* The cruiser *Ajax*, of River Plate fame, helped to sink the battleship *Graf Spee* in 1939 outside Montevideo.

appeared meaningless. 'John has a long moustache.' 'Sabine has just
had mumps.' 'Napoleon's hat is in the ring.' Then a pause. 'Les sanglots
longs des violons de l'automne.' It was enough. Across France 'the
underground' and men of the *maquis* stood by for the next line. It came
on Monday evening, 5 June. 'The weather is hot in Suez.' 'John loves
Mary.' 'The dice are on the table.' 'Blessent mon cœur d'une
langueur monotone.' The message arrived before midnight. Armed
Frenchmen rose instantly to strike at rail tracks and goods yards; telephone lines and cables were attacked everywhere and destroyed. Pipe-
lines carrying petrol on to Sword Beach to set the sea on fire were
cut during the early hours. The response was magnificent and inspiring.

The sea abated as we picked up shelter in the Bay of the Seine.
Thoughts turned to the inner man. Commandos carried forty-eight
hours' rations, and I ordered those who could swallow to eat breakfast.
Hot ship's cocoa and oily sardines (I can still smell them on my
fingers) meant a tin less in rucksacks. No rum ration was issued to my
HQ; it was thought to be bad for the wind in the forthcoming
marathon.

The salvoes from the battleships, left far astern, merged and then
faded against a new approaching sound that throbbed over the flotilla.

We had no experience of air support, but the scene has been well
described by Cornelius Ryan in *The Longest Day*:

> Slowly at first, like the mumbling of some giant bee and then building to a
> crescendo of noise, the bombers and fighter bombers appeared. They flew
> straight across the Channel wing to wing – formation after formation – nine
> thousand planes in all. Spitfires, Thunderbolts, and Mustangs whistled in
> over the heads of the men on decks.
>
> With fine disregard for the naval shelling, they swept down to strafe the
> invasion beaches at low level, swung up and came in again.
>
> Above them, criss-crossing at every altitude, came the 9th Air Forces'
> Lancasters, Fortresses and Liberators.
>
> It seemed as though the sky could not hold them all. Sea-sick men looked
> up and stared with a sudden emotion almost too great to bear. It was going
> to be all right after all.

But carpet bombing and supposed saturation over such a wide front
did not work out as expected. The cloud layer hung so thick above the
Americans at Omaha that 350 bombers, detailed to knock out coastal
defences, flew on in fear of hitting their own troops. Far inland, 15,000
bombs were unloaded wide of target. A mighty orchestration, none-
theless: the jar of falling blockbusters like flat champagne corks; the
popping thuds muted by distance and more noise. Judging by the crumps
on Sword – eight miles ahead – the Germans were taking a pasting;

but, there again, the battery area in Ouistreham – No. 4 Commando's objective – escaped with little damage.

The blitz ended before the coastline hove in sight; desultory firing continued as destroyers closed the range in an attempt to deal with closer gun flashes. Relative silence cast an ominous chill as we blew Mae Wests up, checked safety catches and strapped on equipment. Fur would soon by flying in the opposite direction.

The lowering position (later to be more crowded than Piccadilly Circus) appeared empty, except for one policeman on patrol. It was Admiral Vian, flying his flag in HMS *Scylla*; the gun crews on the cruiser's turrets wore white protective clothing, giving them a curious polar bear appearance. She looked superb – impressive with a massive dignity that came from being almost stationary. Britannia ruled the waves that morning.

The launches passed close to her bows, Rupert's battle pennant snapping in the wind. It is on record that we made a favourable impression: 'As their landing-craft drew level with the Flagship, the commandos of Lord Lovat's brigade – spruce and resplendent in their green berets – gave the thumbs up sign. Looking down on them, eighteen-year-old Able Seaman Northwood thought they were "the finest set of chaps I ever came across".'*

The run-in took forty minutes. At the lowering position we changed formation. The flotilla divided into equal flights, landing ten minutes apart. I did not want the whole brigade boiling on the beach at the same time.

The first wave – twelve motor launches in line abreast – closed to arrowhead formation as we picked up landmarks in the last mile.

No. 6 Commando, the point element in the encounter phase, rode on the port quarter of No. 519 (the command craft), keeping station, a cricket pitch apart. No. 41 Royal Marine Commando (attached passengers in transit) rode to starboard on the right flank.

Ahead, No. 4 Commando and the Free French were landing, passing through the toehold hopefully seized by the 8th Infantry Brigade. They had the benefit of steel landing-craft – slow movers, but proof against the small arms fire expected on arrival. Hutchie Burt's troop went in singing 'Jerusalem', like the fierce Covenanters of old. With them rode the brigade major and his wireless operators. In the event, communication from ship to shore was a wash-out. No. 4 landed at the double, with no help from the first arrivals. The heavy casualties sustained by the infantry on Sword came from mortars and

---

* Cornelius Ryan, *The Longest Day*.

artillery, ranged (as predicted) on the tide-mark and controlled by observation points on high ground behind.

Our own approach, which had the legs of the field, relied on speed rather than protection. The thin skins of three-ply wooden hulls did not stop machine-gun bullets, and we knew it. As a precaution against trouble I had divided headquarters, with Max in charge of the other party. Derek rode within hail, wrapped in a wet oilskin and taking a lot of spray as he searched for landmarks.

Surgeons say good men swear and baddies say their prayers under anaesthetic. Some such nonsense occurs going into battle; approaching crisis is regarded with a certain detachment. I heard later that the troops got a pep talk: on Elizabethan sea captains – Drake, Essex, Howard and Frobisher – the bold men who singed the King of Spain's beard (1596) when they sank the West Indian fleet (thirty-six loaded merchantmen) in Cadiz Harbour. How the English admiral, as he closed the range, ordered each enemy salvo to be greeted by a fanfare from his trumpeters, without exchanging fire with the *Four Apostles*; and how that earlier *Warspite* (Drake's own Flagship) sailed on to warp the *San Felipe* – grapple and board her. Essex that day had doubled his men over three miles of soft sand to enter and sack the town.

The big ships had made an impression on my subconscious mind; but noise spoiled the conversation, and commandos were busy with immediate personal affairs. They looked green after the rough crossing. As we came under heavier shelling Salsbury (who happened to be a mimic) suggested, in an Oxford accent, that the piper should get on with that blank-blank fanfare and find himself a trumpet. From a breastwork of protective rucksacks a voice restored order. 'Watch it, mate – the old man's farting fire this morning.'

Half-seen through palls of smoke, boats were burning to our left front.

Then, with much shouting, Derek picked up his landfall: a conspicuous building, still standing, with a pitched roof. No. 6, who had trained over a stiff course, here took a tot of rum; Curtis made a slight alteration to starboard. A tank landing-craft with damaged steering came limping back through the flotilla. The helmsman had a bandage round his head and there were dead men on board, but he gave us the V sign and shouted something as the unwieldy craft went by. Spouts of water splashed a pattern of falling shells.

Out among the offshore obstacles – heavy poles and hedgehog pyramids with Teller mines attached – we started to take direct hits. Curtis picked his spot to land, increased speed and headed for the widest gap, our arrowhead formation closing station on either side. The quiet orders – a tonic from the bridge – raised everybody's game: 'Amidships. Steady as she goes.' The German batteries mistakenly used armour-piercing ammunition in preference to high explosive and bursting shrapnel. Derek's landing brows were shot away and beyond him Ryan Price's boat went up with a roar. Max had an unpleasant experience when a shell went through his petrol tanks without exploding. Rear headquarters got away with minor casualties. Our command ship took two shells in the stern. It happened in the last hundred yards. There was no time to look back. The impact must have swung round the two boats, Max's and mine, touched down side by side. Each carried four thousand gallons of high-octane fuel in non-sealing tanks aft of the bridge. Had Max blown up we would have gone with him. Five launches out of twenty-two were knocked out, but the water was not deep and commandos got ashore wading; a few took a swim in the shell craters.

The smoky foreground was not inviting. The rising tide slopped round bodies with tin hats that bobbed grotesquely in the waves. Wounded men, kept afloat by life-jackets, clung to stranded impedimenta. Barely clear of the creeping tide, soldiers lay with heads down, pinned to the sand. Half-way up the beach, others dug themselves into what amounted to a certain death-trap. Half left, middle distance, Doc Patterson's No. 4 Commando stretcher-bearers were still struggling, under heavy loads, up to the shelter of the dunes.

'I am going in,' said Curtis. He gunned his engines and bumped over the shallows. 'Stand by with the ramps!' Four able seamen sprang to the gangways. 'Lower away there,' and the brows ran sweetly down at a steep angle. The command craft had a comfortable landing. On these occasions the senior officer, stepping cautiously (rather than attempting a headlong dive), is first off the boat. Surprisingly, it is as safe a place as any. The water was knee-deep when Piper Millin struck up 'Blue Bonnets', keeping the pipes going as he played the commandos up the beach. It was not a place to hang about in, and we stood not on the order of our going.* That eruption of twelve hundred men covered the sand in record time.

* 'At La Brèche (Red Beach) on the Ouistreham half of Sword, men of the 2nd East Yorks lay dead and dying at the water's edge. Although no one will ever know how many were lost during the landing, it seems likely that they suffered most of their 209 D-Day casualties in the first part of the morning. These crumpled khaki forms seemed to confirm the worst fears of follow-up troops. Private John Mason of 4 Commando was shocked to find himself running through "bodies stacked like cord wood. They had been knocked down like ninepins". And Corporal Fred Mears, another of Lovat's Commandos, "was aghast to see the East Yorks lying in bunches. It would never have happened if they had spread out". As he charged up the

And here was Michael – stalking down from the dunes to meet us, with little Johnstone bent double under his pack, looking, somebody said later, like the 'Hunchback of Notre Dame'. In one hand the batman bore a staff with a fluttering Celanese triangle* serving for a banner. As we ran up the slope, tearing the waterproof bandages off weapons, the odd man fell, but swift reactions saved casualties. Ryan Price, sunk at sea, was as good a soldier as he is now a racehorse trainer. After swimming ashore, he re-equipped his troop with the small arms of the East Yorkshire battalion picked off the beach as his men dashed on. Michael delivered good news: the Orne bridges had been seized intact.

My O group did well through the soft sand and flung themselves down behind the nearest pill-box taken single-handed by Knyvet Carr. The job had been the responsibility of the East Yorks; Carr (a skinny but determined subaltern known as 'Muscles' in No. 4) would have won a VC in less demanding circumstances as he bombed his way into the enemy wire.

Others were less fortunate: Max's brother-in-law, Eric Kiaer, was killed before we reached the dunes. Wounded and half drowned in landing, pulled over by the heavy rucksack, then picked out and dragged over the sand – Mullen the artist, like a broken doll, lay with both legs shattered, at the end of a bloody trail. He was beyond speech but out of pain; a glance showed a hopeless case and death was busy with him. A gifted man, Mullen should have landed later with the reserves, but he would not hear of it. No volunteer in the whole brigade was prouder of his beret.

David Wellesley Colley, a Downside boy with a cheerful smile, lay against the pack of a sergeant who had pulled him into shelter. He was shot through the heart. Tears were running down the NCO's face. 'Mr Colley's dead, sir. He's dead. Don't you understand? A bloody fine officer,' he repeated in shock as a field dressing was wrapped round a broken arm. Little Ginger Cunningham, RAMC, with the flaming red hair, had his legs shot from under him as he ran up the beach, but he was picked up by big Murdoch McDougall and got clear. His swearing between lamentations is recorded: 'To think they could miss a big bugger like you, those f—ing Germans, and then blank something well chose to pick on me!' The diminutive Tich Cunningham paraded as the smallest man in the Commando when the order came: 'Tallest on the right, shortest on the left in single rank size! Right dress, eyes front,

---

beach determined to make Jesse Owens, the current Olympic sprint champion, look like a turtle. he remembered cynically thinking that they would know better next time.' (From Cornelius Ryan, *The Longest Day*.)

* A yellow cloth carried by infantry units as a recognition ground signal from army to air force.

number!' Then his voice echoed over the parade ground as the men shouted the count down the line: 'Number four hundred and eighty-nine, Sirr!' Now he hobbled round shot through the knees, both trouser legs soaked in blood. After patching up more wounded and with the help of my wading stick, he set off after the Commando and stayed with them through D-Day. Lance-corporal Cunningham was recommended for the Military Medal: no more was ever heard about it.

While he waited for our arrival, the brigade major lost the signaller in wireless touch with No. 4 Commando, but news came back that Dawson was already wounded and Kieffer, commanding the Free French, was also on the floor. So was Guy Vourch, his second officer and a most dedicated soldier.*

Hard pounding was reported everywhere in Ouistreham, with a resolute garrison fighting from house to house. The landings had gone better than expected; I estimated maybe sixty casualties, but the battle still to come.

---

* His family loved the Tricolore; let France be proud of them. His mother and sisters helped Allied pilots escape out of the country through a network of dedicated citizens. His younger brother reached North Africa to join Le Clerc, whose army crossed the Sahara from Lake Chad. Guy himself was captured before France fell; then escaped to Nantes – where he changed his uniform – and so to Finisterre. He bought a boat and with five others, including his brother Yves (wounded in the retreat), set a compass course for England. The engine failed, the food gave out and a gale force wind blew them off course after leaving Douarnenez. Ten days later they were picked up exhausted off Milford Haven. He lived to fight another day.

CHAPTER TWENTY-ONE

NORMANDY: THE LEFT OF THE LINE

Alors! Tu n'es pas mort, coquin?

Remark passed by a French cavalryman preparing to lance Colonel Ponsonby,
unhorsed and prostrate on the field of Waterloo

There was no time to catch the breath. A hot reception – in terms of
noise. Destructively, it could have been worse. Except for direct hits, the
long-range heavy stuff wasted each shell burst, bunkering deep into the
dunes. During the saturation bombing Monsieur Lefevre, a Resistance
leader, risked death to walk from Ouistreham and cut cables connected
with flame-throwers in the beach defences. He did more than the 8th
Brigade, 3rd Division, who landed to our immediate front. A poor
showing in the last rehearsal was faithfully repeated on the battlefield.*

We passed through them, leaving platoons scrabbling in sand where
the shelling hit hardest, digging holes which would be drowned when
the tide returned. No. 6 Commando led the way. Alan Pyman, commanding the point element with Donald Colquhoun, bombed a path
through the built-up area, widening the gap where No. 4 had previously
forced a passage. Throwing down rucksacks, the troop moved fast,
mopping up pill-boxes and the immediate strong-points with hand
grenades and portable flame-throwers; supporting Bren guns sprayed
lead at every loophole and casement aperture. Houses not destroyed
by bombardment were occupied by a few Germans; firing small arms,
they sniped from roofs and windows. These pockets of resistance were
wiped out by selected marksmen.

The landing was a soldier's battle – total confusion which favours
spirited assault – and it was soon over. Above the crash of shells
and mortars Derek shouted among the buildings like Marshal Ney
leading the Old Guard at Waterloo. Soon a trickle of grey uniforms

* The 3rd Division, in spite of the good record under Monty, proved sticky throughout the landing. They had become muscle-bound mentally and physically, after four years' training in
the United Kingdom.

appeared: bewildered men in shock, their hands clasped behind bare heads. (The captured Germans seemed pleased to throw their hats away.) Some scuttled in a bee-line back into concrete positions already captured. Others hurried towards the beach, where Peter Young and No. 3 Commando were landing in the second flight with No. 45 Royal Marines under Reis. I heard later that, humping some wounded as they ran (including an old friend, Guardsman MacMillan) the demoralized Germans splashed to the boats, crying 'Engeland, Engeland!'*

We were almost through the Atlantic Wall. The immediate defences were laid out with German thoroughness: they warrant a description. They were not, as it were, a continuous row of grouse butts, but rather a system of ingeniously interlocked defence works equipped with every weapon, from underwater obstacles and devices to set the sea on fire to wire and minefields at the water's edge, ranging back through strong-points laced with machine-guns and anti-tank guns to distant artillery and self-propelled half-track cannon – all bearing on the beach. Beyond lay German infantry dug into weapon pits – again with interlocking fields of fire. The tanks and armour were held some distance to the rear. It was a question of how long they would stay there!

Each pill-box was a citadel of reinforced concrete, sunk hull-down and half-buried in the ridges of the dunes. Walls two feet thick stood six feet above ground level, their height made up by a very solid roof giving further feet of concrete head cover. They were certainly bomb – if not blast – proof, and made equivalent precautions at home appear inadequate. Positions sited in depth, 100 to 150 yards apart, were surrounded with barbed wire, with minefields in between. No pill-box faced directly to the front, but each was at an angle to either side, sited to enfilade the wire and deal effectively with approach from the flanks. Each was manned by a crew of half-a-dozen men firing 75 mm cannon and light automatics. In support to the rear were heavy machine-guns

---

* Acting on orders, interpreter Peter Masters (a Sudeten German from intelligence attached first to brigade headquarters and later to No. 6 Commando) interrogated prisoners to find out where the German artillery were sited. Masters recalls, 'My prisoner didn't speak German! He was Obergrenadier Johann Kramarczyk of the 736 Regiment, profession: farmer, from Ratibor in Poland. I know, because I have his paybook still. Remembering that Poles learn French in school I launched a question in my high-school version of that tongue, for although I was "the chap with the languages", my main asset was fluent German with detailed training in matters pertaining to the enemy forces, their organization, weapons, documents and psychology. My first interrogation, in spite of all this, was not a success. Firstly, because Johann, although his face lit up as soon as I asked "Où sont les canons?", didn't know a thing. Secondly, because his Lordship, it turned out, spoke much better French than I. I caught a phrase here and there of their animated conversation before pushing towards Colleville sur Mer.'

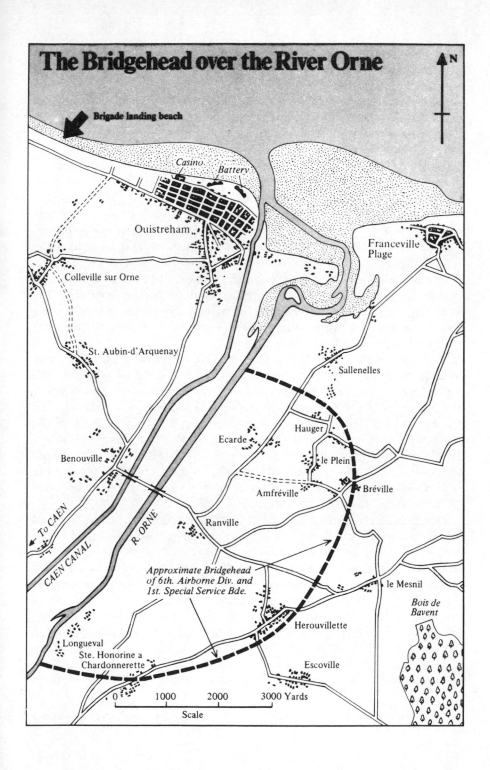

The Bridgehead over the River Orne

N

Brigade landing beach

Casino Battery

Ouistreham

Franceville
Plage

Colleville sur Orne

St. Aubin-d'Arquenay

Sallenelles

Hauger

Ecarde

le Plein

Benouville

Amfréville Bréville

To CAEN

Ranville

*Approximate Bridgehead
of 6th. Airborne Div. and
1st. Special Service Bde.*

le Mesnil

*Bois de
Bavent*

Herouvillette

Longueval
Ste. Honorine a
Chardonnerette

Escoville

0   1000   2000   3000 Yards

Scale

set in less solid foundations equipped with revolving turrets. Some had suffered bomb damage caused by blast.

All this was seen before smoke blinded the last nests of resistance. The furious pace quickened and determination turned the scales. Our team-work went well. Men remembered that 'he who hesitates is lost'. No. 6 Commando moved like a knife through enemy butter.

Bobby Holmes (my demolition expert) hated marching, but he must have been uncommonly strong in the arm. To a slung Tommy gun, a pack of explosives and regulation rucksack (50 lb in weight) he added a motorbike borne across his shoulders. Bobby took cover at my side behind another pill-box. He was wet from head to foot, after disappearing in a shell crater below the churned-up surface. But he held on to his wheels. They provided our only transport for some time to come.

The proximity of the Chief Royal Engineer was an embarrassment as more sappers, laden with gelignite, struggled into the O group. 'Are those bags liable to get touched off?' The brigade major moved hastily round the corner, muttering, 'HQ's going to blow up so fast that nobody will know what hit them.'

Bobby assured us plastic charges were harmless, though detonators (he tapped a hand grenade) were unpredictable. 'And who the hell controls the fuses? Let's get out of here and shake out, you're crowding headquarters.' Even as I gave the order and moved on, a shell pitched among the last REs to get to their feet. Sand partly contained the burst of shrapnel. The 'gelly' behaved correctly, but three of Bobby's men were killed and we left another sitting head down with hands pressed to his middle as blood started to seep through weakening fingers. The poor fellow was beyond speech: nothing could be done for him except the shot of morphine.

Prisoners were forced to hoist packed explosives; here also we collected a windy gentleman in officer's uniform, smoking a pipe at the bottom of a shell-hole and taking no part in the battle. The pipe annoyed me and he was kicked up, told to join the party and consider himself under arrest: Salsbury, who enjoyed any untoward occurrence, put a rifle to his back and we pushed blindly on. Skull and cross-bones marked minefields; with no time to play grandmother's footsteps, the advance went straight ahead, Bangalore torpedoes blasting through wire, which promised the safest route.* A dash across a road, and then a tram-line: we were through.

The smoke rolled away; not far ahead the point element broke from

---

* The idea that the wire was electrified and too hot to handle had spread among the infantry landed to help clear the beach exits.

the built-up area into open country. We were past the immediate defences in the wall – pierced over a three-hundred-yard front. A short pause while Derek regrouped on the first of a series of deep irrigation ditches. Green fields and marshlands lay beyond. There was little cover as the country widened out, climbing to the slope that lay barely two miles ahead. From there the gun flashes were continuous, but shells now screamed over to land behind the advance. After the maelstrom on the coast comparative silence engulfed the perspiring men. Michael looked at his watch. He had been longer in France than the rest of us. 'That round lasted eleven minutes.' The brigade major was a sardonic fellow, but I detected admiration in his voice.

Here an advanced aid post was set up. The walking wounded started to come in, among them, later in the morning, Derek's batman, Corporal Smith, with a bullet through the arm, leading his second-in-command, Bill Coade, temporarily blinded by a stick grenade as he attacked a pill-box.

With a crackle, Derek came through on the set. 'Sunray calling Sunshine. First tasks accomplished. Report minor casualties. Now on start line regrouping for second bound. Moving left up fairway on Plan A for Apple. Still on time. Do you read me? Over.' We had agreed on strict wireless silence; a listening watch required three signals, which affected all concerned. The first: both bridges were still intact; the second: all commandos and brigade headquarters were safely ashore; the third: brigade were through the Wall, leaving two and a half hours to reach the airborne rendezvous on the River Orne.

From this point No. 6 Commando proved its worth. The men pushed forward in Indian file, using cover where possible, avoiding paths or side roads, to slip across the plain. Ryan Price, who had been sunk at sea, took up the running and dealt with the first infantry trenches along the route. Ronnie Hardey, leap-frogging past him, moved farther to the left into scrub country, which provided cover as the new troop fought its way nearer the canal. Here Peter Cruden was ambushed and shot through the shoulder. More prisoners were taken hostage and taken on with the advance, after being offered the choice of being shot by their own side or steering a safe course for the Commando.

Derek was on time when he reached the east end of the straggling houses in St Aubin d'Arquenay. Behind it an Italian battery was surprised and sorted out with no great difficulty. A wrecking party was left to spike the guns. The village itself was supposedly held by German troops. As Derek dealt with one end of the long street, my party came in at the other.

Brigade HQ kept a more direct route before turning left. When

we reached Derek's start line, a bottomless irrigation ditch (it had failed
to flood properly) from which his formation had debouched and spread
like a fan across the plain, No. 6 Commando became an immediate
target for 'Moaning Minnies' (*Nebelwerfer*), a frightening kind of
thermite and petrol bomb that sailed over, six at a time, trailing
black smoke, to burst into flames as they hit the ground, setting the
marsh grass on fire. They were ranging on the advance (half-left) and
looked unpleasant, but in fact did little damage.

Ahead lay the tree-lined hedgerow (seen on previous photographs,
now amputated) that ran at right angles from the coast. Even a
tree-stump offers protection and there were still odd bushes standing
about. A German machine-gunner, as if scenting danger, was firing at
the gaps: I calculated Derek would be up to him by the time we got
started. It looked the route for me. There was a last ditch to cross
before the hedge began and the machine-gun caught us before we got
to ground. Here we lost the pipesmoker; it was not his lucky day, and
there were other casualties, including one of my runners.

In this large drain, which had its bottomless attractions, crouched a
small party of marines from the 4th Brigade: No. 41 Royal Marine
Commando had landed close alongside on the flotilla's starboard
quarter. They had lost their officer and NCOs, but had pushed on gallantly, then missed direction, veered inwards, and brought to a halt,
demoralized by the fire bombs. There were grounds for compassion here. The soldiers were boys who looked absurdly young and
only lacked for leadership. I tapped a pair of boots projecting over the
opposite bank, thinking they must belong to an observer taking bearings. I tapped again, climbed up and found I was talking to a dead
man. We were pinned in mud for two minutes, and I swore foolishly
at the unfortunate castaways – an error, for cursing never helped a
frightened soldier. With irresolute troops, any hold-up can prove fatal
to success: it is hard to get men on their feet again.

A cheery voice sorted us out. It was Robbie Robinson, one of Derek's
best, leading his sixty-man cycle troop. This belated appearance was
due to parachute bicycles (each with a mortar bomb strapped to
handlebars), that had caused balance adjustments climbing in and out
of drains.

They hauled their loads across and struggled on, Robbie remarking
as he did so, 'I'd rather be shot up on a road any day than do this
sort of thing.' Every man in the ditch leapt up and followed him. Such
is the force of example.

The rest was plain sailing all the way. The machine-gun making
trouble had been outflanked. Away out to the right, but pointing in

the proper direction, I spotted a pair of tanks; they were firing inland towards Colleville-sur-Mer. One came to a halt on a track that led to a bridge. It had suffered damage by mine or direct hit, but continued firing at Colleville. The other cut in our direction, as if anxious to find friends.

We got to St Aubin. Tony Smith was told to bring the armour up behind. We required moral support. Brigade headquarters was out on its own. I heard later that No. 45 Commando, who were following, had been counter-attacked in a sortie from the outskirts of Colleville. Peter Young (No. 3 Commando) sensibly bypassed this trouble and was close up, out of sight, steering a parallel course, slowed down by minefields on our left. Max was detailed to form a rearguard, armed with the anti-tank weapon, just in case we had mistaken the identity of our approaching neighbour. We were in good tank country now and a Panzer attack, from any direction, was simply a question of time.

The houses in St Aubin straggled out of sight – a bad example of ribbon development. Strategically, the crest remained of considerable importance: the hogsback gave a fine view of the beaches below. There was rifle shooting ahead. For reasons unknown, we met only token resistance from poor-quality troops as my mixed assortment of signals, sappers, intelligence, medical staff, despatch riders, interpreters, admin. and soldier servants moved up the main street.

Robbie was nowhere to be seen: I presumed he must be in action ahead, having passed through St Aubin unscathed. I disliked street fighting. Scouts were sent out front and headquarters (the motley crew all armed to the teeth) were deployed sufficiently to deal with emergencies, then moved on fast, hugging the pavements and covering windows to either side. I may have grown careless: they say you never hear the shot that kills you; one bullet is usually enough. The moral to this tale: avoid stopping in built-up areas when the presence of snipers is suspected. Officers are favoured targets, and so it proved.

At a crossing, a knot of civilians gesticulated outside a low-roofed edifice damaged in the morning's blitz. The occupants – old people – had been badly injured; the family begged for a doctor and bandages. As I turned to point out the medical team (with armbands) bringing up the rear, a sniper's bullet smacked the wall beside my head with a crack like a whip. A near miss, that showered the relatives with chips and rubble. 'He's over there, top storey,' shouted Bobby Holmes, charging across the street to a house at the corner. Someone tossed a grenade through the window as Bobby kicked the door down and cut loose with his Tommy gun. He looked out at us with shocked surprise. 'This one was dressed in civvies, but he had a Jerry rifle and there's a parachute silk under the bed.' The commentary ended. 'My God, I'm coming down. More Huns are advancing over fields to the south.'* Here was a predicament.

It was the right moment for the DD (swimming tank) to come grinding up the street, but it failed to materialize. Our fire would certainly halt the approaching enemy – who, leaving their horse-drawn vehicles on a side road, were now closing on the village in open order across green patches of standing corn. It was not our job to interfere with hostile strangers, but the Germans would create problems for the inexperienced troops following close behind. The Italians had clearly sent for help. The enemy – a platoon about thirty strong – looked a soft touch: the sun was in their eyes and we were unobserved. Back gardens offered possibilities. I made a quick decision to ambush them. It was a relief to get the rucksack off – straps were already burning shoulders – then tee up behind its bulk and go into action with a light, short-barrelled US army carbine.† Joe Lawrence and Salsbury's brother, a despatch rider, sneaked one of the quick-firing K guns up on to a shed. The Germans were five hundred yards away. We had not been seen. They were in the bag.

The fire order whispered down the ambushing party is not found in training manuals. 'Pick the officers and NCOs and let them come right in.' The ragged volley caused a surprise: dust flew off the back of the fair-haired platoon commander as he spun round and fell; half-a-dozen others, who had bunched in the centre on reaching the buildings, went down in a heap. The rest took cover in the corn. The fight was over before the tank‡ arrived; either the rumble coming closer or the long swinging gun had an electrifying effect on the survivors, who leapt to their feet, throwing weapons away. All shouted their surrender in an unknown language. With shaven heads and smelling to high heaven, they turned out to be Russians press-ganged into service, commanded by German personnel. They appeared delighted to be led into captivity. Schmeisser sub-machine-guns were annexed by brigade intelligence. The horse-drawn transport proved useful: carts were loaded with wireless sets, packs and spare ammunition: Bobby's high

---

* Bobby was a comedian and a good sapper. But he got into trouble shortly before departure, building a demonstration raft at Cowdray on top of which reposed my command vehicle. It was pushed and pulled into the middle of a pond, where it promptly turned over and sank in deep water amid laughter. On D-Day, Bobby Holmes more than regained his damaged reputation.

† This type of carbine was effective only at very short range. With its short barrel, it would have been useful in jungle warfare.

‡ From the 13th/18th Hussars. I never learnt the name of the officer, but heard he was killed in the attack on Bréville later in the week.

explosives kept well to the rear. Preceded by the swimming tank (now an inseparable friend), we moved on to find Derek. Brigade were late and could not raise him on the walky-talky.

At the end of the village, seated on a kitchen chair where the T-junction runs to Caen and Ouistreham, sat a wounded NCO guarding some noisy Italian prisoners – all field gunners. He explained that No. 6 Commando, getting no signal, had crossed the main road and pushed on to the canal, making use of scrub along the bank below. The crack of small arms fire and chatter of automatics traced Derek's progress. Bénouville lay a bare mile ahead. There a battle was raging (a discouraging sound before joining in) supplemented by the vicious reports of self-propelled guns. The approach march had been easy, against moderate antagonists. Now it would be bitter fighting all the way. A counter-attack against the bridges was going on – pressed home by determined Germans from Caen supported by self-propelled guns. The Airborne sounded in trouble, and it was our job to bail them out.

No time to hang about or proceed with caution. But the supporting tank created confidence and the Luftwaffe were elsewhere engaged. (I had not seen a German fighter plane all morning.) Despite lack of cover, headquarters pushed up the road – at best pace, with scouts ahead and thrown wide on flanks through crops red with flowering poppies. Timing and good fortune were still on our side.

The road junction at Bénouville dipped into a hollow; the main road continuing inland, sloped perceptibly up to Caen. Down on the left, close to houses, lay the turn-table bridge across the Caen Canal. Burning transport smoked ahead. A German half-track, upturned in a ditch, provided some protection for wounded men. The dead of both sides sprawled about the hollow, where airborne troops dug deeper into slit trenches. Others brewed up tea: the scene reminiscent of some Indian swoop on the wagon train in a western movie.

The defenders had put up a good performance. They came from the 7th Parachute Battalion, mixed with the Oxfordshire and Buckinghamshire light infantry. The attack had been beaten off: it was not the first that morning. Two anti-tank guns, landed by glider, had helped to stiffen the defences.* One, with all the crew killed, was already out of action. The other, wheeled up on to the bank, engaged the enemy SP guns firing three fields away. The enemy shelling went harmlessly over the hollow.

---

* The airborne solution to landing artillery was the Horsa glider, which, more by luck than prescience (it was designed to carry infantry), could also transport a Jeep, ammunition trailer, 75 mm howitzer and a six-pounder anti-tank gun.

Michael raised a laugh as we got to ground. 'I fear the natives are hostile.' 'The natives are revolting,' I replied. Unbidden, headquarters sprinted to join the hard-pressed square.

The welcome turned to laughter as the carts whipped up to run the gauntlet, came 'Ammer, Ammer, Ammer down the ard igh road': they helped in removing the worst casualties to the *estaminet* (now a show-piece attraction) beside the canal. Here we said goodbye to the tank. Its close support had provided inspiration.

THE BRIDGES

I ran across with Piper Millin, Salsbury and a handful of fighting men. There was a fair amount of mortaring, and a machine-gun up the water pinged bullets off the steel struts, but no one noticed and brave fellows from the gliders were cheering from their fox-holes at the other end.

Soon I was hailing John Howard, a hero in the Ox and Bucks who had crash-landed his ship with its nose almost in the canal. He advised me to keep moving: it was no place to hang about. No. 6 and No. 3 Commandos got across in no time; the marines, close behind, had their CO picked off by a sniper. As the day wore on, Germans infiltrated along the river and canal banks, causing considerable damage.

John Howard's nocturnal *coup de main* was a notable achievement. His fast-dwindling company was to take punishment all day, but they kept the bridge open until darkness fell again. John, a modest fellow, was to be badly shot up later in the war. There were apologies as we doubled over this hot potato. 'Sorry about the mortaring from that ruddy château. The bastards have got the range, but it happens to be a maternity hospital and I have strict orders not to disturb the inmates!'

More apologies for the two-minute late arrival came at the wide river, which flowed four hundred yards of lush water meadows ahead. The grazings provided a nasty spectacle: swollen cattle and horses lay with legs in the air, while others dragged around, tripping on spilled insides, bellowing their agony. Two Schmeisser men under Sergeant Phillot ran over to put them out of pain.

Piper Millin struck into a march and played us across the water.* Peter Wilson, an old friend from early days in B Troop, was shot through the head as No. 4 Commando crossed later that afternoon. Derek had a man killed by a sniper, pierced (curiously) through the nostril. Several other ranks became casualties, but the good music

* I should mention that no piper played on the first occasion while crossing the canal.

drowned the shooting and we managed to stride over in step – almost
with pomp and circumstance!

Pine Coffin of the 7th Parachute Battalion waited in the blessed
shelter of tall trees and a leafy slope, with dead ground behind him.
Also Poett, the commander of the 5th Parachute Brigade, who went
through the 'Doctor Livingstone, I presume?' routine. The appearance
of Gale's G1 'in a pelting heat', as John Bunyan puts it, terminated the
back-slapping.

The news was serious. The Germans (elements of the 21st Panzer
Division) were attacking from the south; their infantry was reported to
be moving along all roads leading to the river. Most of the 3rd Parachute Brigade had dropped miles off course, blown by strong winds;
thin cloud obscured the moon, but in many cases pilots had confused
the river valleys of Orne and Dives. Until the gliders of the Air Landing
Brigade (commanded by Hugh Kindersley) came in that evening the
division was short of troops and already in difficulties.

Gale required a Commando immediately at Le Bas de Ranville to
protect him against attack in the area where gliders were to land.
Parachutists (Colonel Otway, 9th Battalion) dropped overnight on the
Merville battery were also in trouble, reduced to two hundred men
and driven back into Hauger and Le Plein.* Commando assistance
was required without delay; the enemy was moving along both roads
into Bréville. (These place-names became familiar in the days that
followed.) The original plan and other good intentions were cancelled.
In military parlance, 'there was a flap on' and my brigade's rôle was
now restricted through necessity to a more intelligent proforma.

In England, I had tried to forecast the likely course of events, perhaps
taking an optimistic view of our ability to reach the high ground over
the Orne. (There had been shaking of heads among the sceptics.) I
had argued that if the Germans were good – the commandos were better!
The more cautious of my masters looked positively ill on receipt of this
information.

So far, so good. We retained the initiative and events had clicked
into place. The sea crossing, dry landing, rapid break-through and
intact bridges were bonus marks. Shock tactics had proved a complete
success. How would we get on after losing the element of surprise? It

---

* The Merville battery gave trouble, being immediately reoccupied after capture. Thought to
be a four-gun affair, the all-round defences were formidable. The 75 mm cannon were in
steel-doored concrete emplacements six feet thick; two of these were covered by twelve feet of
earth inside a fenced area of seven hundred by five hundred yards of barbed-wire entanglement.
Within lay a double apron of more wire, fifteen feet thick and five feet high. The perimeter,
surrounded by a deep minefield, was covered from within by weapon pits firing light machine-
guns and anti-aircraft weapons. This battery was situated just outside Franceville Plage.

is one thing to go on the warpath; quite another to stay put and take what's coming to you.

These thoughts were uppermost in mind as we climbed out of the river – with two miles to go and dog-tired by this time. To take the ridge above would cost many lives: the enemy had seen us coming. But that was not all. Would I be counter-attacked before we got dug in? The last lines of Masefield's 'Reynard the Fox' in the Ghost Heath run described my feelings exactly.* We'd had a flying start: could we prove our staying power and face the music?

Much depended on the enemy's reaction. The coming battle threatened two alternatives. Having learned the pattern of our army build-up, the Germans would concentrate on first 'containing', then knocking out the bridgehead, by attacking over open country from the south, or rolling up our exposed flank along the coast road and through Bréville (see map), thereby capturing high ground from which to shell the beach below.

This second alternative offered a shorter route for the Panzer divisions, but would slow down the requisite punch. Poor lateral communications and perilous river crossings were less suited to blitzkrieg than the open plain. With Rommel in the ring, however, and one tank available, anything might happen. The field-marshal liked nothing better than a lightning strike from the unexpected direction.

With a modicum of fortune, the counter-attack would come from the south. An advance from Caen seemed the likeliest course of action. That thrust could also head for the Orne crossings: without fire support (and where was that coming from?), we would inevitably be over-run, hopefully to bob up again after help arrived. But that was hardly reassuring. Meanwhile Brigade were out on a limb; retaliation must be expected in the first twenty-four hours. The high ground supplied the answer to the enemy's requirements. There would be no respite. I envisaged probing infantry attacks, looking for soft spots along the defence line. The method was familiar: always effective but lacking originality. Skirmishers, at platoon strength, would come tapping along the forward areas, seeking to draw fire and give positions away. When discovered, defended localities would be systematically destroyed by shelling and mortaring, closely followed by infantry advancing behind tanks to mop up resistance. Fierce pressure

* With feet all bloody and flanks all foam
  The hounds and the hunt were limping home:
  Limping home in the dark dead beaten...
  Dansey saying to Robin Dawe
  The fastest and longest I ever saw.

would be exerted to win the salient and control the beach exits below.*

Derek and Peter Young agreed with my proposed course of action. I had consulted them separately before starting; we then worked out plans together. I think they made common sense. In case of setbacks during early occupation, it had been decided to establish hedgehog perimeter (hill top) defences, well hidden from observation with close-range predetermined killing-grounds, covered by a maximum concentration of small arms fire. I was confident no Germans on foot would get the better of either commanding officer. No brigade had two more resolute leaders. While their combat experience was invaluable, they could read a battle better than most superiors. Both were imbued with a determination to defeat the enemy that made light of the odds: scorning heroics, they rightly considered to throw a man's life away was rank bad soldiering. They were cool customers in a scrap with no illusions.

I had stated these intentions in exalted company. In the planning stages, the airborne forces had seemed preoccupied with their own affairs. I had seen little of General Gale, who, at first blush, struck me as vain and egotistical, with a hectoring manner and a loud voice. I don't think he liked my peculiarities either. (Having challenged and beaten all three of his brigades, first at football and then at athletics, I bet a fiver with the general that we could floor his division in a boxing contest. This was duly done. Pat Porteous VC's soldier servant, Irish Guardsman Moore, knocked out a Canadian paratroop officer in the deciding bout of a great evening's entertainment. These get-togethers may not have satisfied 'Windy' Gale, who, I suspect, was jealous of green berets, but the commandos made friends at all levels with the airborne, who were fine people. The old general was a fine person too, for that matter. The competitive spirit and consuming rivalry can sometimes raise morale too high among soldiers.)

Before departure, the brigade's rôle on D-Day still had loose ends that required clarification. At one briefing I expressed surprise at the vagueness of my remit. 'You will infest this area!' stated General Gale, placing a large hand upon the map and blotting out a land-mass of sixty square miles of territory, including the coast towns of Cabourg

---

*History revealed that 21st Panzer Division lay between Falaise and Caen on D-Day, and that its commander, General Feuchtinger, quickly appreciated the position. He was under command of the Seventh Army, and through the chief of staff made repeated requests that his tanks be permitted to go into action. Feuchtinger realized the importance of the left flank of the Allied bridgehead, which was holding across the Orne, and had actually started to move his division to attack invaders, when he was suddenly placed under command of the 84th Corps and ordered to switch the advance back across to the other side of the river, bypassing Caen.

and Franceville Plage, and stretching back inland to cover enclosed wooded country between the Dives and Orne river systems.

I welcomed independence (since we were under airborne command), but a suspicion had been growing, and I asked, 'Does that mean my four Commandos are considered expendable, and only required to wander about like the *maquis* in no man's land, harrying unspecified targets, without any support or supply system? My brigade has not been trained for bush-whacking.' This caused an explosion. It was a tactless question, and I was hurried out by Gale's G 1. It was clear the general expected blind obedience from subordinates. But I was determined to get the right answer before leaving England. Every commando soldier was going to know his exact duties prior to departure.

My last opportunity to get a hearing occurred at a final conference at Bulford, with General Browning in the chair (his arm in a sling after a glider crash), attended by high-ups in 1st Corps and 3rd, 6th and 50th Divisions or their representatives. Certain officers holding responsibilities on D-Day were given a hearing, or asked questions. I raised the matter again.

'Sir, Phase 1, from the beach to the bridges, presents no serious problem, but I have misgivings about my brigade's subsequent tasks, which remain undetermined. With respect, I wish to offer suggestions for consideration. With no specific targets after crossing the River Orne, four highly trained units, chosen for ability to fight through the West Wall, are apparently going to be dispersed, in order to wage guerrilla warfare. If my brigade scatter they will be lost as an effective formation, and bush-whacking, with only a German battery at Cabourg to destroy, seems a waste of key men. That task can be done by ships at sea, or indeed the Airborne Division. A salient serves no useful purpose. Once dispersed, the commandos will only be able to fight a delaying action, inflicting all the damage they can contrive. Commandos will give a good account of themselves in any situation until their ammunition runs out. But these are shock troops, trained for a man-size job – not beating out bushes. If higher command presuppose there will be no one left after reaching the Orne, they are making a mistake.'

Gale and his planners looked askance; the other generals sat in silence but our corps commander seemed interested. General Browning, a fine soldier (shamefully misrepresented in the film, *A Bridge Too Far*), encouraged me to go on.

'As I see it, the high ground across the river controls the battle, and responsibility must turn immediately to its defence. If one or more commandos push out a salient unsupported, they will simply disappear, perhaps unable to return to the main body. My priority task is surely to consolidate the length of the ridge between Amfréville–Le Plein–Hauger to the sea coast, dug into positions of all-round defence. Could not strong fighting patrols, penetrating deep and sufficiently far, inflict maximum damage to communications and troop concentrations, always returning to previously prepared positions before first light through standing patrols (themselves at all times alert), to give warning of enemy approach? While the defence line has priority importance, it appears essential to disrupt any advance along approach roads which channel traffic to the Orne bridges.

'It is here Brigade can expect to be attacked. I see little sense in beating up the area, or pushing commandos out to Cabourg. May I ask one important question?' Permission was granted. 'This will be difficult to answer, as forthcoming events will be affected by time and circumstance. How long does the army commander, in his overall strategy, expect us to anchor the left flank of the battle? The longer we stay, the deeper we should dig in, and the less we should wander about. Let the enemy find us, until such time as close support is available – I mean armour and artillery. How long will that be? Failing reinforcements, can I please be given ships' guns to break up the Hun concentrations?'

To this I received a favourable reply. Gale wangled the inevitable compromise over Cabourg. It would take a good Commando to march the distance, destroy the battery and then extricate itself, without transport, in daylight from the town. But Peter Young said it could be done. After the meeting General 'Boy' Browning asked me over to the Mess, saying, 'I have got the message and am in full agreement. The idea of wandering about in the blue will get us nowhere, and I liked that crack "to dig in, lie low, and don't shoot 'til you see the whites of their eyes".'

PHASE TWO

The cover ended short of Le Plein, which lay above us a few fields away, crowning the ridge in a long village street which stretched half a mile from end to end. I signalled the brigade to regroup in dead ground for orders, and a breather. There was little respite.

At the six-mile post I called a halt. It was a bad decision. A mortar 'stonk', directed by some invisible observation post, crashed through the trees into the lane where headquarters had sat down. Salsbury and his brother were among the casualties (both walking wounded). Put out of action, they both recovered and survived the war.

The fire fight was thickening, but men were being killed; the Germans had started to react fiercely. We had run into trouble. The enemy had to be dislodged without deláy. Otway's paratroops seemed to have lost the upper hand. I had orders to bail him out; Gale must accept less assistance at Ranville, whither Peter Young was hurrying with his Commando. I stopped one of his officers, Roy Westley, and ordered his cycle troop, which had forged ahead of the main body, to help No. 6 Commando storm the village and find Otway.

Derek was still going strong. His speed during the advance had kept down the casualty rate. A great performance! He had led the way over contested country; but there is a limit to endurance.

No. 45 Royal Marines were largely intact; there was a good man there in Nicol Gray, who assumed command after Ries got picked off. Nicol was told to push up the coast to Franceville Plage (bypassing Sallenelles if necessary) with orders to disrupt communications and contain the enemy long enough for Second Army to establish the bridgehead.

Nicol had an awkward assignment, but it would soon be dusk. I would be well pleased if he could get back into the line the following night, when positions would be prepared for him. In the meantime, the coast road was under heavy fire. The marines waited while Derek sorted out the village.

The mortaring continued as No. 6 Commando worked round for a flank attack. The naval bombardment officer was sent forward with his signal party, with orders to raise any battleship on call. Their feet were in poor shape and ship-to-shore communications were giving trouble. But they pressed on regardless, keen to play a part. Before evening, the FOB's presence was fully justified.

There was hand-to-hand fighting along the ridge. Roy Westley, with a bullet through his wrist, reported the Germans held one side of the street on top of the hill; a wide, leafy 'village green' ran down the middle of Le Plein. Isolated houses on the forward slope still contained small but determined parties of the enemy, firing light automatics at anything that moved below. These were troops who had been in action before – with no intention of giving ground. Roy had contacted Otway, still holding at Hauger, a hamlet (with its château) farther to the left and nearer the sea. The heavy stuff was coming over from the Bréville area. Roy's troop had lost several men, but his Lieutenant Ponsford was dealing with the machine-guns that were doing damage.

Roy Westley was a gallant fellow, the best type of junior leader. His wrist, still numb (before real hurt set in), was bound up tight, and with Max, Joe Lawrence and the few men I could spare, he was ordered back to clear the Germans, get dug in and hold until No. 4 arrived.

18 ABOVE LEFT Brigadier Derek Mills-Roberts, CBE, DSO and Bar, MC,
Légion d'Honneur and *Croix de Guerre*
19 ABOVE RIGHT Lieutenant-Colonel Peter Young, DSO, MC and 2 Bars,
later commanded the Bedouin regiment of the Arab Legion
20 BELOW LEFT Colonel Robert Dawson, CBE, DSO (The Loyal
Regiment), *Légion d'Honneur* and *Croix de Guerre*, at Recklinghausen, 1945
21 BELOW RIGHT A Free French Commando soldier

22 Commandos in training – tin hats were left behind

23 What we were up against – machine guns were dangerous after dark

24 German infantry counter-attacking through blitzed buildings

25 The Atlantic Wall

26 The Battle of the Orchard: the large black cross denotes the edge of the German 744th Infantry Regiment position, the smaller black cross indicates Brigade Headquarters and the white cross marks No. 6 Commando's orchard. Note the shell craters and gliders fore and aft.

27 A broadcast appeal for the disabled after being discharged from hospital, October 1944

As they belted up for departure, Derek signalled he had rolled up one end of Le Plein. Otway was advised that help was on the way. Wishful thinking. I did not like the situation. Isolated driblets of James Hill's disrupted 3rd Parachute Brigade were struggling in behind us, along the coast, after a bad drop in the Dives valley, where bridges had to be destroyed. James landed in the flooded area close to Cabourg – the very resort I had deprecated. He had a struggle to extricate, then reunite, his scattered forces; it seemed the initiative was passing to the enemy.

By mid-afternoon Airborne Division signalled that the coastal battery at Merville, supposedly destroyed, was shelling Sword Beach and shipping lying off the shore. Could I deal with it forthwith? 'Yes,' I signalled back, 'if you return No. 3 Commando.' There was no reply.*

The coast road was wide open, and enemy pressure had suddenly become dangerous. Nicol Gray moved out to plug the gap, his cyclists pedalling ahead with orders to hold Sallenelles, then probe the Merville–Franceville Plage area if possible. Their priority was to block all approaches and hang on. Brigade had scraped the bottom of the barrel. I could field no more of Gale's dropped catches. There were no reserves left.

Credit for the best shot of the war goes to one of General Leclerc's 2nd Division Tank Commanders leading three Shermans – the 'Mort Homme', 'Simoun' and 'Douaumont' on entering Paris some weeks later.† A Tiger tank, prowling round the Obelisk in the Place de la Concorde, blocked any advance from the Arc de Triomphe down the Champs-Elysées. Victory was in sight. There could be no delay. The leading gunner, a Parisian, knew the range, learned as a child from the Almanac: 1,800 metres and dead straight. He kissed the armour-piercing shell, reloaded and set his sights. At extreme range the first shot hit the monster tearing off the tracks. It was a winner.

By way of diversion, I brought off a long shot that afternoon. Waiting for Roy's success signal, I noticed a group of cattle huddled in the corner of an enclosure below and left of our position. Two animals were taking an interest – as cows do, sniffing and throwing their heads – at a thorn bush that overlooked the road. It was a good light, but the

* In the event, on D + 1, Major John Pooley, MC, and Captain Brian Butler, MC, two splendid officers, were sent to Merville to be killed, with heavy casualties among other ranks, reducing two of Peter Young's troops to half-strength in an ill-advised attack across open country in daylight. Although the battery had been reoccupied by hostile infantry, the guns had not, in fact, fired.

† General Leclerc reached Paris on 25 August 1944, just four years to the day after he crossed the Wouri River to start his epic march from the Cameroons in Equatorial Africa with an original force of seventeen men.

carbine did not get the distance. The first round from a borrowed rifle, however, scored a bull's eyes, winging a prone civilian, who jumped up, nicked through the fleshy part of his shoulder. Sergeant Phillott went down to bring him in and paced my shot to better than four hundred yards. The man was a shifty-looking yokel in plain clothes, speaking a French dialect which the interpreters could not break under interrogation. He was clearly some sort of spy, with a wad of money and binoculars, but no signalling aids about his person. 'Looking to the cattle', was his story. I gave him the benefit of the doubt – then a spade, which altered his complexion. When he was told to dig trenches, his relief was all apparent. When the transport arrived, Joe Lawrence handed the wretch over to the prisoner-of-war cage at divisional headquarters, where they subsequently released the collaborator.

The sun had passed the meridian. It was time to dig in brigade headquarters. About now Max, composed as ever, returned from Le Plein with a Camembert cheese taken from the command post captured at the north end of the village. The house appeared suitable for our own requirements: well sandbagged, with deep comfortable trenches in the garden. Max had got a nice room for me, hot and cold water laid on. Joe was clearing bodies from the building. He had garrisoned the newly acquired property against all comers. After a swift 'recce' – shot at en route – on the back of Bobby's motorbike, I decided to take up an overnight position (for better control) in some draughty sheds – astride the main route, with the anti-tank weapon pointing towards Sallenelles and the sea. Who would arrive first? The Germans or No. 4 Commando?

Dog-tired, they were on their way. My old friends, who had been fighting all morning in Ouistreham, had taken a lot of stick, on the beach and through suburbs now called Riva Bella. Dawson and Kieffer were both casualties. The two Allied commanding officers had lost a hundred men in bitter street fighting.

Towards evening, Hugh Kindersley's Air Landing Brigade glided in to relieve the pressure.

For the rest of the week, commandos fought their respective battles, of varying ferocity, which culminated on the night of 12 June. It marked the turning-point in the seven-day encounter. I will leave it to Derek Mills-Roberts to take up the story in 'The Battle of the Orchard'.

CHAPTER TWENTY-TWO

THE BATTLE OF THE ORCHARD
by Derek Mills-Roberts

Hard pounding, gentlemen! But we will see who can pound the longest!

DUKE OF WELLINGTON

We pressed on to a gate on a road which ran through Le Plein. There a soldierly figure, dressed in corduroy trousers and tweed jacket, greeted us courteously; he was the owner of the solidly built farm in front of us. He suggested it would make an excellent observation point.

'You know what that means? It will be badly damaged if used for this purpose.'

He answered, 'I am an ex-warrant officer of the Cuirassiers of the Guard.'

The farm dominated the right half of the village and was an important feature. The farmer supplied useful information. Enemy troops were in Bréville, another village and road junction half-a-mile in front.

Alan Pyman's troop pushed off down the road to seize Bréville. The lanes between were thick and leafy, with snipers active. I turned back to see that a signaller's hair had been neatly parted by a bullet. The extraordinary wound had lifted a flap of scalp and hair from his skull, which he cautiously flattened down before getting on with his work.

We were now on top of the ridge, the final objective: it was to be the left-hand flank of the Allied bridgehead. There was no question of there being a long unbroken line – we were merely a small blob sitting on the disputed high ground. The beach exits were reported closed: we were more or less alone on the ridge. The Germans were reacting swiftly.

As Alan Pyman's troop headed for Bréville we spotted Germans in the wood near the village. The bombardment officer from the royal artillery had a signal link with the navy. To give us direct support was the destroyer HMS *Saumarez*, and I asked the bombardment officer if he could take on the enemy in the wood with this ship. He made a rapid calculation and the shell dropped close in line with us.

'That's no good,' I said.

'No,' he replied, 'she's actually steaming at the moment. But I've got two capital ships on call and they're at anchor.'

An anchored ship has a reasonable chance of hitting a shore target – a difficult task when on the move.

HMS *Ramillies* was armed with fifteen-inch guns which fired a weighty projectile. More calculations, and after what seemed an interminable time there was a tremendous crash as six gigantic shells landed in Bréville Wood. It was a splendid effort which must have shaken the Germans. (HMS *Serapis* also provided twenty rounds of supporting fire.)

Then I got a wireless report saying Alan Pyman's troop were having a rough time and that Captain Pyman himself had been killed. A patrol reported that the enemy were in possession of the château at the far end of the village.

Soon after, we were ordered to evacuate Bréville and consolidate Le Plein. There were still wounded in the village who could not be evacuated. I sent in a fighting patrol: then taking a borrowed Jeep piled the wounded into it and raced back. It was getting dark and twilight is a good time for exploits of this kind – when no one can be sure of his accuracy.

It remained to dig in and fortify our position before night. The farm and its immediate surroundings formed a natural fortress. The Germans were holding Bréville in strength: only four hundred yards separated the outskirts of that place from the orchard. I decided against digging in far out and sited slit trenches along the line of the hedge some seventy yards back within the orchard. If you choose an obvious or suspected place you will inevitably be shelled. In front of these defences I placed standing patrols to give instant information of any enemy advance on the farm.

Screened by vigilant patrol activity, the Commando dug its perimeter defences. Half the men carried miners' picks and the other half general-service shovels, foreshortened to facilitate digging – they were all experts in this digging business. Whatever happened, the camouflage of the area had to be preserved. Two hours later the place was taking shape, except for my own fox-hole, which I would attend to later.

It was getting chilly and wet clothes began to feel soggy. Monsieur Saulnier, the farmer, showed us how to secure the vast lock of his main gate, which was set in an archway. It was massive and old and reminded one of the lock of Hougemont Farm at Waterloo. I went back into the orchard and found that Corporal Smith had rejoined; together we finished the pit which he had almost completed. As we dug he told me he had managed to get Bill Coade to the beach and a dressing station there. He then had his own arm bandaged and immediately made his way back, which, as he modestly explained, 'hadn't been too easy as the route was not yet fully open'. A bullet through the fleshy part of the arm is stiff and sore; digging does not improve the feeling. It required guts to do what he had done.

Lovat came up and joined us in the orchard. He gave me the situation on the map. The size of the blob on the ridge had increased – No. 4 Commando were getting into position on our left. Shimi was annoyed Gale had hived off No. 3 to protect Ranville away below the vital high ground held by our brigade.

We carried rations for forty-eight hours in two concentrated but heavy twenty-four-hour packs. I read the instructions on mine, lit the little spirit lamp and made myself a pemmican stew, adding an Oxo cube which made a messy mixture of the consistency of treacle toffee.

There was cloud over the moon; I expected trouble in the dark. The Germans would make every effort to locate our positions with accuracy, so they could shell and mortar us next day. The night was one of vigil and patrols by both sides. Enemy snipers crawled up in the darkness to the forward edge of the orchard and started shooting, to draw fire and, having succeeded in doing so, to locate our positions. We were determined to prevent them discovering our posts. The snipers had no idea where we were.

Shortly after dawn the farm and orchard were shelled by field-guns and 20 mm shells hit the trees above us – the deflected fragments caused casualties. Our positions were slightly damaged and the heaviest concentration fell on the forward edge of the orchard – the obvious place for our defence line, which I had turned down for that reason.

One of the medium batteries was now ashore and their forward observation officer got it on call: we decided to give Bréville Wood a pasting. When the concentration hit the wood a figure was seen running up the road towards our lines; he was a sergeant in the Parachute Regiment and he explained that he had been dropped the night before and landed almost on top of the Germans. The last artillery concentration, he said, had shaken up the Germans and he thought that if we acted quickly we might take advantage of this.

I chose John Thompson's and Ryan Price's troops and arranged a quick-fire plan. The medium battery would give Bréville a good beat-up, our three-inch mortars trimming the forward edge of the wood; the assault was to attack from the right to capture the battery. It was a great success. The troops went through the wood – after the medium battery had clobbered it – like men possessed and captured four field-guns, two 20 mm guns and five machine-guns in record time. The only casualty on our side was one man killed. Apart from the guns the raid took fifteen prisoners: we hauled the guns back into our own position by means of two Jeeps.

As Ryan came up the road I heard him saying to his batman, 'Why the hell did you fire up into a tree when we were going through the wood?'

'There was a sniper there,' replied his henchman, and then in an injured voice, 'he was just going to shoot you, sir, when I picked him off.'

We were again ordered to draw back into our old positions, with the captured guns.

Shortly afterwards, No. 3 Commando moved up into Amfréville, filling the area between ourselves and No. 4 Commando. The blob on the ridge was now bigger, but there were gaps on each side of us. Two brigades of the Airborne Division were under the same disadvantage.

As long as we remained on the high ground, enemy attempts to occupy it could be dealt with. Infiltration by night had to be accepted. Now we expected a determined attempt to capture the village of Le Plein. If this succeeded the ridge would be lost; at all costs the counter-attack must be held.

The German formation opposite us was the 346th Infantry Division. If the enemy attacked from the direction expected, they would advance from Bréville on the axis of the road and come across the orchard. If this happened, we were not worried, our preparations were complete. We could not extend our defences unless we had a clear field of fire. We had to accept that we had not sufficient men to hold the closely wooded area on the right flank, which offered a good approach for the enemy.

I hoped the Germans would go for us frontally. I envisaged lying low in the orchard during the shelling, which would inevitably precede the attack, and holding our fire while the Germans came into the orchard. Then the Commando would open up with small arms, while the mortars, artillery and capital ships simultaneously smashed up strung-out reserves moving forward. If the enemy concentrated to our front, it was essential to knock out reinforcements and limit the battle to reasonable proportions.

That day and the next, German infantry came searching along the brigade area with fighting patrols, preceded by artillery stonks looking for a soft spot. When they found it – they made trouble. No. 45 Commando were driven out of Franceville, then Sallenelles, and fell back into reserve.

'D' plus three was calm, before the storm, Then came the massive three-prong attack designed to sweep our brigade off the ridge.

THE BIG ATTACK, 'D' PLUS FOUR, 10 JUNE

There was complete silence while we waited for the expected battle. No artillery concentrations were to be fired until the enemy were committed to their linc of advance. Suddenly, all hell broke loose. Patrols forward of the orchard signalled large groups of Germans were approaching; then fell back into the defences, according to plan.

The forward edge of the orchard resounded with the crash of mortar and shell fire – the enemy thought this was the edge of our main position, and they gave it a fifteen-minute hammering – a trying business which strained everyone. Only well-disciplined troops can hold their fire on such occasions. We were out of sight of the enemy until they actually came into the orchard.

I had a quick walk round. The plans were understood by everyone. Each weapon pit contained three men – a veteran of Africa flanked by new boys. They had experienced three days of shelling with no positive action, which is not a good way to start one's active service. The man in the last weapon pit was one of those people who are never surprised. He came from Lancashire.

'I'm telling these lads, sir,' he said as we talked in whispers, 'that this is proper cushy after Africa.'

Whether it was or not was irrelevant. The man in question, faced by the Golden Gates, would not have been at a loss to make some disparaging comparisons. But it was good for morale – the younger men would not give him the satisfaction of getting any change out of them.

The shelling ceased; on came the German infantry. When they reached the orchard they hesitated and came through looking puzzled. This place should have been manned by commandos. Where were they? The leading company entered our defensive rectangle, two sides of which bristled with rifles and light machine-guns. Not a shot was fired. They were well through the hedge and some sixty yards into the orchard before the Bren guns opened up. There was chaos among the apple-trees. The killed and wounded lay still; the rest came on, but the advance of the first wave lost its impetus.

Simultaneously our artillery bombarded the German forming-up position in Bréville and the battleships hammered the roads beyond, where further reserves were being moved up.

The feature of our reverse slope position was that the Germans could not see what was happening till they reached a slight rise in the first hedge.

A second wave came into the orchard without realizing the fate of their comrades. Once more rifles and Brens caused terrible casualties. Now, the Huns knew where we were, and the Commando came under intense mortar and shell fire. Looking along the line of weapon pits I saw ominous craters, and here and there a man sagged listlessly from his slit trench.

German reinforcements now got down along the line of the far hedge and shot us up with light automatics. The shelling and mortaring continued. It seemed they were only waiting until their artillery had knocked us about sufficiently before making the final assault.

We got information from Lovat at headquarters. The enemy were attacking along the whole ridge, but the main effort at present was directed against No. 6 Commando. I knew the danger spot: ground to the right flank – a tangle of houses and gardens, without a proper field of fire.

Soon a party of Germans worked round there and started sniping from various houses. One of our most effective weapons was the K gun. This was an automatic light machine-gun which fired a magazine from a flat drum. It had been used in certain types of obsolete aircraft earlier in the war and was ideal for this situation. We had brought K guns along to thicken up fire power for the counter-attacks expected in the early days of invasion.

Sergeant Wakefield was two weapon pits away on our right. He shouted for a K gun, like a golfer who calls for a particular club: it quickly reached him. Lying behind his weapon, he braved the temporary insult of accurate sniping and got to work on the best bedroom of a house in enemy hands. A stupendous burst from the K caused a lull in the fighting, and I heard from Paul Loraine, my French intelligence officer, spotting on my right, that Sergeant Wakefield had wrecked a mirror but his companion had been wounded. He lay patiently while his sergeant did his stuff. We got him out ten minutes later.

The frontal attack had not died down. After the pounding they had given our defences, the enemy made their full-scale, final assault. After a long burst of fire from spandaus and light automatics, a wave of Germans got up for the third time and came forward. Moving targets required technique and concentration: once, as I was about to fire, Paul Loraine brought down my man.

In the midst of this turmoil, he turned with an apologetic look on his face and said, 'I'm extremely sorry, sir.'

'Don't worry,' I said, 'we're not on a pheasant shoot at the moment.'

The German fire slackened; those who could do so pulled back to the line of the hedge. It was a deadlock – they had made no ground and we, in turn, had sacrificed none.

We could deal with what lay in front of us, but the threat on the right continued. I laid on a counter-attack. A reserve troop pushed across the road and cleared houses from which the Germans withdrew. The enemy made a grave mistake in not putting their main effort into the flank attack; as we hoped, they had concentrated on attacking frontally across the orchard.

The battle started at 8.15 am and it was now 11.30 am. We were again mercilessly shelled and mortared and suffered more casualties. More gaps appeared in defence lines. But gradually the attack on No. 6 died down; soon we heard by wireless that enemy pressure was now concentrating on No. 4 Commando.

As I inspected weapon pits which had received direct hits from artillery fire, killing the occupants, I noticed that a war photographer (unauthorized) – trying to take a photograph of the German attack – had been killed and his camera destroyed.

At the dressing station in the farmyard Peter Tasker, the unit medical officer, and his RAMC detachment were doing wonderful work.

The German attack on No. 4 Commando was a determined effort; soon we heard of enemy infiltration on their right flank. David Style and Hutchie Burt had both been wounded with other officers. I was ordered to send a troop up as counter-attacking reserve. I detailed Captain Robinson and his cyclists. Robbie had done magnificently in Africa and his present troop in Normandy, despite casualties, was a very good one. I was sorry to send them.

The battle was now against No. 4 Commando on the left of the line. If things went wrong there the whole position would become untenable: a breakthrough on the coast would roll up the bridgehead. There was a frightful mill-up at that end; we heard that Robbie had been wounded, with many others. At last came a full report: after fierce fighting the battle on the flank had been stabilized and the Germans finally beaten off.

NO. 4 COMMANDO'S STORY

Lieutenant Murdoch McDougall gives a graphic account of the same engagement. The physical exhaustion that comes from days of fighting without rest is well brought out here (Author):

At the dawn stand-to on the morning of Saturday 10 June, we were at the end of our tether for want of sleep. I walked round my section again and again simply because I dared not sit down. A man would lean back in his trench as I spoke to him and become unconscious before my eyes. Another, standing in his slit trench, fell asleep on his feet and slithered slowly to his knees, to finish huddled in the corner, still holding the mess-tin from which he had been eating as exhaustion overcame him. Others slept in the strangest attitudes: leaning straight-backed against a tree, kneeling with head bowed on the lip of a slit trench, or lying face-down on the ground, still clutching their rifles.

Len Coulson's voice came slowly to me as though from miles away: 'Mac, we've got to keep awake. I think it's bound to come today.'

Nothing happened till ten o'clock when we were heavily mortared. Then the sun struggled through the clouds, and in the brightness of the morning we were astounded to see two German soldiers, one of them carrying a wireless set, strolling blithely along the track to our position. They were obviously going to report on the effects of the mortaring, and were destroyed by half the delighted troop at close range.

Thereafter we were shelled. The noise in the orchards and the wooded park was deafening. The trees resounded to the crash of explosions and rending of timber, while descending shrapnel rattled on buildings behind us. It reached its peak in the early afternoon; the woods reverberated, and casualties in E Troop, who had made the mistake of digging only shallow trenches under trees, mounted rapidly.

Then came the visit of the brigade major. He arrived in F Troop's position, and after grubbing about in various holes finally found the troop leader, who, like everyone except the look-out, was well below ground.

The brigade major, immaculate in green beret and creased trousers, demanded, 'Have the men had a hot meal yet?'

'A *hot meal*, while this is going on?'

'Yes, certainly, they should have had one by now.'

The troop leader leaned from his trench. 'Sarnt-major.'

'Sir.'

'Have the men had a hot meal yet?'

'A *hot meal*? Christ, Nossir!'

The brigade major looked disapproving.

'Well, see they get one right away,' he said, and moved off. He went about twenty yards when there was an almighty explosion; a cry for stretcher-bearers and back he came feet first, whereupon the irrepressible sergeant-major popped his head from his trench and called out, 'It's all right, sir, we'll go without, we're not really 'ungry.'

At about five o'clock we saw the enemy massing directly opposite us at the edge of the wood and at the same time the FOB arrived. An

FOB is a man who imagines he can control and direct the fire of a warship, which generally lies far off-shore. In this case it was *Ramillies*. Len was more than doubtful. 'Those things fire pretty big shells, don't they?' he asked.

'Fifteen-inch; nothing but the best!' said the FOB. 'All I have to do is to get on the blower, give the order and when I say "Shot One" you watch the edge of that wood!'

'All right,' said Len, weakening. The FOB, delighted, manipulated his set, gave some mysterious references, then yelled 'Shot One!' We all turned and fixed our gaze on the wood. For a moment nothing happened. We waited expectantly. Then, without warning, there was an explosion; the ground in the middle of our position heaved upwards; showers of mud and pieces of tree trunks splattered down. We clung to the earth while things subsided. Len turned a baleful eye on the FOB, who gulped and said 'I'm terribly sorry – but the next one'll be over there!' The reply cannot be recorded.

This episode was entirely responsible for the chilly reception which was given to the Airborne FOO (forward observation officer), who arrived on the scene almost as soon as the FOB had crept discomfited away. The task of an FOO is, in principle, the same as that of an FOB except that he 'controls' the fire of an artillery unit somewhere immediately in rear of the troops to be supported. This chap had some 25-pounders to fire and laughed immoderately when told of the black his predecessor had just chalked up. However, as he was from the Airborne camp and seemed a competent sort of chap, he was given a chance.

The system was quite different: as he clambered up on to the wall, he cheerfully took stock of the situation.

'There're quite a lot of the bastards, aren't there? Not much point just giving them one or two. Might as well shove about twelve rounds in right away, don't you think? How far is that from here? About three hundred yards? Well, we'll have a bash.'

While the nearest members of the troop slid into their slit trenches, he gave orders into his set; there was a whistling overhead and the edge of the wood disappeared in clouds of black smoke. The FOO whooped in triumph. The troop look-out in the tree, hung by his toggle-rope, said with satisfaction, 'Just the job. Send some more.'

Another twelve rounds were forthcoming. The FOO climbed down. 'It's no use asking for more,' he said. 'We're rationed down to twenty-four rounds till they get more ammo ashore, so that's the lot for today.'

And he went away.

This breaking-up of the attack annoyed the Germans, who stepped up the mortaring. Casualties were heavy. Hutch Burt, troop leader of E Troop, was wounded, and as both his subalterns and sergeant-major had been wounded on the first day this left the troop without an officer. The senior sergeant was the next to go. The troop was reduced to a third of its original strength.

Then German machine-guns opened up to cover the advance of their infantry across the open field. When they were clear of the wood, our battle began. How long it lasted, I have no idea. I remember Len telling me to take the two-inch mortarmen and shoot up the Germans as they came on. I remember directing fire, and listening to the bullets hissing through the hedge and sighing over our heads. I remember Sergeant McVeigh, DCM, cuddling the butt of his Bren and yelling at the top of his voice, 'Come out, ye square-headed bastards ...' and loosing off burst after burst. I remember Len Coulson, confident and alert now. I remember the German attack melting away, re-forming, then melting away once more. I heard Len say, 'We've done it, boy, they're gone – what's left of them.'

The din of battle died away, and I turned to my slit trench. It had been blown in. I dragged out my rucksack; it was badly holed and wet with the remains of the whisky from my flask, while splinters from my camera stuck out from the shreds of my shirt. I remember little else, for while I was on my hands and knees in the wreckage of my trench, I fell asleep.

This battle of 'D' plus four ended the first phase of our sojourn on the Continent. It was a triumph for the whole unit, but in F Troop we felt that it was a personal triumph for Len Coulson. Prior to that battle, not one of our automatic weapons had been fired. The enemy, probing around, had not drawn any fire other than rifle shots. This may have led them to believe that our machine-guns were elsewhere; in any event, the Germans had chosen our 'killing-ground' to advance over in their attack.

Owing to Len's foresight, casualties throughout the four days of mortaring, culminating in the big attack, were fairly light. Trenches had been dug deep; as many as possible had then been covered over with some sort of head-cover to keep out splinters. No. 4 Commando, as a whole, however, had suffered heavily. Of the 455 British officers and men who had landed on D-Day, there were now 160 left, while only 78 remained of the 200 Free French troops.

As soon as the action was over news from the others filtered in. We heard how David Style had been wounded. In the middle of the battle, a white flag had appeared opposite C Troop's position and a party of about a dozen Germans ran forward. Mindful of Poles wanting to desert from the other side, David ordered his men to hold their fire. The flag party approached. When they were some fifteen paces off, David stood up with a commando on either side and ordered the visitors forward. At once they dropped flat and the hindmost man threw a stick grenade, then loosed off with his Schmeisser. The stick grenade burst beside David's leg, shredded the calf and killed the man next to him. As he fell, David was caught in the chest and shoulder by the burst from the machine-pistol which killed the man on the other side. The Germans were wiped out.

Our existence thereafter changed to stalemate. We were sniped continually, but no longer harassed by sharp attacks. There was no aggressive action except at night, when single men, carrying wireless sets or packages of explosives, attempted to worm their way through to the bridges or where they could lie up and direct fire on targets where it would do the most damage. We installed a system of trip flares around each troop area, which, to start with, were regularly set off by stray horses.

CHAPTER TWENTY-THREE

THE AUTHOR BOWS OUT

Each stepping where his comrade stood,
The instant that he fell.

'Marmion'

We were getting tired. By the end of the fourth day those left in head-quarters showed strain: red-eyed and easily upset. Mental anxiety as it winds up, makes no allowance for snatched rest. I was jumpy and in-creasingly irritable. The shelling seemed continuous: part of the roof had gone; with the crashes of sound one's brains seemed to be blown out as well. Words came slowly from afar, and although the mind raced in mental overdrive it became increasingly hard to concentrate.

There were no laughs any more. When the brigade major limped in half-way through the battle, with a sliver of shell projecting through his bootlaces, we tried unavailingly to pull it out, then cut through the leather with only partial success. In considerable pain, Michael sat on a spartan chair until the boot came off. I suggested it was only bird shot, but nobody thought that funny. He was packed off, sitting upright, to the field hospital for surgery, back across the hard-won bridges. He was driven by Sergeant Rice, who reported that a Black Watch battalion, marching up to the line, had daubed 'HD' (Highland Division) conspicuously on each of these structures. On the return jour-ney Rice noticed that this conceit had boomeranged on the new arrivals: to the divisional lettering 'HD' the figure +4 had been added by a local humorist in red paint! I told Rice to look up a chemist on his way through Ouistreham, to buy a bottle of aspirin. The area was out of bounds and shops were closed. This did not stop my gallant driver. But his French was limited. When he asked the first civilian the way to the *pharmacie*, the citizen replied, 'Tiens – vous voulez une femme aussi?' This witticism took time to sink in, but came back to me in hospital.

Nobody smiled when Bill Millin left his pipes on the grass as he dived for slit-trench cover in a surprise stonk before breakfast. Drones and chanter were shattered beyond repair.

Yet two days before we had laughed like hyenas when a divisional staff officer found his cautious way up the hill to warn of an intended visit from the general; things happened to be fairly quiet and we were sitting pretty, for I discouraged all unnecessary movement. The iron gates of the villa were bolted and barred, with a light automatic covering the street and village green beyond. Accordingly, I advised the back-door approach through a hole knocked in the wall by which a Jeep could enter unseen. Here, concealed in outhouses and lean-tos, we kept petrol supplies, ammunition and stores, with room for Jeeps, a cookhouse and a prisoner lock-up, all on the reverse slope, which was approached unseen from the lane outside. German patrols that prowled by night found our converted residence a tough nut to crack: except for the stick grenade which twirled over on a dark night, we were not interfered with; a vigilant round-the-clock watch saw to that, and every morning at first light a fighting patrol dashed out to catch any infiltrators who had unwisely hung about my end of Le Plein.

This time a snooper had been overlooked. And Gale disregarded the proffered advice: at the appointed hour he arrived at the wrong entrance and sounded his horn. 'I think the Gaul is at our gates,' said Tony Smith. A Bren gunner untied the chains and opened up. With the brigade major I walked down the strip of drive to meet the general. Four immaculate 'hatchet' men in pipe-clayed belts and armed with Tommy guns bounded lightly out of a second vehicle and ran twenty paces to face outwards in all directions, to form a theatrical bodyguard. The general was in a testy mood. 'I cannot make head or tail of your positions. Nobody seems to know where your commandos have got to.' 'No, Sir,' I replied. 'And neither does the enemy.' Even as I spoke the mortaring began. The holes previously dug in the flower-beds again proved useful, and Gale departed down the hill. That all happened in the remoteness of time. The past forty-eight hours had been critical.

On 10 June the brigade stood its ground in desperate fighting which hung in the balance until the very end. To the accounts already given I shall only sum up the final outcome.

The battle rolled all day along the ridge, starting with mass attacks by the enemy from Bréville Wood against Derek's front, where six battalions of the 857 and 858th Infantry Regiments were repulsed with severe losses by No. 6 Commando. Tony Smith and Lieutenant Griffiths (the German interpreter, a fine officer, killed after crossing the Rhine) gathered from Hun prisoners that none had experienced such withering small arms fire or known less about the defences they had been ordered to capture.

Beaten off at that end of the village, the assault switched direction: after an intensive period of mortar and shell fire a massive attack was launched against No. 4 Commando and the Free French in the Hauger area, left of Le Plein and half-a-mile nearer to the coast. Here three battalions of the 744th Infantry Regiment – the remnants of the 857th and elements of the 21st Panzer Division's Reconnaissance regiment, supported by self-propelled guns – were taken on by a weakened Commando and some peerless Frenchmen – perhaps three hundred men in all. Both units had lost their commanding officers on D-Day, and the undermanned troop positions were isolated among small enclosures and spinneys of leafy timber: difficult country to control, without a defensive feature. McDougall has told the story of the ensuing contest. The line was twice broken, but no strong-points in the hedgehog were actually over-run. The enemy were sealed off after pushing forward without the support they counted on – and then systematically destroyed. Every available reserve was thrown in from headquarters, who all welcomed a scrap. With them went a newly arrived, and no doubt bewildered, party of reinforcements from England who had not seen action before. They acquitted themselves well enough. Robbie led the counter-attack and restored control. The artillery support – it was first-class – bombarded the forward areas; capital ships, which fired their broadsides, smashed down among the follow-up concentrations preparing to exploit the breakthrough.

As the battle turned and the Germans broke, Peter Young and Nicol Gray, with strong fighting patrols, routed the withdrawal up to and beyond no man's land, where the Germans had sustained heavy casualties: the bodies of their troops, killed by shelling, lay everywhere.

But if the opposition suffered a bloody nose the brigade had also taken a severe hammering. Nos 4 and 6 Commandos had borne the brunt of the fighting, but mortaring and counter battery fire had taken its toll all through the brigade. Derek Mills-Roberts writes:

There had been a steadily mounting list of casualties since D-Day. Robert Dawson, commanding No. 4 Commando, had been wounded and evacuated; Philip Kieffer had already been knocked out. Colonel Ries, commanding No. 45 Royal Marine Commando, was floored on the morning of D-Day, along with Bill Coade, No. 6's second-in-command; now the brigade major, Michael Dunning-White, was out of action, to mention only senior officers. The Brigadier was soon to follow. Three of Peter Young's best leaders in No. 3 Commando – John Pooley, MC (second-in-command), George Herbert, DCM,

MM, and Brian Butler, MC, were dead. (These last three had been sent to the Depot after the fighting in Italy to act as instructors, for it was felt they had done enough. John had only recently married, but each officer had begged to get back to his troop.) The other ranks' losses had been proportionately heavy.

Philip Kieffer was a big powerful fellow – brave as a lion – and the Free French were wonderful soldiers and fearless to a fault. They had helped to save the day, but at great cost.

Like the still air that follows the impact of a hurricane, the severity of the last attack left a hushed silence over the field. There was no question of a truce when the time came to bury the dead. If there were snipers about we were too tired to notice as we made a round of the forward areas, nor can I remember the names of those who fell: the first elation after victory was outweighed by a deeper sense of loss. The aftermath was desolation. The survivors, as always, rose magnificently to the dark hour: stretcher-bearers moved along the front line; doctors and the walking wounded, helped by medical orderlies, came and went. Burial parties performed their appointed duties: 'The Band of Brothers' were very close that day. The quick and the dead. If I have praised my comrades too highly I make no apology, for they were beyond all praise.

There was a tenderness under the apple trees as powder-grimed officers and men brought in the dead; a tenderness for lost comrades, who had fought together so often and so well, that went beyond reverence and compassion.

And so we dug the commandos' graves, lowering them away in groundsheets in long straight rows; then set about making wooden crosses to mark each spot. Later the men, after repairing trenches, brewed up tea to start again. Funerals in the field, rough and ready though they be, seem less bleak than those performed with formal rites, as though the soldier whose calling deals with sudden death can find a way to stand easy in its shadow.

We buried many friends on 10 June. Senior officers conducted the services. I will mention two that I attended. Peter Young read the lessons over one of the finest commandos of them all. His body was brought in by Donald Hopson. Lieutenant George Herbert, DCM, MM, lately commissioned in the field (he was an original volunteer in No. 3 Commando), fell leading a fighting patrol in pursuit of the enemy and lies buried in the garden of the Château d'Amfréville. The bullet that killed him had cut the ribbon of his Distinguished Conduct Medal. Peter Young, that most dedicated soldier, was not an emotional man and he brushed the tears away as angrily as if a wasp had stung him. For all the barriers were down that afternoon.

René de Naurois, the French padre, who carried a rifle and fought with distinction, prayed a Requiem for Sergeant George Fraser, MM, head stalker at Braulen in Glenstrathfarrer and one of three brothers serving in Lovat Scouts. His decoration, the Military Medal, was awarded posthumously for leading a bayonet charge that drove out the Germans and restored a vital section of the line. There was no time to stand around and only a handful of troops attended. The whole performance was carried out in respectful silence, in keeping with the hush of a golden evening. The files from headquarters cleaned their spades. Then we left the field on tiptoe, so as not to disturb our comrades' sleep.

I called a conference of commanding officers. The situation was grim. Max was busy with casualty lists still coming in. though incomplete, preliminary figures showed losses of 270 men, the majority of which, thank God, were wounded; but *hors de combat*. This could not go on indefinitely; the line was becoming too thin to hold. I suggested to John Durnford-Slater, who came up with congratulations from the army commander, that if the Germans repeated the performance next day, General Dempsey could expect a butcher's bill that spelt curtains for brigade and the certainty that our high ground would fall.

My reactions, I fear, were shrill! 'And where the hell are the reinforcements to cope with this wholesale destruction? You're supposed to have a depot and a training centre full of soldiers – and supply fuck all. Twenty-five lads as replacements for four days' fighting – less than half one troop – committed piecemeal to the battle the moment they arrived. Killed or missing, poor devils, without knowing what hit them.* Fifty more were needed before we cleared the beach. We lost more than this handful, sunk at sea. Who is responsible for the balls-up?'

Old John took the outburst in good part. He was a man of great understanding. The same evening Brigadier Jumbo Leicester, in command of the 4th Royal Marine Commando Brigade, moved into the line with four more Commandos, taking over the open flank on the ridge between ourselves and the Channel coast.

On his way back from the meeting, Derek ran into an SP gun that had found its way into the village, working round from the flank. He was punctured by shrapnel, but after lying up for twenty-four hours took no further heed of his wounds.

---

* In the event Lieutenant Sandy Carlos Clarke got badly shot up as he cheerfully led the reinforcements in the counter-attack.

It was my turn to bite the dust on 12 June. After this Derek took over the brigade in a battle that will be described by spectators, for I was in poor shape at the time. But I remember how, on that last night but one (11 June), everybody's mood changed suddenly. I cannot explain why. The party became hilarious before midnight, enlivened by Arthur de Jonghe, a splendid Belgian who had joined HQ staff as liaison officer. On one of the sorties that afternoon he had captured a large rabbit (described as a Flemish giant) in a cottage garden. This was cooked in red wine by Arthur over the kitchen fire and devoured amid general rejoicing. It made a feast after iron rations. Bobbie Holmes, who was another good scrounger, supplied the red wine. He had been down at the bridges, strengthening their supports, and assured us that heavy tanks would soon come across. Jumbo Leicester's marines and the promise of British armour would put the Brigade back on the offensive.

Emboldened by the Flemish giant, I decided it was the right moment for a message of goodwill to cheer up the local inhabitants. Some must be around, though they were all in hiding. Excepting Monsieur Saulnier at La Grand Ferme and his brave son Bernard, a boy of sixteen, we had seen no French civilians since D-Day.

Before taking off my boots for the first time since landing, a proclamation was composed. Arthur de Jonghe, who seemed to get around, promised to nail up copies at every crossroads in the forward area: '*Les Commandos disent bien le bonjour à tout le monde. Le chêf des Bêrets Verts salue ses amis et veut leur dire que Hitler commence déjà à faire pipi dans sa culotte! Ayez du courage. Nous allons gagner. Vive La France!*'

This task was duly performed by Arthur next morning. Never empty-handed, he returned from beyond the limits of our boundaries leading two valuable brood mares into the lines. He had found them wandering in the next valley.

Here I end the account of my small part in the Normandy Campaign. The final round during that first week's fighting consisted of a night attack – hastily mounted – which turned out to be a very messy affair.

DEREK MILLS-ROBERTS CONTINUES THE STORY

There was to be a conference at brigade HQ in half an hour. I took Paul Loraine with me. There we got full details of the battle for the ridge which had raged all day. The enemy losses had been extremely heavy.

Our exposed flank was now relatively safe. We were pleased to hear that the 4th Marine Commando Brigade had moved up onto the ridge between ourselves and the coast. The same afternoon two self-propelled guns were added to the hard-pressed brigade defence system.

Shortly before returning to our own headquarters, Shimi said, 'You'd better look out, Derek; a German gun is working its way up between 6 Commando and the village square.'

We took a short cut back through the gardens on the right of the street. There was one open place to cross before reaching the farm, covered by spasmodic fire from the SP gun.

I said to Paul Loraine, 'Wait till he fires and then we'll nip across before he has time to reload.'

We had to cover forty-five yards. We were almost across when I felt as if my feet had been kicked from under me, and fell with a bang on the road. I shouted to Paul and Corporal Smith to keep clear and, picking myself up, completed the distance to the farmyard gate. There I met David Powell.

'What sort of bloody place is this,' I said, 'where the old man gets shot up outside his front door?'

David explained that a patrol was already on its way to deal with the nuisance.

I went to the dressing station. I had a shell fragment in my left leg and a smaller one had cut my nose and cheek and wedged itself in the bone of my right eye-socket. The only likely trouble with normal flesh wounds is infection, and when I had been given the routine tetanus injection there was no more to be done.

David Powell walked in. 'We've got the gun, sir,' he said.

I went and had a look at it – the usual version on caterpillar tracks. The crew were dead. It takes courage to drive a self-propelled gun into an enemy area without infantry protection. We buried them in the field where we had made a place for our own dead. The cemetery stretched half-way down the hill.

During the night another self-propelled gun came out of the Bréville area and started shelling us.

'D' plus Six, 12 June

On Monday morning the divisional commander decided to capture the strongly held area of Bréville Wood, which the Black Watch had failed to clear the previous afternoon. Early the same day Colonel Black, the ADMS, told me that my leg was starting to develop gas gangrene and that I would have to have an intra-muscular injection and remain still for some hours after it had been done, or else go to hospital for treatment. I lay on some straw in Farmer Saulnier's cider press – a large stone cistern – feeling rather drowsy after the various drugs I had been given. Corporal Smith came in from time to time to report what was happening. The Germans were well alive to the fact that something was on, and our artillery concentrations on Bréville were reciprocated by German counter battery fire on our farm. The afternoon had been relatively quiet: at dusk a furious battle commenced.

Corporal Smith suddenly appeared and told me that Lord Lovat and Brigadier Kindersley (commanding the Air Landing Brigade) were both badly wounded. 'The brigadier wants to see you immediately, sir.' It was almost dark and the ghastly scene was lit up by the farm buildings, which were on fire.

I got up out of the cider press, feeling rather sick and drowsy, and went to the stable where Lovat was lying. He was in a frightful mess; a large shell fragment had cut deeply into his back and side: Peter Tasker, No. 6 Commando's medical officer, was giving a blood transfusion.

He was very calm. 'Take over the brigade,' he said, 'and whatever happens – not a foot back.' He repeated this several times. And then, 'Get me a priest,' he said. 'Get me the Abbé de Naurois.'

The parachute battalion making the attack on Bréville had been badly hit by mortars and *Nebelwerfers*; petrol bombs from this last insidious weapon had burst, causing desperate injuries and burns among the parachutists. Their colonel was being supported by Paul Loraine, who was holding him in his arms: he was mortally wounded. Other wounded men were on fire and we put out the flames by rolling blankets round them. Farther out lay more wounded. The RAMC were doing magnificent work, while Sergeant-Major Woodcock found every available Jeep to get stretcher cases to safety.

I no longer commanded No. 6 Commando, and, as acting brigadier, had to get on with the wider battle which was reaching round the whole area. I handed over to Tony Lewis and went back into the farmyard, where every available foot of space was covered with stretcher cases, but the Jeeps were making a good clearance. If more shells had fallen into the crowded farmyard the slaughter would have been terrible.

Shimi's Jeep and Sergeant Rice stood under the arch.

I was driven to brigade headquarters under heavy fire. As we went in a clump of mortar bombs landed in the garden. Corporal Smith and I flung ourselves into the nearest weapon pits. I dived on top of an Airborne driver, and Smith on top of the signallers, as more bombs landed: we were covered in dirt and debris.

I went into the house: a shell had come through the roof and made a large hole in front of the fireplace leading into the kitchen. The brigade staff were in their trenches and I joined them and got to work; from the reports coming in I saw all our positions were so far intact. Though we did not know it at the time, the furious fighting had culminated that evening and the seven-day battle was over.

Derek's story covers some close-quarter fighting considered the fiercest that had taken place in France when the Germans still called themselves the Master Race. I give below another view of the action – on the night of the big attack – as seen by a Teutonic soldier.

An intelligence section of German interpreters was attached to brigade headquarters. Each Commando was supplied, as and when required in foreign service, with interpreters from No. 10 Commando. Dudley Lister gave me the best. Some, but not all, were Jewish or Sudetenland subjects. These refugees from the Fatherland hated the Nazi régime in Hitler's Third Reich. After the war they made new homes in different parts of the world. Before being accepted for combat duties they had been carefully screened by the security branch of the War Office and given British names. Their identity was officially forgotten and never referred to. All spoke excellent English – but little French. Those who volunteered for commandos were of a high military calibre and meant business. Their record in Normandy was a most impressive one. The account which follows was sent to me after the war by Peter Masters, now a United States citizen. (Author)

'D' plus Six, 12 June: Peter Masters's Account

No. 6 Commando was to take Bréville, but word came down that an airborne unit, I think the First Canadian Parachute Battalion, were better rested and would do it instead. My friend Gerald Nichols (3 Troop, No. 10 Commando, another good German serving under an assumed name) had done a reconnaissance patrol up to the crossroads outside that village the night before, and seen no sign of enemy armour. So Gerald said to me, 'Let's go out and watch the attack get under way.' I replied that I was not sure that that was a smart thing to do. 'That fellow Dunlop was killed yesterday, when a gun fired short during the preliminary artillery bombardment. How do we know something like this won't happen again?' 'Look,' Nichols said, as we came out of Monsieur Saulnier's farm to the little triangular green. 'Who is standing over there? The brigadier. Now don't you suppose he knows all about that gun firing short? He wouldn't be standing around if he hadn't taken care of it.' 'You have a point,' I conceded. 'Let's watch.'

The airborne soldiers were moving up the road past the farm and the cemetery. We stood on the patch of grass a few feet from Lord Lovat. It was then that the artillery opened up. Shells exploded all around us, and the brigadier was hit – killed, I thought. It was not immediately clear whether it was the Germans firing at the forming-up area, or yesterday's delinquent 25-pounder. We rushed towards cover in the sturdy Saulnier farm, but in the archway entrance the adjutant, Captain Powell, called out, 'We need wire-cutters, quick!' 'I'll get mine from the barn, sir,' I said, and at that moment two shells hit the barn, one on the reverse slope of the roof, and the other high up on the thick stone wall, collapsing it, surprisingly, precisely on to my kit. It took hours to dig out later, removing Norman boulder stones.

The barn did not look attractive as a shelter then, and I ran across the courtyard to dive into a shed opposite. As I opened the door a gaggle of noisy geese, panicked by the shelling, emerged, like white bats out of hell. Hell seemed to be let loose all round. All this had taken only a few seconds, and suddenly Nichols made a wild dash out through the archway, where the brigadier was lying, motionless, and bleeding from abdominal wounds. He picked him up, forked walking stick and all, and, staggering under the weight of the big man slung over his shoulder, re-entered the courtyard. Shells were still bursting all around, and it seemed a foolhardy sortie, at best, to bring in a dead officer. Two of us ran to help him, for he looked as if he were about to fall. He put Lord Lovat down and shouted, 'Where's the MO?' The medical officer was taking prudent shelter diagonally across the long yard, under the huge *cidre* barrels, I believe. Nichols ran across, darting past the captured Sdkfz 251 (half-track) and the people taking cover under it. Cover was hard to come by; every available space was crowded. Someone said, 'Don't get the MO out, we'll need him later!' But Nichols found him, and I saw him pull him by the hand back across the yard.

We kidded Nichols about it later. 'What some people won't do for promotion! I suppose you realized somehow that he was still alive, and saw your opportunity!' But it was a fine thing that he did.

Lord Lovat was given first aid, and loaded on a stretcher, with another general, on a hastily-summoned Jeep, and off tore the driver, no doubt delighted to get away.

Meanwhile the shells were bursting. A Colonel Johnson, of the Airborne, came running into the farm, his right arm dangling by a thread. 'Somebody give me a tourniquet!' he yelled, and I went across and helped hold his arm and tighten a bandage. I carried morphine tablets, like all 3 Troop 10 Commando men active on patrols. I thought I would give him some, but a new series of explosions spilled them – and me – into the mud....

At the other end of the yard lived one Captain Brown, the 6th Airborne forward observation officer. We of No. 10 Commando knew him well, because we always told him what we had seen of the enemy when out on patrol. 'Surely he must be able to do something to stop that battery from misfiring.' We rushed to see him, Nichols and I. 'I know, I know,' he said, 'but I can't do a damned thing about it. All I could do is to stop the whole bombardment, but I can't isolate that gun. Nor can I let the attack go in unsupported. There's only one thing to do and you fellows can help me. The attack seems to have bogged down for some reason, but the faster we attack, the sooner they'll be out of danger back here. Let's go and see if we can't get them going!' He pulled out in his Jeep, his signaller at the wheel, with Nichols running along one side and I the other. We urged the attacking troops, who were lying on both sides of the road, to get up and at them, to move clear from the faulty barrage. But it is impossible to crank up an attack from behind. You have to get to the front of a halted assault yourself. By now we were half-way to Bréville; Captain Brown said this was not our show, and that we commandos should go back. We remonstrated that we wished to stay and help. 'You don't want me to order you back,' he said, 'but there's something you can do – take this chap away who is upsetting everyone. That will be a most valuable contribution. Now go!' Indeed, there was a paratrooper with a particularly painful foot wound, his boot slashed and bloody, hopping around on the other foot hollering. We took him between us, shoved a cigarette in his mouth to take his mind off things, and doubled back. Minutes later the Jeep received a direct hit, and Captain Brown and his driver were killed.

What had slowed the attack, apart from the artillery mishap, was a strange coincidence. The Germans had decided to attack at the same time! They had brought up tanks during the night, a few Mark IVs, and some 75 mm guns mounted on Czech T-38 chassis. The troops were briefed that they were opposite a few *Heckenschützen* (snipers), and the assault was to drive us into the sea, supported by dive-bombing Stukas and surfacing U-boats! All this was romantic fiction, but they did mount quite a spirited attack before our artillery barrage caught them in the open. One German, whom I interrogated, had run all the way from outside Bréville to surrender, with about half a pound of flesh slashed out of his thigh by a shell splinter.

When we got back to the farm, these disillusioned Germans were beginning to come in. I interrogated the first one in the entrance way of our cookhouse and adjacent armoury, when it blew up and burst into flames. My prisoner's back was burned – he smelled like a freshly singed chicken! People thought he had thrown a grenade among the cooking fuel, but he had his hands up on his helmet, and certainly had had nothing on him when I searched him. I realized, as more and more prisoners poured into the yard, that they were adding to the confused scene, which, by now, looked like the Battle of Atlanta in *Gone with the Wind* – an overdone Hollywood tableau, with fires blazing, explosions ripping the air, machine-gun fire rattling in the background, the smell of smoke and gunpowder everywhere. Bodies on stretchers were littering the yard, some still, some stirring. 'Get the prisoners to Brigade Headquarters.' I fell them in, with sharp German words of command. I told them that we would double in formation to their next destination, and that anyone taking cover or even breaking ranks without my permission would be shot on the spot. 'Verstanden? Das ganze links um. Linksschwenken, im Laufschritt, marsch-marsch!' They performed like a drill team. Thus, all thirty or forty of them were moved, with only two of us acting as escort.

I heard later that when my German commands rang out, a British relieving battalion from 3rd Division looked out from under protective cover and saw German boots marching all around; three officers and eleven men surrendered!*

THE END IN NORMANDY –
DEREK FINISHES THE STORY

So the days passed.

The Germans had lost the initiative. It was our turn to take the offensive and contain the Germans in their own positions. Nos 3, 4 and 6 Commandos brought off successful operations at night in which prisoners were taken. The Royal Marines in No. 45 Commando, lacking in experience and perhaps not as steady as the other units, were given a series of strong fighting patrols which greatly increased their confidence. Nicol Gray, who had taken over, impressed me as a purposeful CO of sound common sense.

The sniping programme covered the whole of the brigade front. There were many experts at the game. Camouflaged snipers worked their way up to the enemy positions, with another man acting as spotter. This business required marksmanship and indefinite patience. Two of the

* I cannot vouch for the accuracy of the entire story, or indeed of the last paragraph. The stick referred to was passed on after I waded ashore (see p. 312). But the German section were, without exception, brave, intelligent, and most reliable soldiers, whose assistance was invaluable. (Author)

most successful rifle shots were Guardsman McGonigle of No. 3 Commando and Marine Cakebread of No. 45 Royal Marine Commando, who accounted for over thirty Germans in the Le Plein area. Cakebread was a hairdresser before the war.

The vigorous offensive policy carried out made it difficult for the Germans to mount any fresh attack.

In the earlier flight from Bréville one of the German quartermasters had left a bag behind. This was handed to Tony Smith, who was a fastidious person but also punctilious. He waded through a mass of dirty underclothes and was rewarded by finding underneath them a sizeable wad of French banknotes. When he reported the cash discovery, I felt that the real place for this money was the Commandos' Benevolent Fund. I discussed the matter at length and finally decided to act as a field cashier to all officers in the brigade, giving a slightly higher rate of exchange than normal. This was done and we sent back cheques to the newly formed Commandos' Benevolent Fund. It delighted everyone to think that German money was being paid to the dependants of the officers and men who had been killed in Normandy.

For the past week I had been commanding the brigade as a lieutenant-colonel and I wondered what the permanent position was going to be. I had not heard of any replacement for Lovat, and there was an even chance that I would be given the job myself.

From a purely personal point of view I was not keen; I had brought No. 6 Commando to a fair state of efficiency, and I preferred to go back and command them rather than to take over the brigade permanently.

Lovat was a great loss. Apart from his personal powers of leadership, he also possessed great moral courage. He could stand up to generals when necessary, and in any commander of an independent brigade this is essential. A brigadier who got pushed around would be of little use, and the units he commanded would get pushed around too.

Added to all this, I knew the inevitable irritations and jealousies that any young brigade commander had to put up with. I hoped that a sound brigadier would materialize.

We had a fair number of visitors in those early days – the ridge was still the focal point. General Boy Browning called and we showed him our defences, including No. 6 Commando's orchard.

'All the guns of the German artillery could not get you out of a place like this,' he said. From him this was praise indeed.

General Dempsey again came up to ask about the German dispositions. I gave him the answer in great detail.

'I think you must, indeed, be right,' he said, and left after a few enquiries about the brigade.

Three days later I was promoted to the rank of brigadier, with effect from 13 June, the day I had taken over when Lovat had been wounded.

Attached to brigade headquarters as a liaison officer was Major the Vicomte Arthur de Jonghe, of the Guides Cavalry of Belgium. Arthur, a tall, lean, wolf-like fellow, had been a prisoner of war of the Germans when he was aged fifteen – a thing he well remembered. In 1940, when Belgium capitulated, Arthur operated as an agent for his country; when his face became too well known he joined the commandos. His knowledge of Europe, and of those who lived there, was of great use to us. He spoke with a slight accent, protesting the while that he was more English than the English – in many ways this was true. He had the charm and manners of a light cavalryman, which indeed he was.

One day he came into my headquarters and announced that we were faced with a problem. A man had turned up in our area with a number of women and suggested to Arthur he should run a brothel for the 1st Commando Brigade.

Arthur laughingly pointed out to me, 'He tells me many men may be dealt with most quickly by these ladies, provided the Military Police are there to see there is no delay or messing about outside.'

Despite this reassurance, I told Athur I was afraid public opinion in England was against such things.

'Of course,' he said, 'nor do I approve. And I think I will go and chase these women out of the area.'

Later that day he told of an embarrassing incident which had occurred.

He had gone down to the place of ill repute near Bénouville, and the general, seeing his car outside, asked 'Where is Major de Jonghe?'

Arthur's driver unwittingly replied, 'He's upstairs, sir, with the three prostitutes.' This caused much ragging.

On the banks of the Orne there was a large factory at Colombelles; Harry Blissett, the new brigade major, heard that it contained a store of German white wine in casks. He suggested a lorry could be sent with specially cleaned petrol tins to fetch it; then bottling could begin. I agreed and an excellent job was done.

Later the mess corporal informed Harry that the wine was finished. He was told to get more, but unfortunately Harry overlooked the swilling out of the petrol cans. The wine came back and was duly bottled. The new consignment looked better than the first – crystal clear, with inviting labels which Harry had specially designed.

After a harassing day, a very senior officer called upon us.

Harry suggested he take a glass of wine and, producing a splendid bottle, whipped out the cork with an air of pride and poured the amber liquid. The general raised the glass on high, surveyed it with delight, then took a sip. His other hand went to his stomach and he hastily put the glass down. The most casual observer could see that vitriol would have been more palatable.

Harry was not the man to panic in a crisis: he rang the bell for Max Harper Gow, who came in. 'Max,' he said, 'try some of this.'

Max took one sip and put it down. 'High octane,' he said, with the assurance of a connoisseur.

At the end of July the brigade moved from Le Plein down to the Bois de Bavent, where we took over from the 3rd Parachute Brigade. It was a dull period of waiting.

Just holding ground, without making physical progress, has a depressing effect on soldiers who have been trained in launching violent assault.

At last came a forward move. The brigade led the way across the valley of the River Dives, with its wide marshes, deliberately flooded by the Germans, who had opened the irrigation sluices. Varaville and Robehomme, two island villages in the watery plain, were captured without difficulty. The enemy appeared to be pulling out and fighting a delaying action as they went.

On the far side lay a ridge of hills. I pointed them out to Tony Smith. 'We can expect trouble up there. That position looks very easy to defend.'

But at first light the next day we crossed broken bridges unopposed – taking advantage of early morning mist – and got well up on to the other slope.

That night the whole brigade infiltrated through prepared enemy defences to capture the high ground at Angoville. It was the main tactical feature on the line of advance. In bitter fighting, the commandos held against repeated infantry counter-attacks throughout the following day. Once darkness fell the shelling ceased. We were undisputed masters of the Angoville Heights.

During the night the Air Landing Brigade came through and we were united with 6th Airborne Division. Once more our supplies of ammunition were replenished.

Next morning we were once more in the lead. We had one sharp skirmish near a small *estaminet*. Shortly afterwards we were put into reserve for a day's rest.

I walked into the restaurant and asked the proprietress if she could manage four for lunch.

'Certainly, we have an excellent lunch going here,' and she smiled.

Paul Loraine then told me, 'German officers ordered their lunch about twenty minutes ago – and have also paid for it. The proprietress will look after us well.'

A genial figure poked his head into the window of the hostelry: one of the local *gendarmes* who had looked after several parachutists when they had been dropped in this area on D-Day. His information we found most interesting.

I went into a little orchard by the road and came across some Parisians who had moved into the country when Paris fell; they also had been looking after parachutists. One couple, who looked like Darby and Joan, both being aged over eighty, knelt down and said their prayers. Part of France had been liberated and they thanked God for it.

We moved to a place called Drubec, spending one night in the priest's house, which was next to the old village church. Ernest Hemingway, in *For whom the Bell Tolls*, wrote a description of the smell of death and I remembered this when I opened the church door. The whole fabric of the building seemed to be permeated with the indescribable smell. Near the altar lay a coffin. They told me a stout Norman farmer had died and, owing to the fighting, they had been delayed in burying him. Next day the funeral took place. The hot August sun beat relentlessly down on the roof of the little church; inside stood the packed congregation, wearing their thick broadcloth. No windows were open, nor was there sign of ventilation.

I turned to Paul Loraine. 'These Normans of yours,' I said, 'aren't squeamish.'

'They're the salt of the earth,' said Paul.

Next day we were on the move again. Beyond Pont l'Evêque we did an ambitious night flanking operation behind the German positions and captured a long sector of road. As the Germans started to withdraw next morning they ran into trouble.

When the Allied advance reached the Seine the 6th Airborne Division and the 1st Commando Brigade were withdrawn to England, to train for operations elsewhere.

We pulled out at short notice, and drove back to Mulberry Harbour at Arromanches, the embarkation point for England, passing all the old familiar ground: the café beyond the Heights of Angoville and then the narrow lanes blocked by guns and dead horses, which had not been moved; the last few weeks had made them stink abominably.

At the café I found the *gendarme* we had met before. I asked about the family whose son had been in the French Legion of the German

Army. He shrugged his shoulders; they were now wanderers on the face of the earth, hunted through France and terrified to return home.

We passed over the Caen Canal, then by our old landing beach, and so on to Arromanches. This artificial harbour of huge caissons and sunken ships looked lonely as darkness fell upon us – its work was still not done but nearly so. Somehow it had thrived on the noise and turmoil of landing fighting men. The embarkation of troops for England, after that task was finished, seemed tame by comparison, for such an instrument of war.

Arrangements for our return had been muddled by the transport people: there were no ships to take us back or accommodation ashore for the brigade. What was to be done?

I sent off Harry, the brigade major. 'Get your facts right about the muddle,' I said, 'and any relevant evidence concerning it all.'

'I've got that, sir,' said Harry, 'and we're in no way at fault.' This was so.

I then told each commanding officer to get his Commando back to England. There were a number of empty tank landing ships ashore. I hailed the captain of one of them: he said that he sailed with the tide. We sailed with them. So we came home.

The brigade had been in the fight for eighty-three days without being rested. Of the 146 officers and 2,452 other ranks who had taken part, 77 officers and 890 other ranks had become battle casualties, killed, wounded or missing.

POSTSCRIPT

Then he said – 'In yonder forest there's a little silver river. And whosoever drinks of it – his youth shall never die.'

ALFRED NOYES

Field-Marshal Wavell, who never wasted words, can supply a worthy postscript:

In Military Service we have been inclined to believe that our armed forces are both excessively professional and unimaginative.

This war has shown as others have done before it that the British make the best fighters in the world for independent and irregular enterprises. Our commando soldiers, to whom a special memorial was recently unveiled in Westminster Abbey, have proved that where daring and initiative and ingenuity are required in unusual conditions – unrivalled leaders can be found among the British race.

The spirit which had its most renowned expression in the Elizabethan adventurers perhaps lived before them, but most certainly lives on today. It will surprise other foes in other wars – if wars are still to be.

The 1st Commando Brigade fought its way to Neustadt on the Baltic, leading the van of the general advance through Holland and Germany, always maintaining an important rôle which included the assault crossings of five great rivers: the Maas, Rhine, Weser, Aller and Elbe. Then they were disbanded.

I did not hear the victory bells which rang out across the Jutland Peninsula to proclaim the end of the war in Europe. The shell wounds that had driven clothing through my body, back and side took some months to heal, being too extensive to sew up. I lost two stone in hospital. But I was on my feet again in time to visit Russia in uniform as one of the Parliamentary Delegation paying their respects to Stalin and the Praesidium of the Soviet Union early in 1945. On my return, Mr Churchill sent me on a lecture tour of the United States, after which I became Under-Secretary of State for Foreign Affairs in the 'Caretaker' Government.

Within the year the military curtain rang down on Special Service. Old statistics are as dry as dust, but I will wind up this story on a proud note. Army Commandos had fought with distinction in 147 separate actions, including engagements in the Far East. In five years a force of five thousand volunteeers had been reduced by casualties in different theatres to a quarter of their original numbers. The soldiers who had stepped forward since 1940 from different regiments had won 8 Victoria Crosses, 35 Distinguished Service Orders, 162 Military Crosses, 32 Distinguished Conduct Medals and 218 Military Medals. There were battle honours besides. Now they have gone their various ways and the mantle has fallen on the Royal Marines. May they worthily uphold a new army tradition. But the commando spirit has not passed away. Seven hundred old comrades attended the reunion dinner in 1976, while the numbers who turn up for the D-Day pilgrimage to Normandy are equalled only by friends in the Airborne Division. Thirty years on, the feelings are the same. It is a moving experience, and when the pipes play down the rows of crosses behind the beach, few dare to look each other in the face. The graves are well kept and the simple setting recalls Thermopylae: 'Go, tell the Spartans, thou who passest by, That here obedient to their laws we lie.' The average age of commandos who fell in France was only twenty-three. 'The flowers of the Forest – that aye fought the foremost': they are not forgotten.

In Scotland we have raised a lasting memorial. Three steadfast figures, a rugged group in raiding order, emerging from the sea; alone but not afraid. Cast in bronze, they stand proudly in Lochaber, their keen eyes lifted to the hills. The plinth is plain.* But on it I would have written the message that nobody grows old by merely living their allotted span. People grow old by abandoning their ideals. You are as young as your faith and as old as your doubts: was it all worthwhile? The causes of war are falsely represented: its purpose dishonest and the glory meretricious. Yet we remember a challenge to spiritual endurance and the awareness of a common peril endured for a common end. That is the only way to come through a shattering experience – and enshrine the memory of those brave men who did not return. The spirit of comradeship still burns undimmed. It is a strong emotion. If we had been individuals in battle we should have run away. But we were a team, held together by our training and our shared experience. No other form of endurance, except great danger at sea or on the mountains, could forge the same links of iron determination. We were bound to one another – men on the same rope. War is a bloody business, but it united a nation in its hour of need.

What happens now? I am willing to believe in a paradox: the

---

* It only bears the formal words, 'United we conquer', and lower down, 'This country was their Training Ground'.

future is best built on good things which have emerged from the past. In the 1940s this country's record and performance were not inglorious. In the 1970s we seem to have lost our way, but surely not the map as well. It is another generation's turn to put things right.

I shall end this soliloquy with a quotation from the historian, Hillary Saunders. His tribute, which flatters my old comrades in No. 4 Commando, may not be out of place, for this book is written for them: 'When in the course of time these former soldiers grow old and die the memory of great deeds, greatly performed, will remain with the glory. And that same sun which gilded the bayonets of Lovat's men as they charged the guns at Varengeville will shine upon a new generation as strong and as all-compelling as their fathers.' Who knows?

But that is how it was 'on commando' in the brave days of old.

> When all the world is young, lad,
> And all the trees are green;
> And every goose a swan, lad,
> And every lass a queen.

APPENDICES

LA CHASSE AU CERF
by Lord Ribblesdale

I arrived at the Hôtel de France in ample time for a capital breakfast. The waiter recommended a Graves, which he assured me was
much esteemed by the gentlemen of the hunt. The meet was at the
Belle Croix, within half an hour's ride of Fontainebleau, and at the
telephone-appointed hour I was informed that my horse was at the
door. I have never yet got over the pleasurable feeling caused by this
familiar announcement – I hope I never shall – so I rushed out full of
curiosity. 'Le cheval de Monsieur' proved to be a fine bay Prussian. A
brand-new bridle with many buckles caparisoned his lusty neck; a glossy
ultramarine frontlet adorned a pensive brow; a burly saddle surmounted his thick withers. The manager of the livery stable had
courteously ridden round in order to show me the way, and we looked
him over together, before mounting. I remarked with deprecation upon
his very German appearance, at the same time paying a flattering tribute
to the nice horses bred in France. 'Now this horse,' I said, 'was never
bred in the Nivernais' – this happened to be the only district which
occurred to me at the moment – 'a heavy horse like this can only be a
Prussian.' I saw at once that I had made a point. 'Ça se voit que
Monsieur est connaisseur. Dame, c'est un cheval qui vient d'un peu
partout.' From that moment the manager and I understood each other,
and we set off together, excellent friends. The manager himself, who
sat as he should, was riding a little brown horse with a cock-tail, all
quality, and quite a stag-hunter. As I watched his gay movement, and
the vein tracery starting into relief on his clean neck, I made up my
mind to swap horses, and so to be quit of the cosmopolitan. After
some more slighting remarks upon the Prussian, I explained, within the
limits of my French, that he was riding my sort of horse, and that he
ought to have sent him for me instead of the the Teuton. 'But,' objected
my companion, 'you said you weighed over one hundred kilos.' This
was the telephone guess at my weight made by my representative at
the Continental. 'That is so,' I admitted, 'but blood and action carry

weight. Can you expect either from "un cheval qui vient d'un peu partout"?' He knew quite as well as I did that it was not to be done. And after loyally citing the merits of my mount as a *cheval de retraite*, an equine type that needed explanation, but which turned out to be a good hack horse, he agreed to change at the Belle Croix, after he had shown the brown horse to a client. The client was not out, so we changed at once. I had a charming ride, which, as he assured me the brown horse was 'parfait pour les dames', was not to be wondered at.

D-DAY: A SECOND OPINION
A MEDICAL REPORT OF THE MORNING'S BATTLE
by Dr. J. H. Patterson, RAMC,
of No. 4 Commando

For a tomb they have an altar;
For lamentation, memory;
And for pity, praise.

SIMONIDES (written after the Battle of Marathon)

Patterson's contribution shows a sensitivity that blends without heroics with the spirit which won the Normandy beaches. Equal determination were shown by the navy – the crews of the Silent Service, who gunned their engines and drove the boats up on to the sand under withering fire. The medical officer's account also illustrates the foul weather conditions. (Author)

On Wednesday 31 May we drew our rations, cooking equipment, etc, and the next day moved down to the *Princess Astrid* and the *Maid of Orleans* in Southampton docks. The French troops and the rest of the brigade remained in camp: they were all landing from LCIs, which are small and cramped and cannot keep men aboard for more than twenty-four hours. We were the lucky ones, as we knew the ships well, and settled in quickly. I distributed my final items of kit and lectured the troops on the use of morphia, etc. Soon we were absolutely ready.

An extra six bottles of fluid plasma suddenly arrived, sent by mistake. I was very glad of it and we managed to persuade the ADMS to get the 'giving sets'. This plasma proved its worth in the early stages of the battle. It appeared later that the advice of the people who were in the Sicilian landings (in which each RAMC man carried a bottle of plasma and a giving set) was not taken by the medical units following us, with disastrous results on the beach-head. The blood and plasma stocks were sunk, and none was available until late on D-Day.

On 1 June we pulled out from the quayside and lay off Cowes. The whole of Cowes Roads and Southampton water, as far as the eye could reach, was packed with shipping. Thousands of ships and landing-craft of all kinds filled the sea; in the sky were serried ranks of balloons, hundreds and hundreds of them, the farthest mere pin-points in the sky.

No sooner were we aboard than the weather started to break. It blew hard, up to gale at times, from then until and including D-Day. As the weekend wore on we watched the sky. It was clearly impossible to start in such weather, and we got more and more depressed. The prospect of postponement for fourteen days appalled us; it would have meant imprisonment in camp again. On Sunday 4 June the spindrift was flying in scuds across the roadstead, and people coming aboard in duty craft arrived soaked to the skin. The 5th was to have been D-Day, but by eleven o'clock on Sunday the 4th the 'postpone' signal came through.

Apart from boat drill and one or two conferences, the days on the ship passed quietly enough. We were crowded, what with the brigade major and extra liaison officers and FOBs. They were fitted into the landing-craft somehow on boat drills. I slept on deck each night in David Haig Thomas's sleeping-bag, which kept me warm in spite of the wind. It was during the small hours in the blackness of the nights that I felt chicken-hearted, but the foul weather persuaded me each morning that 'it couldn't be today'.

It was tea-time on 5 June when I noticed the minesweepers beginning to creep out, and I realized the show was on. At lunch-time I had been offering heavy odds that it would be cancelled again. I felt appalled when I looked at the sea. It was blowing half a gale from the south-west, and banking up black and beastly for what promised to be a dirty night. The 'Operation On' signal came through; we were issued with real maps and real place-names. Our job was to land at zero plus thirty minutes on Red Beach (Sword) – that is to say, on the extreme left flank of the whole invasion.

The East Yorks were going in at zero hour to clear the beach defences and make gaps in the wire and minefields for us. We were to follow thirty minutes later, pass through the East Yorks, dumping rucksacks, then crash into Ouistreham, destroying beach defences up to the canal mouth from the rear, clearing up strong-points round the Casino and the six-gun battery at the eastern end of the town. The French troops were given the Casino job – a very tough one, as it turned out. We knew now that 140 German guns could train on our beach, to say nothing of machine-guns and mortars. We were comforted with the assurance that the RAF and the Royal Navy would give a lot of attention to all these guns. The six-gun battery at Ouistreham we just didn't like to think about – it was a bare mile from our beach. Disturbing news of panzer divisions at Caen came in at the last minute to cheer us up, also of a new battery of four guns just east of the Orne and a mile or two inland.

The rest of the brigade, Nos 3, 6 and 45 Commandos, were to land at zero plus 75 and make their way straight to the bridges over the Orne canal, cross them, and establish a link with the 6th Airborne Division. Some gliders were to be landed right at the bridges to prevent them from being demolished by the Germans. This was the part of the operation I felt most doubtful about. In the event of the Orne bridges being blown, we would have had to cross in rubber boats. No. 4 would follow the brigade out of Ouistreham and take up a defensive position with the other Commandos and the 6th Airborne Division. The Canadians were on the right of the British forces, and the Americans on their right again, though how far the beach-head was to extend we had no idea.

We hoped that the R E chaps were going to make a job of the mine-fields. A demonstration on an exercise of A V R E (Assault Vehicles R E), with flails and other gadgets, had been most unimpressive. On LCTs we could see the 50 lb Bangalores, the bridging sections and the huge faggots for filling anti-tank ditches, all mounted on flail tanks.

Now all the shipping was quietly sailing out through the booms. The rest of the brigade was in LCIs, and our French troops ranged up alongside us in theirs to give us a hail – they had gone aboard only that morning. It was wonderful to watch the steady stream of craft slipping out past the Portsmouth forts, silent and orderly, with no sirens or fuss, their balloons marking those already hidden by intervening ships.

The wind howled and it rained in vicious scuds. The skipper said in his speech, 'The High Command must be counting heavily on surprise, for the Germans must surely think that not even Englishmen could be fools enough to start an invasion on a night like this.'

The wheels were turning and nothing could stop them now.

The battle fleet was steaming to join us, *Warspite*, *Ramillies* and many others, and, blow as it might, we were to go through with it.

Feeling small, I set about my final packing. I spliced my identity discs on a new string and gloomily hung them round my neck. I dished out sea-sick tablets and morphine, and talked to all the troops on final medical plans. I had a heated tussle with the brigade major on the subject of the rum issue, finally settling that the troops could have it thirty minutes before landing and not prior to getting into the LCAs, with cold and sea-sickness ahead; this would have been folly, but I had to be firm to gain my point.

I had a bath ('washing off the *B. coli* in case you stop one', as the naval doctor said), ate a huge dinner of I don't know what, soaped my socks, dressed for battle, and, after midnight, rolled utterly weary into my bunk.

Breakfast was at 0400 hours. The wind howled and the ship rolled, and the adjutant, sharing my cabin, was sick in the basin, which was depressing. It was grey dawn when I emerged, and I could just see the shapes of the ships. There was some ack-ack in the distance and an occasional red flare. Then came a distant metallic thump. I went out in time to see a Norwegian destroyer on our port beam roll over with a broken back, and heave its stern and bows out of the sea. No craft seemed to take the slightest notice, and, like chaff on a stream, the ships drifted on. But before long a little tug was alongside the destroyer, picking up survivors.

We were going slowly now; the wind had lessened a bit. As the dawn came up the sky filled with the roar of aircraft. We began to heap on our equipment, everyone very preoccupied with his own arrangements.

The anchor went down with a rattle, as the *Princess Astrid* anchored in her appointed place, eight miles out. It was quite light and the ship rode steadily at anchor, though a big swell and the white horses made the assault craft look unsteady. I was in S4, the last boat on the starboard side. The *PA* carries eight LCAs. We stepped gingerly over the gap with our heavy loads and began to pack in. As always, the last man couldn't sit down. The last view as I sat down was of the bow of the Norwegian destroyer, sticking out of the water a mile astern. Soon 'Lower away' sent the LCAs down in turn into the water, to bump and wrench on their davits as the swell took them; but quickly the shackles were cast off and we rode free – very free – in that sea.

The LCA flotillas, like tiny corks afloat, were lifted into view and then sank between the waves of the huge swell. With their camouflage, they were hard to see against that white, grey and blue sea. The various assault craft were forming up, the first wave already on its way towards the shore. The sky was clearing and full of aircraft, big formations of them tearing south. Far out on either beam we could see the warships, with the big orange puffs from their guns and the flashes lighting up the smoke. All landing-craft were moving now, with the LCAs forging ahead and the LCIs close on their heels. The LCTs crept in slowly, allowing us to pass them. The artillery and tanks aboard were firing through camouflage nets, and the din grew absolutely shattering as the guns closed their range. The rocket ships loosed their streaming sheaves of flame in volleys at the beach. All along the coast were flashes

of bursting shells and answering coast artillery, and over it all
drifted a thick pall of smoke, streaked here and there with streams of
tracer.

It reminded me of coming up to the line at the start of a sailing race.
We were rolling heavily in a big south-west swell, which broke continually over us, drenching us to the skin. My hands grew numb and
dead, and my teeth were chattering with cold and fright. I had a look at
my batman, sitting on the thwart. He was looking awful, but gave me a
big grin through his green. It was zero hour, and the first infantry were
going in. We passed round the rum, and those were not sea-sick took a
good swig. The sea was dotted with 'bags vomit' and I could see the
boys on the LCIs rushing to the rail.

I took a look around the boat. Private Hindmarch beside me, polite
as ever and looking surprisingly pink. Lieutenant Kennedy – I always
remember him as a sergeant in No. 5 – looking grim, but enjoying his
rum. Just as well too, as he was never seen again after leaving the
boat. Little Sapper Mullen, the artist, as grey as a corpse, who died of
wounds later that day. Gordon Webb and Peter Beckett in the bows,
peering forward, alert and tense.

The chaps in the other boats were passing round the rum, and I
could hear snatches of song through the hellish din. Hutchie Burt's
boat went in singing 'Jerusalem'. We didn't sing in our boat. My mouth
was bone-dry and I was shaking all over; I doubt if I could have produced a note.

The shore was obscured by smoke, but I made out the fountains of
shell bursts, and the rattle of small arms fire cut through the roar of
the heavy shells. Something was hit on our starboard bow, and a huge
cloud of black smoke went up, with orange flame licking against the
murk of battle. With four hundred yards to go, we were late. Men
struggled into their rucksacks – an almost impossible feat. I gave it up
until we grounded.

Bullets rattled against the craft and splinters went whining overhead.
'Ready on the ramp.' We cowered down. The explosions were very near.
'Going in to land.'

We touched, bumped and slewed round. This was no true landing.
Then the order: 'Ramps down.' The boat began to empty, and, being at
the stern, the medics were the last to leave. I seized a stretcher. No
one seemed ready to take the other one, so I picked it up too, staggered
to the bows and flopped into the water. It was thigh deep, as the craft
had grounded on some softish obstacle, probably a body.

The next stretch of time is muddled in my memory. I have no idea
how long it took from the boat till I reached the enemy wire. There

was thick smoke over the beach, and the tide low but flooding. There were many bodies in the water; one was hanging round one of the tripod obstacles. The shoals were churned with bursting shells. I saw wounded men among the dead, pinned down by the weight of their equipment.

The first I came to was little Sapper Mullen, the artist. He was submerged to his chin and quite helpless. Somehow I got my scissors out and with my numb hands, which felt weak and useless, I began to cut away his rucksack and equipment. Hindmarch appeared beside me and got working on the other side. He was a bit rattled, but steadied when I spoke to him and told him what to do. As I was bending over I felt a smack across my bottom, as if someone had hit me with a big stick. It was a shell splinter, as appeared later, but it hit nothing important and I swore and went on. We dragged Mullen to the water's edge at last.

The Commando were up at the wire and clearly having trouble getting through. Hindmarch and I went back to the wounded in the water. I noticed how fast the tide was rising, and wounded men began to shout and scream as they saw they must soon drown. We worked desperately; I don't know how many we pulled clear, though it wasn't more than two or three.

Then I saw Donald Glass at the water's edge, badly hit in the back, and we went to him, and started to cut away his equipment. Doing so I became conscious of a machine-gun enfilading us from the left front. In a minute I was knocked over by a smack in the right knee and fell on Donald, who protested violently. I cautiously tried my leg and found that it still worked, though not very well. Donald was too bad to walk, so I got Hindmarch to open a stretcher. I looked for help, but the only standing figure anywhere was my batman, who was working on his own with the drowning wounded in the water. He smiled and I left him to get on with it. I tried my leg again and took one end of the stretcher.

Hindmarch is a big strong fellow and between us we began to carry Donald up towards the wire. I had to have one rest, and at the finish I was beat to the wide, and just lay and gasped for some minutes. We took the stretcher from Donald and left him in a bit of a hollow in the sand, where he had a certain amount of cover. The troops had got through the wire, and I stumbled after them across the minefield to the demolished buildings, which the air photographs had shown so clearly and which were our assembly area. I am a little vague about the minefield, but I remember thinking that it might be wise to walk in footprints.

I should explain that the stretcher-bearers were in another LCA with the sergeant. One of my RAMC lance-corporals was with each troop, and I had only my batsman, Hindmarch, and Cook with me in my LCA. My orders had been to move up to the assembly area with the troops as soon as possible, and not to linger on the beaches attending wounded, but to be in readiness for the assault task. Following waves of the landing would include medical units to deal with early casualties. I had overlooked the factor of the rapidly rising tide, but can be excused for not anticipating that the East Yorks would fail to cut the wire and clear the path for us. The men themselves had definite orders to push on, and not to help the wounded. I think these orders were sound.

I found the unit assembling among the buildings. Someone gave me a swig of rum, which did me good, and Lance-Corporal Cunningham put a dressing on my leg. It turned out to be a lucky wound, through the muscles and tendons behind the knee-joint, which had missed the popliteal artery very narrowly. The shrapnel in my buttock made me stiff but was not worth bothering about. I could find no sign of my batman Smith, but later found him, badly wounded in the legs, at the beach dressing station awaiting evacuation. The last I heard of him was a letter from his widow, asking how he came to die on D-Day, so I am afraid he must have eventually been killed by air strafing during the night.

We were safe in the demolished buildings, though shells were going close overhead, but the Germans were ranging on the water's edge, where troops and craft were packed together in the shallows. The rest of my section turned up, minus Marine Porter, who had been killed, among others, when a mortar bomb landed right in their LCA. We moved off to the main east–west road, which leads straight into Ouistreham. The road was under heavy mortar fire, and I came on six of our men lying dead along the way. One was my Lance-Corporal Pasquale. I was very lame, but Marine Boyce carried my rucksack and gave me an arm.

Farther on we passed two more of our chaps, one dead, and the other almost gone, with his head smashed. I pushed his beret over his face and went on. Then we found my Lance-Corporal Farnese, quite dead. The same mortar bomb had killed another man (an old Frenchman who had come out to welcome us) and severely wounded another commando in the shoulder. A bad case: his shoulder was practically severed, with blood gushing from it. With some trouble we got the bleeding under control with finger pressure in the neck, and, holding on to this point, we put him on a stretcher and carried him the rest of the way to the point where the assaulting troops left their rucksacks, three hundred yards farther on.

Here I set up my RAP, on a patch of grass under some pine-trees. The attack on the battery and the coast defences was well under way, but I was preoccupied with my badly wounded man, and packed the wound as best I could before giving him a pint of plasma, which brought back his pulse. Now casualties were pouring in; I was hard at it until late in the afternoon.

A lot of officers were hit. The CO had a split temple and some fragments in his leg, but refused to be evacuated. The Germans persistently mortared the road junction beside us: the bombs were clearly visible as they sailed overhead, skimming the roofs of the adjacent houses. Later we put as many of the wounded as possible inside, under cover. Casualties kept coming in all the time, but luckily some Jeeps turned up and I began to get the severely wounded back to beach-head, which cleared the congestion. At about 1200 hours the fighting died down, and troops began to reassemble for the move inland up to the Orne bridges.

Troop leaders came and reported dead and missing. We had suffered heavily. The French doctor was among the dead, killed by a burst of LMG fire from the Casino as he was tying someone up.

The rest of D-Day I spent in and around Ouistreham and on the beach-head. It was a day full of incidents, though I have forgotten their chronological order. It stayed fine, though windy. The arrival of the glider troops (6th Airborne) in the afternoon was a wonderful sight. For over an hour the planes towing the gliders came streaming very low over our heads; then the air was full of multi-coloured parachutes. I saw a number of them 'candle', to my horror, but found out later that they were only dropping supply containers. The gliders were cast off and circled down to land over to the south-east in the Orne salient, where the commandos were reinforcing airborne troops.

I talked French with the locals as I drove around the town collecting more wounded in a Jeep, getting cautious as it became clear that the only troops left in the town were probably Germans. I drank 'à bas les Boches' with the white-bearded local doctor in the little hospital, across the bodies of civilians and soldiers. I had a cup of hot sweetened milk and a wash at the cottage across the road. I flattened to the ground as we were straddled by a stick of small armour-piercing bombs and practically disappeared when a rocket-firing Focke-Wulf strafed the streets. More than once I looked out my wounded on the beach. Nothing was being done for them: there was no plasma or blood, and they lay there being bombed and machine-gunned all night long. Very few were taken off that day.

I learnt that Sergeant Garner and two marines had been taken prisoner by some Germans still holding out underground in a strong-point by the battery; they sent the sergeant back, asking me to attend their wounded and some of ours being held there. This struck me as foolish, and I sent out a Red Cross Jeep with orders to bring the wounded to me. To my surprise, about fifty Germans came doubling down the road, escorted by my Red Cross Jeep. The strong-point had surrendered. I took two German medical orderlies prisoner and used one as a guide and shield to explore the battle areas for more wounded.

By dusk, most of the wounded were back at the beach-head. I moved in and sheltered in some trenches the Germans had dug in the dunes. As soon as it was dark the enemy started attacking the coast from the air, and all night it went on. I was exhausted, but my wounds were so stiff and sore that I got no rest. At about 0400 hours I took my section back to the old RAP in the town. I was amused to see new troops in the ditches advancing very cautiously into Ouistreham along the road we had used to attack the morning before. I was feeling very sick and stiff, and found a Jeep awkward to climb in and out of. I couldn't walk, so it was the only way.

I set the men on to getting themselves some food and went off to locate the Commando. This I succeeded in doing, across the Orne bridges. The landing areas presented a wonderful sight. The gliders in their hundreds lay close-packed everywhere. I found Lovat shaving and got a warm reception. I set up shop in the billiard room of a château (at Hauger), conscious of the fact that it must be a clear ranging mark for the German artillery and mortars when they got going again.

The section of medics was collected together. We dug ourselves foxholes in case of trouble. Very soon it was dark, and we began to get a stream of casualties, which kept coming in all night long. Soon the ground floor was covered with bodies, hurt in varying degrees. Light was the biggest problem, as the blackout was very sketchy, and most of the work was done with the help of a night light; there were several of these lying about, which the Germans had left behind. Throughout the night there was continual small arms fire, and a lot of enemy air activity (not directed at our hamlet but near enough to keep one interested). The German, good soldier that he is, made full use of local knowledge and small parties penetrated right into our area. Some of the firing was between our own troops, and one sentry was shot dead by another as he approached with a message.

Somehow the night came to an end, and I began to get the wounded away; by mid-day we were pretty well clear. There were still the odd Germans about, and I got sniped at as I was hobbling into the bow window of my RAP. After that I used the door, and kept the shutters closed on that side of the house. The ADMS appeared and battered on the shutters, asking why I wasn't using the window as an entrance. He hopped in pretty smartly when I explained.

I told him about the wounded being drowned by the rising tide, and he made me very angry, saying that it was entirely due to lack of training. He had come in comfortably three or four hours after us, when the beach defences had been silenced. 'A guardsman,' he told me, 'will stand to attention with his guts hanging out until he is told to stand easy.' I was in no mood to argue, but decided there and then just how much notice to take of him in future.

The afternoon was quiet and I began to get worried about my leg, which was swelling and showing a regrettable tendency to get bubbly. The night brought more firing and some casualties, but I got some sleep, and remember dozing off, listening to the quick whoomp-crack whoomp-crack of SP guns and thinking how nice it was that some of them had got over the bridges to join us. Later in the day I discovered that the Orne bridges were not strong enough to carry SP guns, and that the Germans were hitting a pontoon bridge which the REs were trying to put up. The SPs I heard were German. There is not much else to tell. In the afternoon (D plus three) the ADMS appeared again and told me that I must get evacuated, and the doctor of No. 3 could look after the rest of the unit (which was now considerably smaller) until a replacement arrived. I wasn't really sorry, for I had had about enough, and the CO came with me, as he was all in too. Lovat ordered Kieffer to go with us.

The journey home was long and tedious, with many incidents before we felt safe. We got down to the beach in good time and eventually set off in a DUKW to a hospital ship.

The DUKW was making rather heavy weather of the seas. Then the hospital ship upped its hook and sailed away. So back we came to the beach-head. Twenty-four hours later we got off in an LST from Arromanches. For the night we were taken back to the 16th Casualty Clearing Station, which had opened in a village three miles inland. This was where they shot two women snipers out of the church tower.

Next day (10 June) there was a flood of casualties, with a lot of red and green berets among them. I heard from some of our men how the Commando was being heavily attacked and though the Germans were being repelled, our casualties were heavy, and No. 4 (which had lost two-thirds of its men) was now back in reserve.

We got heaped on a truck that evening and driven back to Arromanches, where we went aboard an LST just in time to catch the tide. It was full of wounded, and for two nights and a day we lay in that huge cavern of a hold before entering Portsmouth harbour. I was lucky, being slightly mobile, and was able to have a look around and to scrounge some food. The poor devils on stretchers had nothing but bully and bread and tea, though some had bad belly wounds.

In the Channel we picked up twelve survivors from an American tug, torpedoed by a U-boat while towing a section of 'Mulberry' harbour. The sub had biffed the harbour section with two torpedoes, but it remained afloat, though anchored to the tug, which was sitting on the bottom. Nine out of the twelve men had fractured thighs, due to the deck coming up under them when the torpedo hit.

We spent two more days in hospital trains, where we got M and V for supper and soya 'links' for breakfast.

At long last I arrived at the Queen Elizabeth hospital, Birmingham, on the night of D plus eight and troubles were at an end.

THE FREE FRENCH COMMANDO
by Robert Dawson

Lieutenant-Colonel Robert Dawson, CBE, DSO (Loyal Regiment), commanded a detachment of Frenchmen from No. 10 Commando. Robert was biligual and much respected by his Allied men – never more so than when he gave Kieffer a half-length lead so that they could be the first to touch down in La Belle France on D-Day. (Author)

The French consisted of two troops, 1 and 8 (commanded originally by Kieffer and Trepel respectively). Trepel disappeared on a small raid in the winter of 1943, and his place was taken by Lofi – a small, stocky, immensely muscular man, a regular in the French navy who had come up from the lower deck and eventually risen to the highest rank open to men promoted in this way: 'Officier en Chef aux Equipages', equivalent to, but below commander (Capitaine de Fregate).

Kieffer was a naval reserve officer, and had spent a large part of his life in the West Indies. He was far older than anyone else in the Commando. He heard about Special Service and agitated through the French naval headquarters in London to raise a French Commando unit. He had his way in the end; No. 1 Troop was mainly raised from volunteers in the Free French forces who had either come out through Dunkirk, or had subsequently escaped from France – many of them by fishing boat from Breton ports. There were many Bretons among them, and when Kieffer (an Alsatian) joined No. 4 Commando he became my senior Officer of French Troops; he also formed a very small staff and Guy Vourch (now a senior Professor in the Medical Faculty of Paris University) took over the men. No. 8 Troop was mainly composed of sailors, *fusiliers marins* and soldiers of *infanterie coloniale* – which does not mean 'native troops' – many having served in West Africa, where several French territories declared for de Gaulle in 1940 and 1941, others having seen action in Syria, and yet others who had come from Koenig's Free French brigade, which won fame in the desert at Bir Hakeim.

These were tough, self-reliant soldiers, quick in action and very brave indeed. Even before D-Day a measure of mutual respect had developed between the British and French, but after the initial phase, the Commando became so firmly welded together at every level that it seemed entirely natural that we should fight and live side-by-side till the end of the war, as indeed we did. I cannot speak too highly of those Frenchmen, whose homes and dreams were for many years beyond their reach, behind the enemy defences. It is true they had everything to fight for, but they gave all they had; in addition they gave us, and we gave them, a comradeship that has never cooled – as anyone who has visited Ouistreham and Amfréville on the anniversary of 'The Day' can testify. We shared everything – the wonder and excitement of Paris just after the liberation, saddened for poor Philippe Kieffer by the news of his son's death in the Resistance, as a boy of seventeen, just before the city fell. So much springs to mind: the 'rough house' which exploded in a rest camp on the Belgian coast after Walcheren, when some enterprising Frenchmen stole the local girls from the stolid British gunners who had invited them to a dance; the occasional wine ration which the French were able to claim from their supply services; above all, the leave-taking when the French, who had been with us right into Germany, were recalled to France in June 1945. A large detachment of No. 4 Commando, including all those who had been awarded French decorations, travelled to Paris, and were invested with their medals by the French Minister for the Navy, M. Jacquinot, in the courtyard of the Ministry, followed by the traditional 'vin d'honneur'. Then we received perhaps the most gracious compliment the French nation could pay: we marched as a unit and alone (apart from the French naval band that led us) from the Ministry in the Place de la Concorde to the Arc de Triomphe, where Philippe and I jointly laid a wreath on the Unknown Soldier's tomb and relit the flame. I doubt if any other British Regiment has ever been honoured in this way.

REGIMENTS, CORPS AND OTHER FORMATIONS FROM WHICH THE OFFICERS AND MEN OF THE 1ST COMMANDO BRIGADE WERE DRAWN

Royal Horse Guards
Royal Dragoons
8th Hussars
Lovat Scouts
Royal Armoured Corps
Royal Tank Regiment
City of London Yeomanry
Royal Regiment of Artillery
  (Field, Heavy Anti-aircraft,
  Coastal and Searchlights)
Welsh guards

11th Hussars
12th Lancers

Corps of Royal Engineers
Royal Corps of Signals
Grenadier Guards
Irish Guards
The Buffs
King's Regiment (including the
  King's Liverpool Irish)
Royal Fusiliers
Devonshire Regiment
Somerset Light Infantry
East Yorkshire Regiment
Green Howards
Duke of Cornwall's Light
  Infantry

Hampshire Regiment
Dorsetshire Regiment
Royal Welsh Fusiliers
Life Guards
Scots Greys
King's Dragoon Guards
Inniskilling Dragoon Guards
9th Lancers
Berkshire Yeomanry
Lanarkshire Yeomanry
Ayrshire Yeomanry
Coldstream Guards
Scots Guards
Royal Scots
Royal Warwickshire Regiment
Royal Norfolk Regiment
Suffolk Regiment
West Yorkshire Regiment
Bedfordshire and Hertfordshire
  Regiment
Lancashire Fusiliers
King's Own Scottish Borderers
Cameronians
Gloucestershire Regiment
East Lancashire Regiment
Black Watch
Oxfordshire and Buckingham-
  shire Light Infantry
Loyal Regiment

Welch Regiment
Essex Regiment
Sherwood Foresters
Royal Berkshire Regiment
King's Own Yorkshire Light
  Infantry
King's Royal Rifle Corps
Highland Light Infantry
Royal Ulster Rifles
Royal Army Service Corps
Royal Army Ordnance Corps
Liaison Regiment
Free French No. 10 Commando
No. 45 Royal Marine Commando
Royal Inniskilling Fusiliers
Worcestershire Regiment
Easy Surrey Regiment

Border Regiment
South Staffordshire Regiment
South Lancashire Regiment
Queen's Own Royal West Kent
  Regiment
King's Shropshire Light Infantry
Durham Light Infantry
Liverpool Scottish (Queen's
  Own Cameron Highlanders)
Rifle Brigade
Royal Army Medical Corps
Royal Army Dental Corps
Honourable Artillery Company
US Rangers (attached)
Gordon Highlanders
Royal Marines

THE SPIRIT OF THE REGIMENT
by Andrew Maxwell, MC

Andrew Maxwell served as a lieutenant in the Scots Guards during the North African Campaign. I asked him to send me an account of his experiences during the fighting on Rigel Ridge in June 1942. The following text is taken from the letter which he wrote in reply. (Author)

War experiences can be boring unless they are your own; then they are fascinating!

The 2nd Battalion Scots Guards, with whom I was serving as a newly arrived junior officer, was strung out behind Rigel Ridge, not far from Tobruk, in an area called 'Knightsbridge' – a misnomer and not to be confused with London, SW. Our task, during those mid-June days of 1942, was to engage and hold up an enemy attack during a critical period while the Knightsbridge 'box' was being evacuated.

On 11 June, we were most heavily shelled throughout the morning. It did not sound too bad because the sequence of bangs was reversed. First, the crack as the 88 mm shell – hopefully – went over us. Secondly, the explosion as it hit the ground, and only thirdly the noise from the gun. This was heartening as it gave the impression that these were our shells going towards the enemy! During lunch I was punctured by shrapnel – which brought my morale back to normal. It was only a small fragment and in the back, but no conclusions should be drawn from this. A medical orderly bound it with the army equivalent of Scotch Tape and the only real casualty was my shirt, and I did not retire hurt as the company commander had been killed and his second-in-command was also a casualty.

Unfortunately a telegram went out which caused my family some concern. Two days after this, on 13 June, at about mid-day, a highly alarmed forward observation gunner came hurrying through our line in his armoured car and informed us in the briefest possible terms that a heavy enemy tank force was approaching.

We had adjusted ourselves to this new situation and before long a great number of ponderous grey tanks came into view on our exposed flank. Our anti-tank gun hit a few before being put out of action. It must have been rather frightening for them, as they were bound to be demolished within a few moments of firing their first shot. During the heavy shelling which accompanied all this action, I was impressed by the nonchalance of the desert larks, which flew around in full song, showing no slightest sign of alarm. The other incongruous sound came from our portable gramophone, which kept repeating a currently popular (desert) song called 'Sand in my Shoe'. All too soon I heard a grinding sound in front of me and over the ridge came the largest tank I had ever seen – no doubt the same size as all the others, but it looked impressive against the sky at about thirty yards' range. The monster had a look of confidence, power and majesty, and in a fleeting way reminded me of Queen Victoria. It also presented a wonderful camera shot and I reached for the Minox camera which I always – improperly – kept in the breast pocket of my shirt. It was not there and I remembered that when I had last changed the old blood-soaked garment, I had forgotten the Minox – a piece of very bad luck. All unrecorded, the tank lumbered through our position and two hundred yards behind us slowly turned round with grunts and groans to face our rear.

Its gun ended pointing straight at my pick-up truck – understandably, I suppose, because it had an antenna sticking out of its roof and was obviously the company command vehicle. At this late stage, all my modest military training came to my aid and it seemed imperative that I send word back to battalion HQ that our company had been over-run. Causing as little commotion as possible, I crept to the pick-up and managed to get the switch on and pull out the headphones. Then I snaked under the truck and made contact with HQ. Thoughtlessly, I passed a message, 'Enemy tanks have over-run our position and I have surrendered.' This seemed a perfectly straightforward and informative message, until I remembered with a shock that officers and men of the Brigade of Guards never surrender. At considerable risk I had to return to the truck and go through the whole process again to cancel the last message and replace it by 'Enemy tanks have penetrated our position and we are pinned down and surrounded'! The joke being that when I later asked battalion HQ what had been the reaction to my signals, I was told that nobody had any recollection of receiving any message at all.

But to return to the battle. In a few moments worse was to come. The whole area was suddenly full of German infantry. They looked small and purposeful but quite unlike a war image on the silver screen.

There is a moment of euphoria when one realizes that one is still alive but that one's war is over. I found it interesting that during a period of action, when risks are high and confusion great, one's various senses become saturated with unaccustomed noises and, particularly, smells – maybe some of them are human fear smells. Certainly feelings are dull in some areas and acute in others. For instance, one discovers all sorts of abrasions, bruises and cuts on elbows and knees, but has no recollection of when these occurred.

The euphoria did not last and while I still had the chance I stowed my oil compass under my hat. It is hopeless trying to move long distances in the desert without a compass, especially at night.

I thought this an appropriate moment to send a final message to HQ, but was dissuaded by the new arrivals, although I pointed out that this was 'for information only'.

There were some packages of cigarettes lying about on the ground and a German NCO indicated that I could pick them up if I wanted them. This was nearly a sad accident. As I bent down, the heavy compass jumped forward and nearly took my hat off. Fortunately, the strange movement in my cap went unnoticed.

We were herded together and, as some of the men were walking wounded and, inevitably, thirsty, I stopped a great eight-wheeled armoured car and asked for some water. The German officer handed me a jerry-can full of liquid and told me not to use more than necessary, as this was part of his crew's ration. I observed this strictly and as we parted he gave me a very guards-like salute, which I returned with all the parade-ground skill I could muster.

As it got dark a loud, most ugly-sounding voice called all prisoners to line up. This imminent loss of liberty had a profound, if belated, effect on me – also on my brother officer, Ian Calvocoressi, and we immediately called a conference on the subject of escaping before it was too late. We could not agree on the bearing to follow, so we decided to go our own ways independently. We each took a deep breath – at least, I suppose Ian did – and ran like the devil out into the dark. The last word I heard was a shout from Ian. Later he told me that I was running directly at a sentry. He seemed to see better in the dark than me. His warning proved to be right when at full speed I passed within a few feet of one of the surrounding guards. He took no action whatever and I can only suppose that he had had a wearing day and did not want to start anything. After running about two hundred yards without any lights going up or alarm given, I knew I was out of trouble, and this was the most exhilarating moment I had in the war.

Later that night, as I was passing back through our old position, I saw a crouching figure moving in somewhat the same direction as me. He was small and looked like Ian, but, having got so far, I did not want to take even the slightest risk, so we did not make contact until the following day at our battalion HQ. With my oil compass, I felt very secure. It was rather a long walk, nine hours in all, because I decided not to take a short cut by going through our minefield, but round it. They were all anti-tank mines, so could not be detonated by the weight of a body. But you can never tell, can you?

Days like this made you thirsty. At one point during the night I passed some old earth-works where there had been earlier fighting. In the hope that I might find an abandoned water container, I went into a deep dug-out, but suddenly I had such a strong feeling of people all around me, and the atmosphere was so spooky, that I resisted the need for drink and kept going.

The next surprise was to see a light in the distance. I thought of every possible solution to account for it, and eventually, out of curiosity, crept up and found the source was the luminous dashboard of a crashed aircraft. In due course I got back to the battalion and was greeted by 'Why the hell did you take so long?' At least I made it several hours before Ian, which proved my superior desert craft! Ian later had a good story that during the night he had seen a crouching figure moving and peering in all directions, which he rightly took to be me. He was able and willing to demonstrate my action in a way which was considered humorous to anybody who cared to watch!

VI

ABBREVIATIONS AND SHORT TITLES

ADMS	Assistant Director Medical Services
CCO	Chief of Combined Operations
CIGS	Chief of the Imperial General Staff
COHQ	Combined Operations Headquarters
COSSAC	Chief of Staff Supreme Allied Command
CRE	Chief Royal Engineer
DA	Deputy Assistant
DD Tank	Swimming Duplex Drive Tank
DUKW (Duck)	Amphibious 2½-ton lorry
E-boat	German equivalent of MTB
FOB	Forward Observation Bombardment (officer)
Folbot	Folding boat
FOO	Forward Observation Officer
G1	General Staff Officer: Grade One
GOC	General Officer Commanding
IO	Intelligence Officer
ISG	1st Scots Guards
LCA	Landing-Craft Assault (carried four crew and thirty-five men)
LCF	Landing-Craft: Flak
LCI	Landing-Craft: Infantry
LCT	Landing-Craft: Tank
LMG	Light Machine Gun
LSI	Landing Ship: Infantry
LST	Landing-Ship: Tanks
ML	Motor Launch
MO	Medical Officer
MT	Motor Torpedo
MTB	Motor Torpedo Boat
O Group	Order Group
QMG	Quartermaster General

RAMC	Royal Army Medical Corps
RAP	Regimental Aid Post
RNVR	Royal Navy Voluntary Reserve
RSM	Regimental Sergeant-Major
SAS	Special Air Service
SBS	Special Boat Section
SNO	Senior Naval Officer
SOE	Special Operations Executive
TCV	Troop Carrying Vehicles
TSM	Troop Sergeant-Major

INDEX